Water Perspectives in Emerging Countries

Water in Agricultural Practices:
Training the Trainers

Jose Araruna Jr., Aluisio Granato, Julia Stuchi, Rodolfo Silva (Eds.)

September 15-21, 2019 - Rio de Janeiro, Brazil

Water Perspectives in Emerging Countries

Water in Agricultural Practices: Training the Trainers

Jose Araruna Jr., Aluisio Granato, Julia Stuchi, Rodolfo Silva (Eds.)

September 15-21, 2019 - Rio de Janeiro, Brazil

Issue Editors

Prof. Dr. Jose Araruna Jr.
Pontifical Catholic University of Rio de Janeiro, Department of Civil and Environmental Engineering,
Rua Marquês de São Vicente, 225, Rio de Janeiro, Brazil; araruna@puc-rio.br

Prof. Dr. Rodolfo Silva Casarin
Instituto de Ingeniería-UNAM, Circuito Escolar s/n, Edif. 17 Ciudad Universitaria 04510 Coyoacán,
México, DF; RSilvaC@iingen.unam.mx

Prof. Dr. Aluísio Granato de Andrade and Julia Stuchi, M.Sc.
EMBRAPA SOILS, Rua Jardim Botânico, 1024, Rio de Janeiro, Brazil;
aluisio.andrade@embrapa.br, julia.stuchi@embrapa.br

Exceed Chairman & Editor-in-Chief

Prof. Dr.-Ing. Norbert Dichtl
Technische Universität Braunschweig, Institute of Sanitary and Environmental Engineering,
38106 Braunschweig, Germany; n.dichtl@tu-bs.de

Publishing Editor

Prof. em. Dr. mult. Dr. h.c. Müfit Bahadir
Technische Universität Braunschweig, Leichtweiss Institute, Exceed Office,
38106 Braunschweig, Germany; m.bahadir@tu-bs.de

ISBN 978-3-7369-7100-4   © 2019

Bibliografische Information der Deutschen Nationalbibliothek

Die Deutsche Nationalbibliothek verzeichnet diese Publikation in der
Deutschen Nationalbibliografie; detaillierte bibliografische Daten sind im Internet
über http://dnb.d-nb.de abrufbar.
1. Aufl. - Göttingen: Cuvillier, 2019

CONTENT

PREFACE

The International Workshop on "Water on Agricultural Practice", held in Rio de Janeiro, Brazil on September 16-20, 2019 involved 30 participants from 14 different countries that made the event very broad in participation and expertise. Their findings on water-related issues dealing with sustainable agricultural practices covering erosion control and monitoring, use of reclaimed wastewater for irrigation, and environmental services and education are set in chapters grouped around these 3 themes.

This proceedings document is an outcome of the EXCEED SWINDON Project termed "International Network on Sustainable Water Management in Developing Countries". This initiative is a venture for capacity building through higher education and joint research based on a network of 35 partners from Latin America, Middle East, South East Asia, and Sub-Saharan Africa, coordinated by the Technical University of Braunschweig. This project is carried out within the framework of the DAAD Program "Excellence Centers for Exchange and Development".

Professor Dr. Ali Müfit Bahadir reviewed all papers submitted to the workshop in terms of their scientific content and also made the final publishing review. As the issue co-editor of this proceedings book, I cordially acknowledge his invaluable contribution while publishing this book.

I believe that the wide range of global interests on sustainable agricultural practices combining with the issue of this proceedings book should contribute to attaining the Millennium Goals set by the United Nations.

Prof. Dr. José Araruna, co-editor
Department of Civil and Environmental Engineering
Pontifical Catholic University of Rio de Janeiro

EROSION CONTROL IN AGRICULTURAL PRODUCTION SYSTEMS IN BRAZIL

Aluísio Granato de Andrade, Julia Franco Stuchi

Brazilian Agricultural Research Corporation, Embrapa Solos, Rua Jardim Botânico, 1024, Jardim Botânico, Rio de Janeiro – RJ, Brazil, CEP 22460-000; aluisio.andrade@embrapa.br; julia.stuchi@embrapa.br

Keywords: Public policies, Soil restoration, Sustainable practices, Water conservation

Abstract

With the increase in demand for agricultural and forestry products, the pressure on soil and water resources intensifies, leading to extreme food, nutritional and socioeconomic insecurity in the world. This paper aims to present the political and technological strategies to control erosion in agricultural production systems in Brazil presented during the Workshop - *Water on Agriculture: Training the Trainers*. Erosion has been causing economic damage and degradation in different regions of the country, mainly over sandy soils and areas used since the first agricultural cycles in the 16th century. The largest land use is occupied by pasture with different levels of degradation. Radical changes in soil management and throughout the production system are required to control erosion and to prevent soil and water degradation. The continental expression of Brazil, housing six biomes with different types of vegetation, relief, soils and climate, in addition to the contrasting socio-economic and cultural aspects, requires that the construction of a national policy for sustainable soil and water development should include adaptations considering territoriality and regional specificities. It is necessary as well to develop information on the potentialities and limitations of soil for agricultural production. A strategy to start solving these problems is being initiated trough the National Soil Mapping Program (PRONASOLOS) with detailed information on the country's soils over the next three decades at the most appropriate scales in order to plan agricultural production in a sustainable manner and to prevent water and soil degradation. Zero tillage systems and agroecological production systems can also be considered as examples for erosion control. To further reduce soil and water losses due to erosion in Brazil, it is necessary to expand the use of more sustainable agricultural production systems considering the different limitations and potentialities of the soil and the socioeconomic and cultural conditions of the country.

1 Introduction

The conversion of natural areas into productive areas has made it possible to increase the supply of food, fiber, energy and other raw materials for the development of humanity. However, many lands that generate wealth through livestock are currently degraded or in the process of degradation. It is estimated that some 33% of land in the world show some kind of degradation [1]. In Brazil these areas occupy approximately between 16.5% [2] and 22% [3] of the territory with varying levels of degradation.

Water erosion accelerated by inadequate land use and management is already considered one of the most serious environmental problems nowadays, as it causes the removal of the richest layer of the soil with the consequent loss of nutrients, and creates pollution and siltation of water resources, floods and a sharp drop in agricultural productivity [4, 5].

In a scenario of growing world population and increasing demand for agricultural and forestry products, pressure on soil and water resources intensifies further and could lead to a world of extreme instability as well as socioeconomic, food and nutrition insecurity [1]. This issue worsens during water crises either due to lack of water during drought periods or excess water during the rainy season, causing severe losses to the productive sector [6-8]. Lack of application of conservationist practices causes a reduction in soil infiltration and water storage. Consequently, in periods of less precipitation there is less water availability in degraded areas compared with well-managed ones. This results in decreased tolerance of crops to periods of water stress due to less soil water retention. In situations of excessive rainfall, degraded soil exposed to erosion agents produces more sediment, causing greater siltation of water resources [6]. This situation leads to an increase in the occurrence of floods and reduces reservoir life, increasing costs for hydroelectric power generation and water treatment for urban water supply [7].

With climate change, the problems caused by the degradation of soil and water resources can become worse. Increasingly frequent extreme events such as high-intensity storms worsen erosion processes, and rising temperatures generate higher water consumption for agricultural production, impacting negatively a wide range of crops which creates risks to food security [9]. Erosion has been causing economic damage and degradation in different regions of the country mainly over sandy soils and areas used since the first agricultural cycles in the 16th century (Figure 1).

In the semi-arid climate region of northeastern Brazil, erosion contributes to the expansion of desertification (Figure 2). This paper aims to present the political and technological strategies to control erosion in agricultural production systems in Brazil presented during the Workshop – *"Water on Agriculture: Training the Trainers"*.

Figure 1: Degraded areas due to agricultural exploitation using inappropriate practices with laminar erosion and gullies in different regions of Brazil.

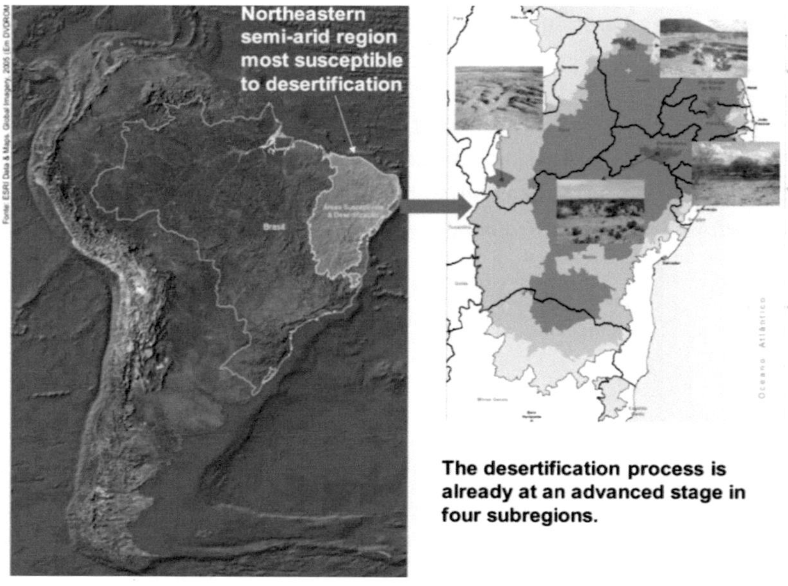

Figure 2: The northeastern region of Brazil with semi-arid climate is the most susceptible one to desertification in the country [10].

2 Land Cover Use in Brazil

In Brazil, 61% of the territory is protected by law (Figure 3). The largest land use is occupied by pasture with different levels of degradation. Although they occupy only 5% of the land, annual crops cause large erosion losses when conventional planting is used [11].

Figure 3: Land cover in Brazil

3 Economic Losses from Erosion under different Land Uses in Brazil

According to Polidoro et al., 2016 [12], considering the different land uses in Brazil, it is estimated that the financial losses caused by erosion amount to over US$ 6.3 billion/yr. Areas with annual crop planting without soil conservation practices and intensive tillage are the most problematic ones due to erosion (Figure 4).

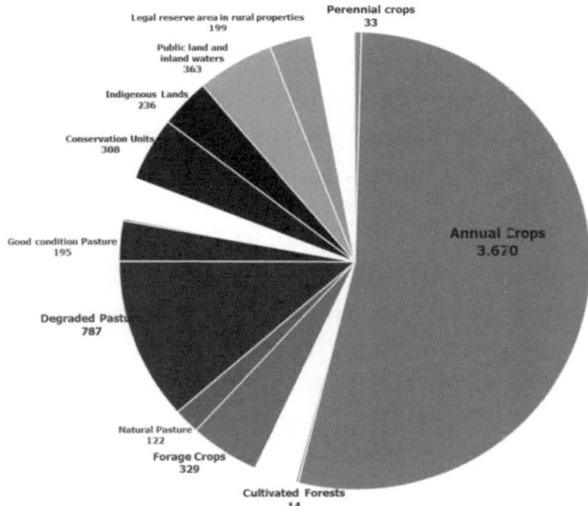

Figure 4: Economic losses from erosion under different land uses in Brazil (billion/yr)

4 The Green Revolution and the Consequences on Soil Erosion

During the 1970s' green revolution, the models of agriculture in Brazil were not efficient to control soil erosion, mainly due to the intensive use of mechanization (Figure 5). With the advance of erosion it was necessary to make changes in the production system. Initially only terracing was performed as a strategy to prevent erosion. However, this technology was not enough to significantly reduce the damages caused by erosion. Radical changes in soil management and throughout the production system are required to control erosion and prevent soil and water degradation.

Figure 5: Unstructured soil due to intense tillage through plowing and harrowing

5 Controlling Erosion in Brazil

5.1 Laws and Policies related soil and water conservation

In Brazil, people are becoming increasingly aware of the importance of environmental preservation and the restoration of rivers, steep slopes and other permanent preservation areas provided for in the environmental legislation. However, the importance of properly managing soil and water for agricultural production and their consequences beyond gates of rural properties still have not reached all the regions of the country. In the socioeconomic dimension, the problem of erosion and loss of fertile soils has a strong influence on the reduction of job offer and income in rural areas [13]. The devaluation of rural property, the abandonment of previously productive land and the rural exodus with migration to large urban centers worsen the social problems related to urban agglomerations.

However, the country still has a major socioeconomic and environmental liability caused by soil degradation in areas of agricultural production, bequeathed by the various agro economic cycles and a lack of governance and planning in the countryside. To prevent this degradation with the advance of agriculture in the country, the National Soil Conservation Plan was created in 1975. Later in 1987, it was replaced by the National Program of Watershed. Currently, only a few states participate in this program or develop independent programs. The absence of a national soil and water conservation policy has been shown in Brazil's Soil Governance Audit Report [2] as one of the main risks to sustainable management of these resources. Although there are legal instruments governing the theme, such as Law 6.225/1975, which calls for discrimination by MAPA in regions for the mandatory implementation of soil protection and erosion control plans, and Law

6.938/1981, which establishes the rationalization of subsoil, water, restoration of degraded areas as one of the Principles of National Environmental Policy, protection of areas threatened with degradation and grants special incentives to producers, who adopt conservation practices, there is still a considerable risk for the country as soil and water resource degradation advances [2]. On the other hand, new initiatives point to efforts to reverse this situation through policies such as the National Policy on Agricultural Land Use, Management and Conservation bill and the recent Decree No. 9,414 of June 19, 2018 establishing the National Soil Survey and Interpretation Program of Brazil (PRONASOLOS). It will make it possible to detail the information of soils in the country at scales more compatible with the planning for sustainable use of soil and water resources.

Considering the large underused areas with varying levels of degradation that currently exist in Brazil, the possibility of their reintegration into the high yield chain is an opportunity to increase national agricultural production on a sustainable basis without the need to expand agricultural frontier by converting new areas of natural vegetation.

5.2 Strategic Guidelines for formulating integrated soil and water conservation policies
The continental expression of Brazil, housing six biomes with different types of vegetation, relief, soils and climate, in addition to the contrasting socio-economic and cultural aspects, require that the construction of a national policy for sustainable soil and water development should include adaptations considering territoriality and regional specificities. Through the survey and systematization of information on the existing institutional, legal and technological structure on soil and water governance in the country, training of multiplier agents of the five Brazilian regions, strengthening interpersonal networks and management tools based on participatory approaches and an update of the National Soil and Water Conservation Plan is being proposed. Discussion processes and propositions will be promoted involving actors from the governmental, productive sector and civil society organizations with influence on soil and water governance for sustainable agricultural production and the provision of other ecosystem services. To update the National Soil and Water Conservation Plan, six Strategic Guidelines are proposed:

(1) Legislation - Compliance with the bill: National policy for soil and water conservation in rural areas for watersheds, states and municipalities. It proposes: (a) strategies for the approval of the bill (PL); (b) mechanisms for adaptation and implementation of the PL, as necessary, at the Country, State and Municipal levels; (c) the development of monitoring instruments and systems to ensure the use, management and conservation of soil and water in different regions of the country; (d) alert mechanisms, fines and integration with the Rural Extension to correct the deficiencies detected; and (e) the creation of a fund to receive fines and other financial contributions from other sources of financing for the use of funds in the priority Work Plans defined by the management committee of the Plan.

(2) Prevention - Prevention of agricultural land degradation aimed at new areas with high environmental vulnerability and to curb and promote the replacement of degrading practices It proposes: (a) mechanisms to improve the use of existing information to formulate public policies that

promote the use of new areas only in accordance with their agricultural suitability, indications of agro ecological zoning and adoption of soil and water conservation practices; (b) strategies to prevent the exploitation of areas of high environmental vulnerability; (c) incentive mechanisms for conservation farmers; (d) strategies to prevent erosion and soil and water degradation processes; (e) the management of multiple demands for land and water use, emphasizing the use of techniques and procedures aimed at sustainability; and (f) strategies to curb the use of degrading practices.

(3) Conservation - Conservation of arable land: to promote greater productivity and sustainability of agroecosystems. It proposes: (a) strategies to conserve soil and water quality and promote the development of more sustainable agricultural production systems in areas with good productivity; (b) mechanisms for valorization and traceability of agricultural products that use good agricultural practices; (c) strategies to ensure for current and future generations the qualitative and quantitative availability of soil and water; and (d) improvement of soil and water conservation actions that promote the integrity of aquatic ecosystems as well as the functions of forests and areas with natural vegetation and conservation units to improve and maintain the water regime.

(4) Recovery - Recovery of degraded lands aimed at restoring areas of ecological interest and/or reinserting them into sustainable agricultural production. It proposes: (a) strategies for the development of tools for the characterization and mapping of lands with different levels of degradation; (b) mechanisms for the application of the most appropriate technologies for the different levels of degradation, environmental conditions and production systems of the country; (c) a strategy for the development and implementation of low-cost practices for erosion control and recovery of degraded areas; and (d) the provision and payment of environmental / ecosystem services to the recovery areas, taking into account the benefits for agricultural production and the environment in general.

(5) Monitoring - Monitoring, evaluation and adaptation of the Plan's actions. It proposes: (a) mechanisms and protocols for the diagnosis of the current situation of land degradation in the different biomes of Brazil; (b) creation of tools to predict the situation of land degradation without the intervention of the NAP - Soil and water, with the aim of establishing scenarios to take preventive measures to avoid the worsening of critical situations, such as the growth of areas in desertification, salinization and ravine; (c) a strategy to create a support database, through which technicians, producers or citizens themselves can feed the database from local and georeferenced observations; (d) mechanisms for the establishment of a management information system based on indicators and statistics related to the goals of the Plan; and (e) strategies and mechanisms to adjust and to redirect the Plan's actions as well as an estimate of the socioeconomic and environmental benefits delivered to Brazilian society.

(6) Integration - Integration of teaching, research and extension actions for soil and water conservation and the recovery of degraded areas. It proposes: (a) the insertion of the topics involved in the issue (soil and water) in the primary schools' program with appropriate regional

adaptations; (b) mechanisms to standardize the concepts related to the use, management and conservation of soil and water in rural and urban areas for the various higher education courses with adherence to the subject, such as: agricultural engineering, forestry engineering, environmental engineering, zoo technics, hydrology, geography, biology, among others; (c) the alignment of the research demands identified in the subject under discussion, with the R&D projects of the research institutes and the definition of priority topics for the development of undergraduate thesis and master's and doctorate dissertations; and (d) strategies to raise awareness of the importance of natural land and water resources at all levels and for all public, urban and rural.

Given the complexity of the proposed plan, it will be necessary to establish joint strategies to ensure the participatory building processes are efficient. In this sense, adjustments will be made to the proposed methodological strategy, the available financial resources and the priorities indicated by the team, as well as outlined efforts to obtain complementary resources.

The set of intended actions of the present proposal is in alignment with the Voluntary Guidelines for Sustainable Soil Management [6] and the Sustainable Development Goals (SDGs) established by the UN, especially the SDGs 1, 2, 3, 6, 7, 8, 11, 12, 13, 14, 15, 16, and 17. It is also expected to contribute to the improvement of other public policies, such as the National Action Plan to Combat Desertification and Mitigation of Drought Effects, National Water Resources Plan, Low Carbon Agriculture Plan and National Agroecology and Organic Production and Rural Credit Programs.

Through the expansion of legal provisions that enable greater incentives for innovation to address the problem of degradation, it is expected that an increase in the adoption of conservation technologies with regional adaptations to stimulate development of new technologies will occur, thus contributing to reverse the degradation scenario, enabling more sustainability of production systems and resilience to climate change.

5.3 National Soil Mapping Program (PRONASOLOS / BRAZIL)
Brazil has been developing its Soil Classification System for over 50 years [15]. It was necessary to develop information on the potentialities and limitations of soil for agricultural production. A National Soil Mapping Program (PRONASOLOS / BRAZIL) is being initiated. The objective is to detail information on the country's soils over the next three decades at the most appropriate scales in order to plan agricultural production in a sustainable manner and to prevent water and soil degradation. In Figure 6, one can see the soil map of the same area at different scales. The more detailed the map scale (the more soil types appear) allows the area to be divided into more homogeneous management zones, increasing the efficiency of inputs and ensuring greater production and protection of the soil.

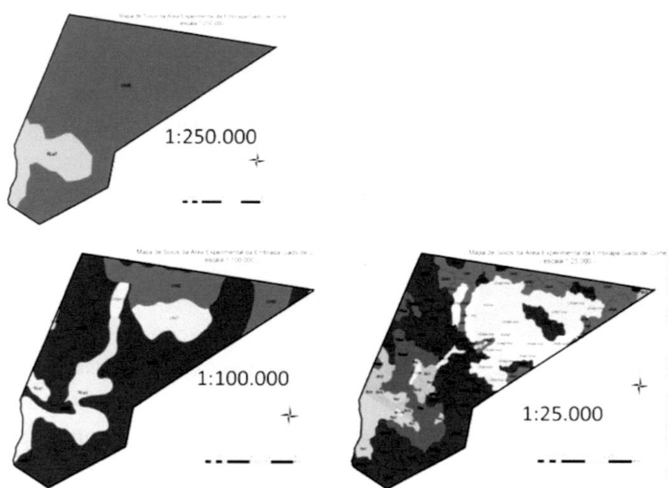

Figure 6: Soil map of the same area at different scales (Source Embrapa Soils)

5.4 Land Use Planning in Watersheds and Farms

By mapping the soils in more detailed scales in different regions of the country already prioritized by PRONASOLOS, it will be possible to carry out more appropriate land use planning. This information, along with the level of soil degradation through different technologies such as proximal sensors, satellite and drone image interpretation, makes it possible to monitor changes in soil quality in real time and to prevent soil degradation. By interpreting this information, land use planning can be carried out in watersheds and rural properties, including:

- Division of the area into management zones according to relief, soil type, degradation level, use and crop yield;
- Indication of crops and cover crops according to agricultural and climatic suitability of the area;
- Recommendation of soil and water management and soil conservation practices.

The division of the rural landscape into management zones enables a more efficient management of the soil, the use of inputs, labor and equipment, and the development of more sustainable production systems for large and small farmers. This makes it possible to use the soil with the most appropriate management system for the different positions of the rural landscape, either in Precision Agriculture or Agroecological Systems.

5.5 Zero Tillage System and Integrated Crop-Livestock-Forest

Among the existing production systems in Brazil, the no-till system is one of the most important for erosion control. This system began to involve not only the lack of tillage, but also crop rotation, permanent soil cover and traffic control [13]. This system is evolving in Brazil. It already occupies more than 50% of the area with annual crop production over 30 million ha.

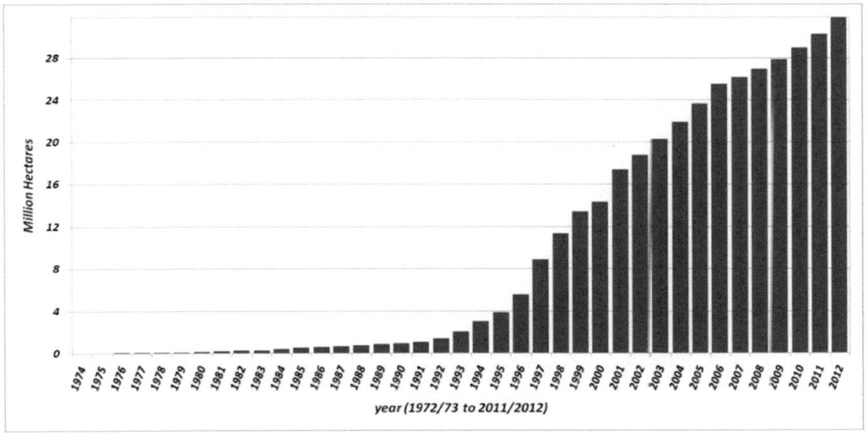

Figure 7: Evolution of planting area with no-till system in Brazil
(Based on *https://febrapdp.org.br/download/area-PD-Brasil-e-estados.pdf*)

The evolution of the zero tillage system combines annual crop, livestock and forest activities in different ways. Integrated crop-livestock-forest management systems are shown in Figure 8.

Figure 8: Integrated crop-livestock-forest systems

This system occupies over 11.5 million ha, in different parts of the country (Figure 9) [13]. Currently, most deforested areas in Brazil are being used as pasture (about 173 million ha). Only 10% of these areas adopt less impacting pastoral systems, such as fallow period, rotations and integrated crop-livestock-forest. Degraded pasture recovery is part of the voluntary commitments of Brazil in the 2009 United Nations Climate Change Conference in Copenhagen (COP15). These commitments were ratified in the National Policy on Climate Change, in which the Plano Setorial de Mitigação e de Adaptação às Mudanças Climáticas para a Consolidação de uma Economia de Baixa Emissão de Carbono na Agricultura (Sector Plan of Mitigation and Adaptation to Climate Changes for the Consolidation of a Low Carbon Emission Economy in Agriculture) (ABC Plan) was established for agriculture. Its targets include the recovery of 15 million ha of degraded pastures by 2020. This scenario makes the recovery of degraded pastures for sustainable agricultural production one of the greatest opportunities to increase national agricultural production with no need for converting more natural vegetation areas to the advancing agricultural frontier. Also, in addition to generating income, these degraded lands, when properly recovered, will also provide ecosystem services, such as erosion control, regulation of groundwater recharge, increase of carbon stock on soil, and, consequently, mitigation of the effects of greenhouse gas emissions [16].

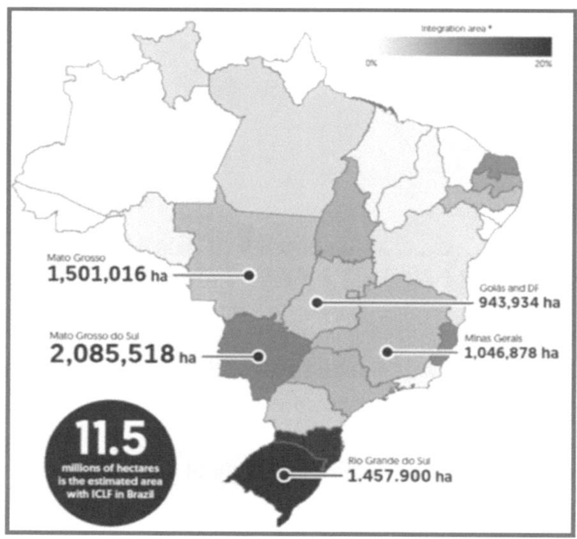

Figure 9: Areas with Integrated crop-livestock-forest systems in Brazil [13].

5.6 Agroecological Systems

Agroecological production systems can also be considered as examples for erosion control. These systems employ greater rotation and intercropping of crops and animal husbandry, recycling organic waste and stimulation for biological nitrogen fixation with varied combinations and agroforestry models (Figure 10). The important point is that the soil is always kept covered, either with plants or with mulch. These systems still occupy a small percentage of the Brazilian territory, mainly small farmers, but the demand for agroecological products has grown a lot. It is a very interesting option to make the agroecological transition of conventional production systems. In addition to helping to reduce erosion, these systems also enable the production of pesticide-free food.

Figure 10: Agroecological systems in southeastern Brazil (Embrapa Agrobiology)

The area with organic production in the country is growing, but still occupies a small part of the territory (approximately 750,000 ha). According to the Coordination of Agroecology (Coagre) of the Secretariat of Development and Agricultural Cooperatives (SDC), linked to the Ministry of Agriculture, Livestock and Supply (Mapa), this type of cultivation in the field is already found in 22.5% of Brazilians counties, and the outlook is even higher for the next years. This indicates that there was an increase from 6.7 million units in 2013 to approximately 15.7 in 2016, more than double the growth in three years. The Southeast is the region with the largest area of organic production, totaling 333,000 ha with 2,729 records of producers in the National Commission of Agroecology and Organic Production (CNAPO), also linked to the Mapa. It is followed by the North (158,000 ha), Northeast (118,400 ha), Midwest (101,800 ha) and South (37,600 ha) regions [17].

5.7 Integrated of Mechanical, Edaphic and Vegetative Practices
For areas with high degradation (mining areas, gullies, hills, cutting slopes, and landfills) and advanced erosion processes, the use of the integration of mechanical, edaphic and vegetative practices along with the use of organic residues for use as mulch or as organic fertilizer has been shown to be efficient (Figure 11).

Figure 11: Integration of practices: slope re-conformation, terraces and water basins, planting of nitrogen-fixing tree seedlings, mulch application [18].

Mechanical practices aim to reduce the speed and quantity of runoff and to promote water infiltration and sediment retention. These include terraces, basins, level cultivation, subsoiling of compacted areas, etc. Soil practices related to fertility management with proper application of organic and mineral fertilizers, soil improvers and conditioners promote greater availability of soil, nutrients and water for vegetation in periods of water stress. Vegetation practices correspond to plant selection and management for production purposes, soil protection, biological nitrogen fixation, organic matter supply, nutrient cycling, biological decompression and soil structure. This set of practices promotes soil structure preservation and/or improvement, water and organic matter infiltration increase, nutrient cycling, and maintenance of soil coverage, which increases its resistance against erosion. No-tillage systems, agroforestry systems, and integrated crop-livestock-forest are good examples of systems that contribute to soil conservation.

To prevent further degradation of soil and water resources in Brazilian agricultural areas, a set of methodologies is available to analyze potentials and land limitations for agricultural production at different geographical scales, such as: the Soil Analysis Methods Manual, the System Land Classification System, the Land Crop Aptitude Assessment System, the Brazilian Land Classification System for Irrigation, the Agroecological Zoning, the Conservation Planning of the Agricultural Land and the Integration Participate in Knowledge on Soil Quality Indicators, among others [16].

6 Conclusions

The following conclusions can be drawn from the presentation of state-of-the-art in Brazil above.

- To further reduce soil and water losses due to erosion in Brazil, it is necessary to expand the use of more sustainable agricultural production systems considering the different limitations and potentialities of the soil and the socioeconomic and cultural conditions of the country.
- The mapping of Brazilian soils in more detailed scales will allow to plan even more the use of these areas and to avoid still further the erosion advance.
- Increasing the use of technologies for more efficient land use planning can prevent major economic and environmental damage caused to the country by erosion and promote increased agricultural production on sustainable bases and other ecosystem services.
- The empowerment of society for sustainable soil and water management must be increased.
- Brazil is updating its National Soil and Water Conservation Program. It is expected to increase the use of conservation practices through public policies that meet the different needs of the country's regions.
- The replacement of degrading practices with sustainable practices should be increasingly encouraged through public policies that encourage conservationist farmers.
- The integration of mechanical, soil and vegetative practices together with waste utilization has proved to be efficient to control erosion even in areas with a high degree of degradation. Research must continue to mitigate the negative impacts caused by erosion in different regions of the country.
- Reinsertion of degraded lands into the sustainable agricultural production system can promote significant increases in national agricultural production and other ecosystem services and prevent severe erosion damage, avoiding the deforestation of new areas for agricultural production.
- The agroecological transition of conventional production systems should be encouraged as a way to prevent soil and water degradation and to increase the production of organic foods, which have higher market value.

If one has so many technologies for erosion control and the development of sustainable agricultural production systems, why is erosion still a threat to the survival of humanity?

7 Acknowledgements

Authors thank the Exceed Swindon project and DAAD for promoting the workshop on *"Water on Agricultural Practices: Training the Trainers"* from 15 – 21 September 2019, as well taking charge of participation of the authors in the workshop. The authors also thank Embrapa Solos for the support during the event and PUC-Rio, especially to José Araruna.

8 References

[1]   Food and Agriculture Organization of the United Nations (FAO and ITPS): Status of the World´s Soil Resources (SWSR) – Technical Summary. Food and Agriculture Organization of the United Nations and Intergovernmental Technical Panel on Soils, Italy, Rome, 2015, 77 p.

[2]   Tribunal de Contas da União (TCU): Relatório de auditoria: Governança de solos em áreas não urbanas. TC 011.713/2015. Brasília, DF, Brazil. 2015, 36 p.

[3]   Bai, Z.G., Dent, D.L., Olsson, L., Shaepman, E.: Proxy global assessment of land degradation. Soil Use and Management, 2008, 24(3), 223 – 234,

[4]   Feng, X.M., Wang, Y.F., Chen, L.D., Fu, B.J., Bai, G.S.: Modeling soil erosion and its response to land-use change in hilly catchments of the Chinese Loess Plateau. Geomorphology, 2010, 118, 239–248.

[5]   Zhao, W.W., Fu, B.J., Chen, L.D.: A comparison between soil loss evaluation index and the C-factor of RUSLE: a case study in the Loess Plateau of China. Hydrology and Earth System Science, 2012, 16, 2739–2748. FAO & ITPS. Soil erosion: the greatest challenge to sustainable soil management. Rome, 2019 100 p.

[6]   ANA - Agência Nacional de Águas (Brasil): Os efeitos das mudanças climáticas sobre os recursos hídricos:desafios para a gestão / Agência Nacional de Águas. – Brasília: ANA, 2010. http://www.ana.gov.br.

[7]   OECD: Governança dos Recursos Hídricos no Brasil. OCDE - Organização Para a Cooperação e Desenvolvimento Econômico. OECD Publishing, Paris, 2015. http://dx.doi.org/10.1787/9789264238169-pt.

[8]   World Bank Group: High and Dry: Climate Change, Water, and the Economy. World Bank, Washington, D.C., 2016. Available at: https://openknowledge.worldbank.org/handle/10986/23665. License: CC BY 3.0 IGO. Consulted in 06/12/2016

[9]   PBMC - Painel Brasileiro de Mudanças Climáticas: Impactos, vulnerabilidades e adaptação: contribuição do grupo de trabalho 2 ao primeiro relatório de avaliação nacional do Painel Brasileiro de Mudanças Climáticas, Brasília. 2013.

[10]   Anonymus: Instituto Brasileiro do Meio Ambiente e dos Recursos Naturais Renováveis. Atlas das áreas susceptíveis à desertificação do Brasil / MMA, Secretaria de Recursos Hídricos, Universidade Federal da Paraíba; Marcos Oliveira Santana, organizador. MMA, Brasília, 2007, 134 p., ISBN 85-7745-048-X.

[11]   Manzatto, C.V., Freitas Jr., E., Peres, J.R.. (ed.): – Rio de Janeiro: Embrapa Solos, 2002, 174 p.

[12]   Polidoro, J.C. et al. (24 authors): Programa Nacional de Solos do Brasil (PronaSolos). Rio de Janeiro: Embrapa Solos, 2016. 53 p. (Documentos / Embrapa Solos, 183).

[13]   Michellon, E., Reydon, B. P., Chicati, M.L.: Impacto econômico do manejo de solo e água em microbacias hidrográficas paranaenses. Agropecuária Técnica, 2014, 35(1), 54-61, Available Online: ISSN 0100-7467http://periodicos.ufpb.br/ojs/index.php/at/index

[14]   Andrade, A.G. de, Freitas, P.: Prevention of advancing degradation and recovery of degraded lands. In: Life on land: Contributions of Embrapa / Gisele Freitas Vilela ... [et al.], technical editors; tranlated by Paulo de Holanda Morais. – Brasília, DF : Embrapa, 2019. PDF (120 p.): il. color. – (Sustainable development goal / [Valéria Sucena Hammes, André Carlos Cau dos Santos] ; 15) Translated from: Vida terrestre: contribuições da Embrapa 1st edition. 2018. ISBN 978-85-7035-919-3

[15]   Washner, S.: Produção orgânica mais que dobra em três anos no Brasil. 2017. In: http://www.organicsnet.com.br/2017/01/producao-organica-mais-que-dobra-em-tres-anos-no-brasil.

WATER EROSION IN MOROCCO

<u>Abdelmalek Dahchour</u>[1], Souad El Hajjaji[2]

[1]*Hassan II Agronomy and Veterinary Institute, BP 6202 Rabat-Instituts, Morocco;*
abdelmalekdahchour@gmail.com

[2] *LS3MN2E- CERNE2D, Mohamed V University, Faculty of Science, Av Ibn Batouta, Rabat-Agdal, Morocco, <u>hajjajisouad@yahoo.fr</u>*

Keywords: water, aeolian, erosion, soil, Rif Mountains, Morocco

Abstract

Erosion phenomena could be due to natural phenomena such as heavy rainfall, wind, climate change, or results of human activities. Erosion by water is the dominant process in 55% of the affected area estimated to be between 11.4 and 32.5 million km^2 worldwide. They are non-renewable on a short time scale and very expensive to reclaim. The specific environmental conditions prevailing in Mediterranean Region are in favor of erosion because of low annual precipitation, high evapotranspiration, intense rainstorms, drought occurrence and steep slopes. Wind erosion is a common cause of land degradation in the arid and semi-arid grazing lands. Soil erosion could also be subsequent to chemical and physical processes occurring in the soil. In Morocco, soil erosion leads to a decline in reservoir capacity of dams by 0.5% per year. The impact of erosion varies according to climate zone and it impacts mainly mountains of Rif, where the degradation exceeds 2000 t/km^2.yr. Various studies have been conducted to assess the current and future soil erosion risk in vulnerable zones using different models integrating environmental parameters and soil properties. Others have focused on human activities and their role in the erosion of soil. Little has been dedicated to erosion by wind and coastal erosion, which was the main goal of this study.

1 Introduction

Erosion phenomena stand for the removal of matrial of soil due to natural phenomena such as heavy rainfall, wind, climate change, or results of human activities. This leads to deposition of sediments in areas with reduced slope and reduced velocity of water. Loss of top soil and its redistribution changes the potential of soil to act as source of nutrients for the plants and causes shift to less desirable plants such as shrub species, in plus to breaking soil structure.The main factors that govern erosion include the power of running water on soil surface, slope of the land and climate. Erosion rate depends on the type of soil, sandy and clayey soils being less erodible. The ability of any soil to be eroded could be assessed by measuring surface and agregate stability, infiltration, compaction and organic matter content. Erosion by water is the dominant process in 55% of the affected area worldwide. The size of the affected surface is estimated to be between 11.4 and 32.5 million km^2 worldwide. In Africa, 65% of the agricultural area, 31% of the permanent pastures and 19% of the forests and woodlands suffer from soil degradation. They are non-renewable on a short time scale and very expensive to reclaim once that are subject to soil degradation. The specific environmental conditions prevailing in Mediterranean Region are in favor of erosion because of low annual precipitation, high evapotranspiration, intense rain storms, drought occurrence and steep slopes[1-3]. In Morocco, soil erosion leads to a decline in reservoir capacity by 0.5% per year.

Wind erosion is a common cause of land degradation in the arid and semi-arid grazing lands. Serious impact of wind erosion occurs, when strong winds blow over light-textured soils that have been heavily grazed during periods of drought. Due to the loss the plant nutrients from the top soil, fertility is reduced as well as reduction of soils capacity to support productive pastures and to sustain biodiversity.

Soil erosion could be subsequent to chemical and physical processes occuring in the soil. Soil is subject to a series of human-induced degradation processes including displacement of soil material and internal soil deterioration [4], salinization, compaction, reduction of organic matter and non-point source pollution[5]. Globally, it has been estimated that nearly 2 billion hectares of land are affected by human-induced soil degradation [6].

In MENA Region, more than the half of the total area recieves less than 150mm anual rainfall leading to reduced or absence of plant cover. The situation is made worse by high land use pressure from human and animals. Wind erosion is the most common environmental problem in the region and accounts for approximately 60%(135 million ha) of soil degradation [7].Therefore, the aim of this study was to investigate the significance of soil erosion through water in Morocco.

2 Erosion in Morocco

In Morocco, soil erosion leads to a decline in reservoir capacity by 0.5% per year that threatens the availability of irrigation water and, thus, food security [8].Total land area is estimated to 71.085 million ha, including 5.8 million ha of forest, 9.2 million ha of agricultural land (13%), and 46 million ha of pastures. It counts 4,500 plant species with 537 endemic plants[9]. Soil vulnerability to degradation processes is cosely linked to climate context in Morocco. Close focus on the distribution of soils according to the main climatic zones reveals that 93% of soils is subject to aridity as it is shown in Table 1.

Erosion of soils impacts mainly mountains of Rif, where the degradation exceeds 2000 t/km^2.yr. This is accompanied by siltation that has been estimated at 125 million m^3 [10] coresponding to a reduction of storage capacity of 0,5% per year [11]. In some cases, it has achieved 100% such as for the dams, e.g., «Ali Thelhat Dam», which is completely silted after having a storage capacity of 30 million m^3 in the past [12].

Table 1 : Distribution of soils in different climate zones[10]

Climate zone	Surface area (ha)	% of total surface area	Precipitation (mm)
Desert and arid	56,000,000	78	< 250
Semi-arid	10,000,000	15	250-500
Subhumide to humide	100,000	7	> 500

2.1 Water erosion

The water erosion is very intense with a specific degradation exceeding 2000 t/km^2.yr in the Rif catchment to the north of Morocco. It varies between 1000 and 2000 t/km^2.yr in the pre-Rif, and 500 and 1000 t/km^2.yr in the middle and high Atlas, while less than 500 t/km^2.yr in the remaining regions.The main catchement are presented in Figure 1 [13].

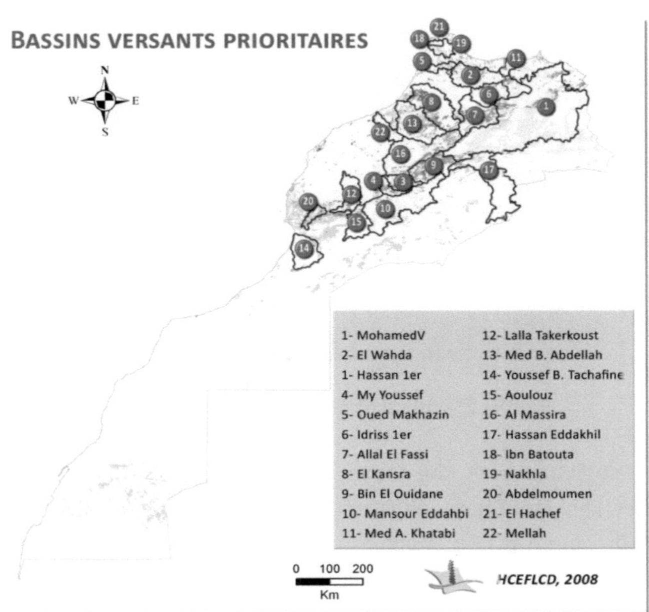

Figure 1: The main catchements in Morocco [13]

More than 2 million ha of agricultural lands are water eroded with an average soil degradation varying from 2.1 to 20 t/ha.yr and higher in northern and north-western basins. In 1988, an estimated 700 million m^3 storage capacity was lost. The actual annual siltation amounts to 50 million m^3 and is expected to reach 150 million m^3 in 2030 [14]. This represents a loss of 2% in storage capacity per year, compared with 0.5-1% per year in the Mediterranean circumference. Most of studies indicated that the Rif region is the most affected one by erosion with rates sometimes exceeding 60t/ha.yr. Attempts of mapping the risk of erosion were conducted in the Rif area, central Atlas and Smir Dam, using the USLE (Universal Soil Loss Equation) model of Wischmeier and Smith, geomatics approach based on the processing of satellite images and interpretation of spectral indices, such as the Form Index (FI), the Coloration Index (CI), the Brightness Index(BI), and the Normalized Difference Vegetation Index (NDVI) and GIS to model and to assess potential erosion and the amount of sediment carried out by either runoff or through rivers [15-17].

Other studies in semi-arid upper and middle Draa catchment using the physically-based, distributed soil erosion model, PESERA (Pan European Soil Erosion Risk Assessment) were conducted in order to assess the current and future soil erosion risk for five periods between 1980 and 2050. Climate change scenarios were approached with the regional climate model REMO, together with the scenarios of socio-economic change (Figure 2).

Figure 2: A view of eroded soil in the Draa Valley

In the region of the Draa valley, the mean erosion rate was estimated at 19.2 t/ha.yr. Erosion culminates in the high mountain zones, more precisely in the western (Tizi-n-Tichka), central (Skoura Mole) and eastern (M'Goun chain) part of the Central High Atlas. This is aggravated by rainfall reduction and higher temperatures, leading to a decrease in vegetation cover. Intense precipitation events would cause an increase in soil erosion up to 31%. This was aggravated by the behaviour of local population forced to satisfy their energy demand by enhanced extraction of firewood that increases the erosion rate of 27% [8].

The GIS can manage in a rational way a multitude of data on various factors of land degradation, which allowed to conclude that the essential factors in the Khmiss watershed are slope and consists of erodibile soil and vegetation cover. This has led to the first map of erosion risk in the Khmiss watershed. It shows that the basin loses 36 t/ha.yr in average, which is explained by moderately high erosion. This loss is a combination of factors including high soil erodibility (83% soil shows a K factor between 0.47 and 0.67) and degraded vegetation cover (76.9% of the total area are occupied by agricultural activities) (Figure 3).

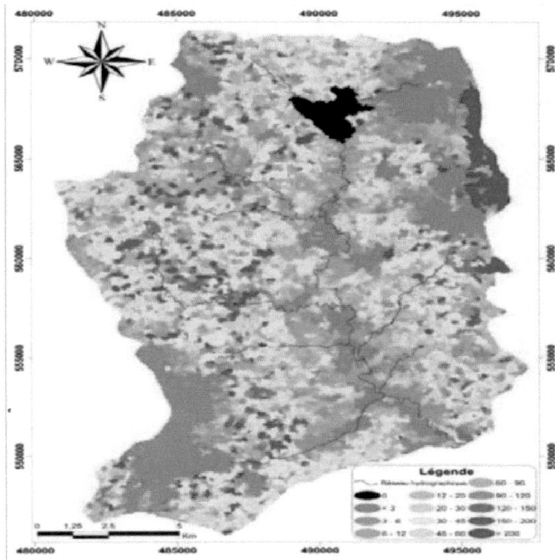

Figure 3: Erosion risk map of Wadi Khmiss watershed

Another approach using GIS combined with empiric models allowed mapping and modeling water erosion risk at the Smir Dam watershed and identifying areas, which require the highest priority interventions for the protection of soils and, equally important, the reduction of sediment transportation at the Dam closures. The resulting map was obtained by overlying different thematic maps, representing the calculated results of the five USLE factors and showing an average rate of 45.45 t/ha.yr.

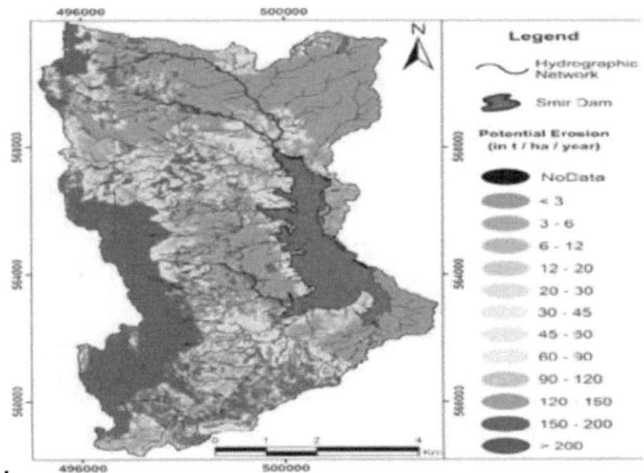

Figure 4: Erosion hazard map in t/ha.yr in the Smir Dam watershed

Studies on siltation have been also reported [18]. The risk of erosion was also under focus by some workers [13], who used the Universal Soil Loss Equation (USLE) coupled with GIS. The model integrates five erosion factors: the erosivity of the rain, the erodibility of the soil, the inclination and the slope length, the vegetation cover and the anti-erosion practices applied in the area of Moulay Bouchta watershed (7,889 ha), located in the western part of the Rif Mountains. The model revealed an average erosion rate of 39.5 t/ha.yr, 87% of the basin has an erosion rate above the tolerance threshold for soil loss (7 t/ha.yr). Soil losses per subbasin range from 16.2 to 81.4 t/ha.yr. The amount of eroded soil is estimated at 311,591 t/yr, corresponding to a specific degradation of 12.1 t/ha.yr. In the absence of any erosion control, 25% of the soil losses would reach the new dam located upstream of the basin outlet, reducing its water mobilization capacity to 59,625 m^3/yr.

The impact of climate change was not neglected. Thus, Choukri et al. [19]have reported a suty on the of climate change on erosivity by raifalls in Rif Mountain.The application of this model shows a weak evolution of erosivity on an annual time scale, but a very strong evolution of the latter depending on seasons. R factor was reduced during winter and spring, and a pronounced increase during summer and autumn was noticed. This discernable change of the seasonality of rainfall erosivity is very useful for adjusting the evolution of agricultural practices and for selecting appropriate soil protection measures.

2.2 Wind erosion
Wind erosion constitutes a non-negligible part of soil degradation. According to Abahusain et al. [7], a total of 600,000 ha would be affected by wind erosion, compared with other counries in MENA region such as Saoudi Arabie (50 million ha), Algeria (12 million ha), and Sudan (71 million ha). Morocco is affected by wind erosion at a medium range. This situation is aggravated further by high land-use pressure both from human and animals, which ultimately causes severe topsoil disturbance. The resulting wind erosion is the most common environmental problem in the region and accounts for approximately 60% (135 million ha) of soil degradation. Wind erosion has resulted in detrimental effects on land quality by removing the fertile top soils. In addition, the accumulation of eroded materials in irrigation canals, agricultural fields (sand encroachment) and water harvesting points affects the cropped areas in the region severely [7].

2.3 Overgrazing
Overgrazing is the the dominant cause of soil degradation, accounting for 49% of the impacted soils, followed by agricultural activities (24%), deforestation (14%), and overexploitation of vegetative cover (13%). All these types of degradation cause a decline in the productive capacity of the land and reduce potential yields [20]. The need for tillage and the usage of heavy machinery and agricultural practices has been questioned in the last decades, partly because of the excessive erosion from tilled farmlands [21].

2.4 Other factors

Other factors are contributing to the degradation including over exploitation of the water table, water pollution resulting from untreated domestic and industrial waste waters, and salinity. Approximately 500,000 ha located mainly in the coastal areas are threatened by salinity. The oases are also threatened by salinity and sanding. 35% of the Tafilalet palm grove soils were salty (4-6 g/L) and 18% very salty (> 16 g/L) [14].

To cope with this issue, one promising solution was proposed by the joint FAO/IAEA division that used two isotopic techniques: fallout radionuclides (FRNs) and compound specific stable isotopes (CSSIs) expressed as FRNs-CSSIs to pinpoint erosion sources in the Tetouan watershed area. Non-tillage practice and planting cereal crops together with fruit trees and shrubs such as "Atriplex" that have root systems known to hold soil were suggested as the best solution. As a result, the Tetouan area saw its soil losses reduced from 54 t/ha.yr to 32 t/ha.yr and Oued Mellah watershed from 10 t/ha.yr to 3.5 t/ha.yr [22].

3 Estimated costs of degraded lands

Attempt to assess the cost of degradation agricultural land was made on the basis of lost of agricultural production in the case land that are mainly planted with cereals [23]. Degradation is expressed by a temporary or permanent decrease in land productivity due to human activity. Four classes of land degradation are identified (Table 2).

Table 2: Classes of degradations of land [23]

Degree of degradation	Effects
Slight	Low reduction in productivity
Moderate	Considerable reduction in productivity
Strong	Biological functions of soil are considerably destroyec, no potential for rehabilitation and use
Extreme	Biological functions of soil are considerably destroyed; not recoverable

Combination with five range intervals of degradation, FAO obtains four ranks of degradation severity (slight, moderate, severe, and very severe). Application to the case of Morocco leads to the following scenarios:

- 10 to 25 percent of the land is severely degraded;
- 25 to 50 percent of the land is moderately degraded;
- 50 to 100 percent of the land is lightly degraded.

The costs of degradation ranged from 84 to 168 million USD, averaging to 126 USD (or 0.36% of Moroccan GDP).

4 Management actions

Nowadys, the total costs of land degradation in Morocco amount to around 0.7 billion USD, which is 1.7% of the GDP. According to the High Commissioner for Water and Forests, 93% of Morocco's dry-weather areas (arid to sub-humid) are vulnerable to desertification due to overexploitation of natural resources. Annual loss is estimated at 30,000 ha of forest land. Widespread damage is the cause in mountainous regions, while in the plains, the most frequent problems are siltation of canals, rivers and ports, and flooding of major riverbeds [24].

Chronologically, the oldest project dealing with desertification problems dates back to the 1960s. It aimed at contributing to the rural development of Western Rif and controlling erosion risks threatening this region. Actions undertaken within the framework of this project included plantation of fruit trees, land development projects, herd management development, construction of earth roads, rehabilitation of springs and erosion control.

Nowadays Morocco has launched a national strategy to limit desertification (LCD) in agreement with the United Nation convention against desertification. It aimed at promoting income actions toward local population , imroving knowledge and information about this issue. The main axis of the actions considers the folowing items:

- Improvement of storage capacity of water. This amounts to 17 BCM in 140 dams,
- Supplying drinking water, which has achieved 94% in rural areas, and
- Generation of energy to cover 100% of rural zones.

These actions are accompanied with prorgams to improve lands in poor regions, forest program aiming at reforestation of 530,000 ha and the restoring 4,000,000 ha. Pastural lands program has identified and improved several areas via the constitution of co-operatives [14].

5 Conclusions

Morocco is under arid climate in favor of intense erosion phenomena. Water erosion is an important phenomenon affecting the degradation of soils in Morocco. This impacts the fertility, the structure of the soils, reduces the infiltration and enhances sediment transportaiton at the dams. The different approaches reported, revealed an effective erosion and have assessed some models to evaluate flow of water and siltation in some vulnerable catchments. The effect of climate change was also focused on the evaluation of erosivity factor, mainly in the Rif area. The data obtained are very usefull to orient decision makers towards the priorities of soil protection. Actions undertaken by governemental authorities are heading toward limitation of the impact of erosion, taking in consideration the improvement of live conditions of local populations.

6 Aknowledgements

The authors would like to thank EXCEED Swindon project and DAAD (German Academic Exchange Service) for support to participate at the *"International Expert Workshop on Water inAgricultural Practices: Training the Trainers"*, 15 – 21 Sep, 2019, Rio de Janeiro, Brazil.

7 References

[1]   G. Verstraeten, J. Poesen, J. de Vente, X. Koninckx: Sediment yield variability in Spain: A quantitative and semiqualitative analysis using reservoir sedimentation rates. Geomorphology 2003, 50, 327–334

[2]   J.C. Gonzalez-Hidalgo, J.L. Pena-Monne, M.de Luis: A review of daily soil erosion in Western Mediterranean areas. Catena 2007, 71, 193–199

[3]   J. de Vente, R. Verduyn, G. Verstraeten, M. Vanmaercke, J. Poesen: Factors controlling sediment yield at the catchment scale in NW Mediterranean geoecosystems. Journal of Soils and Sediments 2011, 11, 690–707

[4]   R.S. Dwivedi: Spatio-temporal characterization of soil degradation. Tropical Ecology 2002, 43(1), 75-90.

[5]   V.A. Kavvadias: Soil degradation. Soil Science Institute of Athens-National Agricultural Research Foundation, http://www.prosodol.gr/sites/prosodol.gr

[6]   UN: UN Secretary General's report A/504/2000 Chapter C. "Defending the Soil" (2000)

[7]   A.A. Abahussain, A.Sh. Abdu, W.K. Al-Zubari, N.A. El-Deen, M. Abdul Raheem: Desertification in the Arab region: Analysis of current status and trends. Journal of Arid Environment 2002, 51, 521–545

[8]   A. Klose: Soil characteristics and soil erosion by water in a semi-arid catchment (Wadi Drâa, South Morocco) under the pressure of global change. Rheinische Friedrich-Wilhelms-Universität Bonn, Germany (2009).

[9]   S.M. Elyousfi, O. M'hirit: La conservation et l'utilisati on des ressources génétiques foresti ères marocaines : les régions de provenances. In : Bani-Aameur F, ed. L'arganier et les plantes des zones arides et semi-arides. Colloque international sur les ressources végétales, 1998, 23-25 avril 1998, Agadir, 1998, 11-8

[10]   Evaluation Du Portefeuille De Pays: MAROC (1997 – 2015) Volume I – Rapport D'Evaluation. GEF/ME/C.50/Inf 01 19 mai 2016

[11]   W. Remini, B. Remini B: La sédimentation dans les barrages de l'Afrique du nord. Éditeur: Larhyss Journal, Algérie, 2003, 45-54

[12]   A. Lahlou: La dégradation spécifique des bassins versants et son impact sur l'envasement des barrages, in Recent Developments in the Explanation and Prediction of Erosion and Sediment Yield (Proceedings of the Exeter Symposium, July 1982). IAHS Publ. no. 137, 163-169

[13]   A.Zouagui, M. Sabir, M. Naimi, M. Chikhaoui, M. Benmansour: Modelisation Du Risque D'érosion Hydrique Par L'équation Universelle Des Pertes En Terre Dans Le Rif Occidental: Cas Du Bassin Versant De Moulay Bouchta (Maroc). European Scientific Journal 2018, 14(3), 524-544

[14]   M. Ghannam: La desertification au Maroc – Quelle stratégie de lutte? 2nd FIG Regional Conference Marrakech, Morocco, 2003, December 2-5, 2003.

[15]   Anonymous : La production agricole en climat aléatoire : acquis et possibilités de régulation. Commission de réflexion sur la sécheresse. Rabat: INRA, (1995) p. 31

[16]   L. Khali Issa., A. Raissouni, R. Moussadek, A. El Arrim: Mapping and Assessment of Water Erosion in the Khmiss Watershed Western Rif, Morocco. Current Advances in Environmental Science, 2014, 119-130

[17]   A. Raissouni, L. Khali Issa, K. Ben Hamman Lech-Hab, A. El Arrim: Water Erosion Risk Mapping and Materials Transfer in the Smir Dam Watershed (Northwestern Morocco). Journal of Geography, Environment and Earth Science International 2016, 5(1), 1-17

[18]   W. Nouaim, S. Chakiri, D. Rambourg, I. Karaoui, A. Ettaqy, J. Chao, M. Allouza, B. Razoki, M. Yazidi, F. El Hmidi: Mapping the water erosion risk in the Lakhdar river basin (Central High Atlas, Morocco). Geology, Ecology, and Landscapes, 2018. DOI: 10.1080/24749508.2018.1481655.

[19]   F. Choukri, M. Chikhaoui, M. Naimi, Y. Pepin, D. Raclot: Analyse du fonctionnement hydro-sédimentaire d'un bassin versant du Rif Occidental du Maroc à l'aide du modèle SWAT: cas du bassin versant Tleta . Revue Marocaine des Sciences Agronomiques et Vétérinaires *(in press)*.

[20]   F. Choukri, M. Chikhaoui, M. Naimi, D. Raclot, K. Lafia: Impact Du Changement Climatique Sur L'évolution De L'érosivité Des Pluies Dans Le Rif Occidental (Nord Du Maroc). European Scientific Journal 2016, 12(32), ISSN: 1857 – 7881; (Print) e - ISSN 1857- 7431

[21]   R. Lahmar, A. Ruellan: Dégradation des sols et strategies cooperati ves en Méditerranée: la pression sur les ressources naturelles et les stratégies de développement durable. Cahiers de l'Agriculture, 2007, 16(4), 318-232

[22] IAEA: Erosion in Moroccan Watersheds Can Be Reduced up to 60 Percent Through the Use of Iso topic Techniques; www.iaea.org/newscenter/news,Octobre 2016

[23] FAO: Land Resources Potential and Constraints at Regional and Country Level. World Soil Resources Report 90, Rome (2003)

[24] R. Mrabet: No-Tillage systems for sustainable dryland agriculture in Morocco. INRA Publication, Fanigraph Edition, 2008, p. 153

[25] R. Dahan, M. Boughlala, R. Mrabet, A. Laamari, R. Balaghi, I. Lajouad: A Review of Available Knowledge on Land Degradati on in Morocco OASIS — Combating Dryland Degradation. 2012, Series Oasis country report 2. ICARDA (2012)

HYDROGEOCHEMICAL NITROGENOUS BEHAVIOUR IN AN ANDEAN AGRICULTURAL BASIN

<u>A. Berbesí-Jaimes</u>[1], M. Casamitjana-Causa[2], J.C. Loaiza-Usuga[1], S. Cardona-Gallo[1], G.A. Correa-Londoño[3]

[1]*Universidad Nacional de Colombia, Sede Medellín, Facultad de Minas, Av. 80 # 65 – 223, Bl. M2. Medellín, CP 13114, Colombia; afberbesij@unal.edu.co*

[2]*Corporación Colombiana de Investigación Agropecuaria – AGROSAVIA, Centro de Investigación Tibaitatá – Km 14 Road Mosquera - Bogotá, Cundinamarca, Colombia; PhD student Universidad de Girona, Spain.*

[3]*Universidad Nacional de Colombia, Sede Medellín, Facultad de Ciencias Agrarias, Departamento de Agronomía, Medellín, CP 13114, Colombia*

Keywords: Agricultural basin, andisol, hydrological nitrogenous balance, water dynamics

Abstract

Water dynamics in watersheds are related to the interaction and variability of soil moisture, vegetation, temperature, and precipitation, among other factors. Changes of those elements can be used as control tools to simulate hydrological process and proposals intended for ecosystems conservation. Soil is the key component of these systems especially in reference to mountain wetlands behavior, where most processes associated with water levels are regulated. Anthropogenic activities can alter the constant interaction of these elements and generate physical, chemical and biological imbalances. Agricultural and livestock activities have the greatest influence on the ecosystem alteration. The most important factor associated with these activities is nitrate, polluting ground water and surface water reserves. In Colombia, the use of ammonium nitrogen has been increasing during last years but is not as abundant as in European or North American countries. This research had as results that water regimes mostly constrains soil moisture in forest soil use due to factors such as interception; crop and grazing soil uses have high water retention. Environments and polluting nitrate loads do not become significant to the risk to human health at concentrations less than 50 mg/L NO_3^-. High levels of soil moisture, high rainfall, infiltration ranging from 70 to 90% of total precipitation and low runoff (less than 1%), acidity and high redox potential (typical properties of soils derived from volcanic ashes such as andisols) condition the environments of mobility systems of these pollutants give high resilience to these ecosystems.

1 Introduction

Prevention from water pollution in the world water reservoirs is one of the key points of current environmental policies. To integrate knowledge by means of research is essential to improve continuously those policies. One of the main causes of water deterioration in a catchment is agricultural activities and the use of fertilizers to improve production yield. Nitrogen is the most common fertilizer used in the agriculture in developed countries, characterized by high use of chemical supplies with low technological level. It is estimated that the use of nitrogen fertilizers reached 119,400,000 t/yr in 2018, of which 7% of the implementation is attributed to Latin America. Colombia is one of the main users [1].

Nitrate is a chemical compound with broader interest for its great use in agricultural activities. It is present in mineral deposits, soils, fresh or marine water bodies, atmosphere and even the biota of an ecosystem. Ammonia and ammonium are nitrogen compounds important due to their role in the formative processes of nitrification [2] with important effects on human health. In Colombia, it was found that there was an epidemiological response to the availability of nitrates in wells water, causing stomach cancer [3].

The understanding of hydrological dynamics associated with water movement in basins is important because of allowing prediction and prevention of events that can affect anthropic and natural activities of any region in the world. One of the major drawbacks of understanding water movement focuses on the physicochemical process interaction, correlated with biomass and nutrient production and presence of compounds of health interest. Additionally, these processes are conditioned by factors such as land use (forests − crops − pastures), and the insertion of anthropogenic chemical compounds to the natural systems as agricultural supplies, that modified soil and water quality and in turn affect the health of nearby populations [4].

NO_3 is a compound associated with oxidative processes, and thus water interaction is essential as a medium for its presence. The concentrations of this compound will be defined by the mass quantities and volumes of infiltrated water. Hydrological variables play an important role as formation medium and transport mechanism. These mechanisms are related with precipitation, evaporation changes, percolation and lateral flows. Some researchers are focused on analyzing nitrate surface behavior and the effect on surface and subterranean reservoirs, in conjunction with the dynamics of hydrological components such as precipitation and evapotranspiration. For example, [5] studied the impact of climate change on the transport and accumulation of nitrate in soils, founding that land use types play an important role in the movement and impact of nitrate. [6] investigated forested areas finding significant levels of NO_3^-. This behavior is in accordance with leaching in relation to surface or groundwater sources. On the other hand, growing areas with use of nitrogen fertilizers and intensive livestock are considerable sources of nitrate generation and transport [7, 8].

Soil moisture is a parameter that allows determining accurately the stored water into the soil profile according to timescale, which involves flow dynamics. There are variables such as precipitation and evapotranspiration resulting from the integration of evaporation and transpiration processes, which generate a water phase change from liquid to vapor phase.

To study these processes can help to understand the total water amount available in soil, vegetation and atmosphere. After heavy events of rain or irrigation, soil tends to drain water until reaching its field capacity, or the amount of water in a well-drained soil, when these sources of direct water supplies are scarce. Plants reduce the water content in the soil by the effect of absorption of the roots, reducing the potential energy of water movement [9]. In order to understand precisely the behavior of any contaminant in the water system-ground, it is necessary to analyze at multiple scales the soil moisture dynamics in order to avoid any data masking.

Andisols are characterized by high water mobility associated with high porosity, low bulk density, high content of organic matter and humic acids [10]. Studies under low precipitation conditions showed low nitrate leaching rates, which can take years to generate impacts on underground sources [11]. The effect of high precipitations has not been studied thoroughly. Andisols cover 8.5% of Colombian territory [12]; studies of nitrate mobilization through this soil and interactions have not been addressed.

The aim of this research is to find the relationship between hydrological variables (superficial and sub-superficial flows) with geochemical flows of nitrogenous compounds by means of monitoring and control periodically those flows under different land uses present in the study catchment area: forests, pastures and potato crops under andisols presence.

2 Materials and Methods

Study area
This research was carried out in the Las Palmas basin, municipality of Envigado, Department of Antioquia, Colombia. This basin belongs to the integrated system of Basins of Rio Negro in the upper rural area of Las Palmas basin, with the latitude of 6°9'7.27"N and longitude of 75°31'18.40". The geological unit is made up of ultramafic rocks (dunite and serpentinite), in which there are land uses such as stubble, plantations, transient crops, secondary forests and pastures [13, 14]. The development of hydrological monitoring and measurements on the basin scale were carried out in accordance with methodologies developed by [15-18], collecting runoff, infiltration, and percolation flows on hydrological plots under three land uses (forest, pasture, potatoes cultivation), instrumented at the upper part of the catchment (Figure 1).

Since March 2017, hydrological dynamics were studied in this mountain basin using 6 experimental complete hydrological plots and 6 piezometer plots. Two plots were implemented for each land use (forest, grass, potatoes cultivation) (Figure 2), where the following parameters were measured weekly: precipitation, runoff, soil moisture and percolation at 20 cm and 50 cm depths.

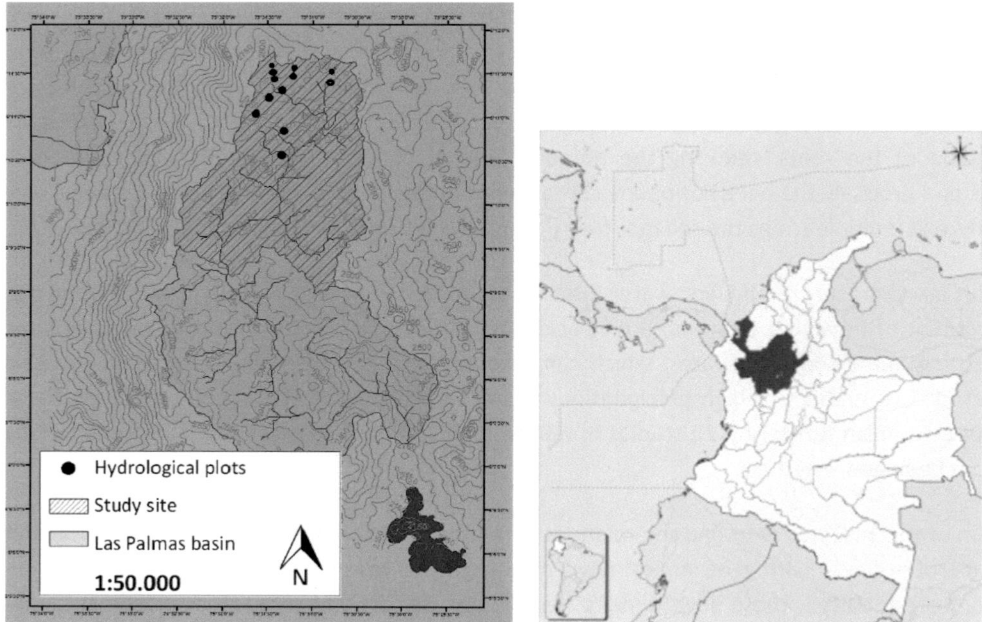

Figure 1: Las Palmas basin, study site, Envigado municipality, Colombia, South America

Figure 2: One of the experimental plots of this study under potatoes land use,
with measurement of rainfall, soil moisture, infiltration, percolation, and runoff

Water table was measured from 1 m to 20 m by means of 6 piezometers installed along the catchment (see Figure 1). The accumulated flows were measured weekly in rainy seasons, and every 15 days during the dry periods along 7 months in 2018. Chemical variables were measured on each collected flow in the Soil Science Laboratory, Universidad Nacional of Colombia (Campus Medellin). The measured variables were nitrate (NO_3^-), nitrite (NO_2^-), ammonium (NH_4^+), ammonia (NH_3), bicarbonate (HCO_3^-), dissolved oxygen (DO), pH, redox potential and electrical conductivity (EC). This hydrological approach has been applied to monitor and to simulate hydrological variables at different scales. In case of hydrogeochemical variables, the methodologies used were ion selective electrode for NH_3 (multiparametric HACH HQ40D), and spectrophotometer for NO_3^-

and NO_2^- (DR 3900 Hach Lange) [19-21]. This research also takes into account elements associated with the nitrification process of production. To analyze the chemical variables associated with nitrification behavior in a chronological path, Biplot graphics were realized to understand, which type of nitrogen compounds predominate in the different sampling times. The analyses were divided in three periods: first rainy period (March to May), second drought period (June to August), and third the rainy season (September).

3 Results and Discussion

Soil moisture and evapotranspiration on hourly scale
Forests soil use has the lowest soil moisture with an annual average value of 75.41 mm/month. This value does not exceed the values for field capacity in these soils (111 mm/month). High precipitation periods in forests results in soil moisture of 89 mm/month, and recharge times ranges from 5 h to 1 d. The soil moisture in dry periods reaches values of 82 mm/month with recharge times between 1 to 2 d maximum, loss water storage quickly to gradually.

Pasture and potato growing areas show different soil moisture behavior. For the moistest month (May), the soil moisture presents an average of 114 mm/month with recharge times ranging from 15 h to 3 d. In the driest month (July), the average soil moisture is 145 mm and the recharge times up to 20 h. These values are not representative, because the refills are minimal compared with the total water inside of the system in May. The areas under potato cultivation register the highest soil moisture in the study site during the months with high precipitation, having soil moistures of 173 mm/month with recharge times ranging from 15 h to 2 d. The soil moisture in low precipitation periods amounts to 132 mm/month; recharge times take up to 4 d. Under potato, the soils have high moisture stability and water retention with a trend to retain more water than the other soil uses. This soil behavior in relation to soil water recharge and retention is associated with soil porosity and infiltration changes due to alteration during cultivation and ploughing. However, in spite of high water retention (50% more than in forests) in culture soils in comparison with forest, potato regulatory functions are scarce, because water remains a short time in the systems and interception through flow does not exist.

Analyses of hydrological variables at annual scale
In the study catchment, the hydrological behavior is bimodal with two rainy and two drought periods. The rainy periods in Colombia are March to May and September to November; the drought periods are December to February and June to August. The high precipitation in the basin in the studied period is 319 mm/month for the rainy season and 37 mm/month during dry period.

The changes in the soil moisture are an important parameter to understand the water flow dynamics in high mountain systems, especially in relation with mountain wetlands behavior. The water balance components bring information about parameters like behavior in relation to rainfall intensity and volume. The response of soil moisture in relation to soil properties plays important role in Andean volcanic ashes soils. Figure 3 and Table 1 show the behavior of the different variables associated with water balance in different soil uses related to rainy and dry periods.

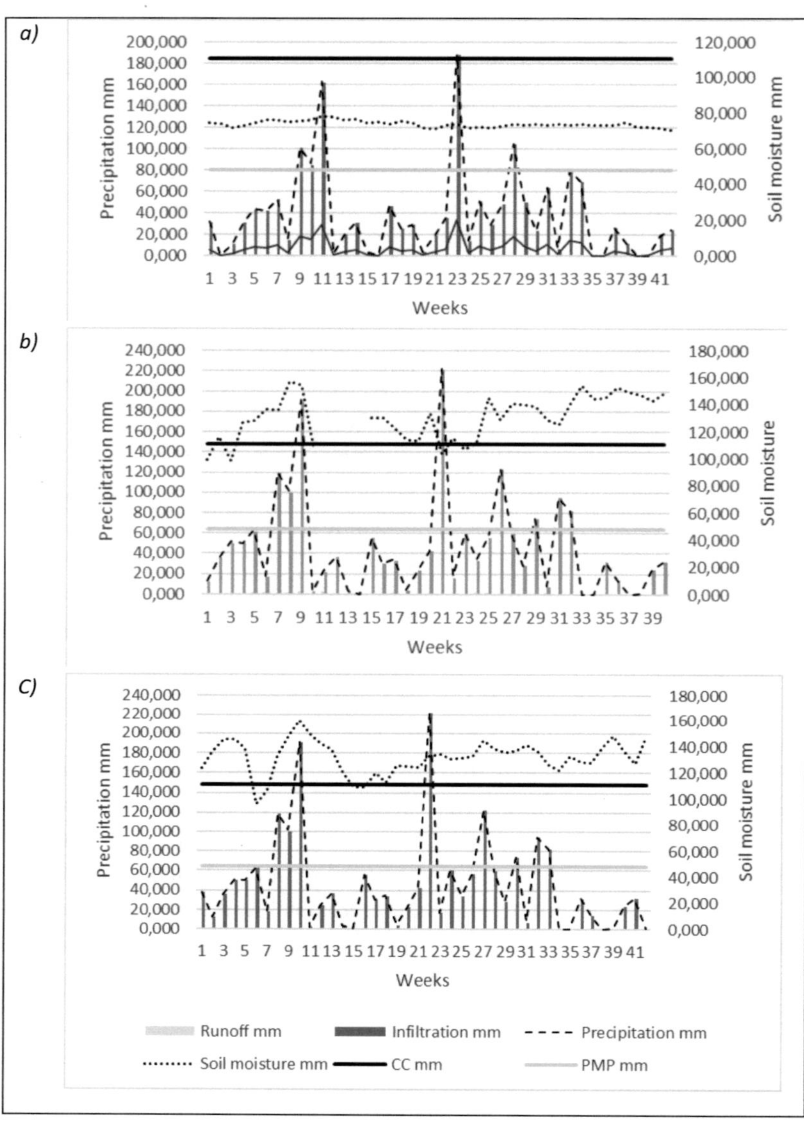

Figure 3: Water balances during the studied period under (a) forest, (b) pasture, (c) potatoes.

Table 1: Water balance in different soil uses on Las Palmas basin

Soil use	Precipitation (mm)	Infiltration (mm)	Interception (mm)	Runoff (mm)	LF 0.2 (mm)	LF 0.5 (mm)	SW (mm/mm)
Forest	1,876	1,580 (90.1%)	289 (21.1%)	7 (0.37%)	333	110	74.5
Pasture	1,862	1,838 (78.5%)	-	34 (1.8%)	2,522	967	132
Potatoes	1,905	947 (99.5%)	-	9 (0.62%)	580	486	131

LF: lateral flow at 0.2 m depth and 0.5 m depth, SW: soil moisture

Forested areas have the lowest runoff values (3.7% of the total precipitation). These values can be due to water interception by the canopy, which reduces precipitation inside the forest. The presence of litter favors high soil porosity, infiltration capacity, and increased subsurface flow. Runoff values in pastures and potatoes are less than 18% and 6%, respectively. This response is mainly conditioned by soil structure, texture and hydraulic conductivity. The erosion capacity on soils is related to cultural practices especially under anthropic interventions. In the wet periods (March to May), runoff response is related to high soil moisture conditions that conditioned water flow into the soil (saturated porous systems). The dry periods (June to July) register the lowest runoff values, because the rainfall is used to recharge water into the soil, since these soils rarely reach permanent wilting point values and do not develop hydrophobicity problems. Infiltration ranged between 78 to 99% of the total rainfall by event. Forest is the land use with the lowest infiltration values due to rainfall interception, which achieve 21% of the total precipitation. For all studied land uses, the largest infiltrations values are registered in the rainy period and the lowest in the dry period.

Figure 4 shows the concentration of nitrogenous compounds (ammonia, nitrate and nitrite) related to ions. These values are lower than the levels for contamination established by international standards [22]. The values reported in this study are at acceptable levels for water consumption NO_3^- (50 mg/L), NO_2^- (0.2 mg/L) and NH_3 (< 3 mg/L).

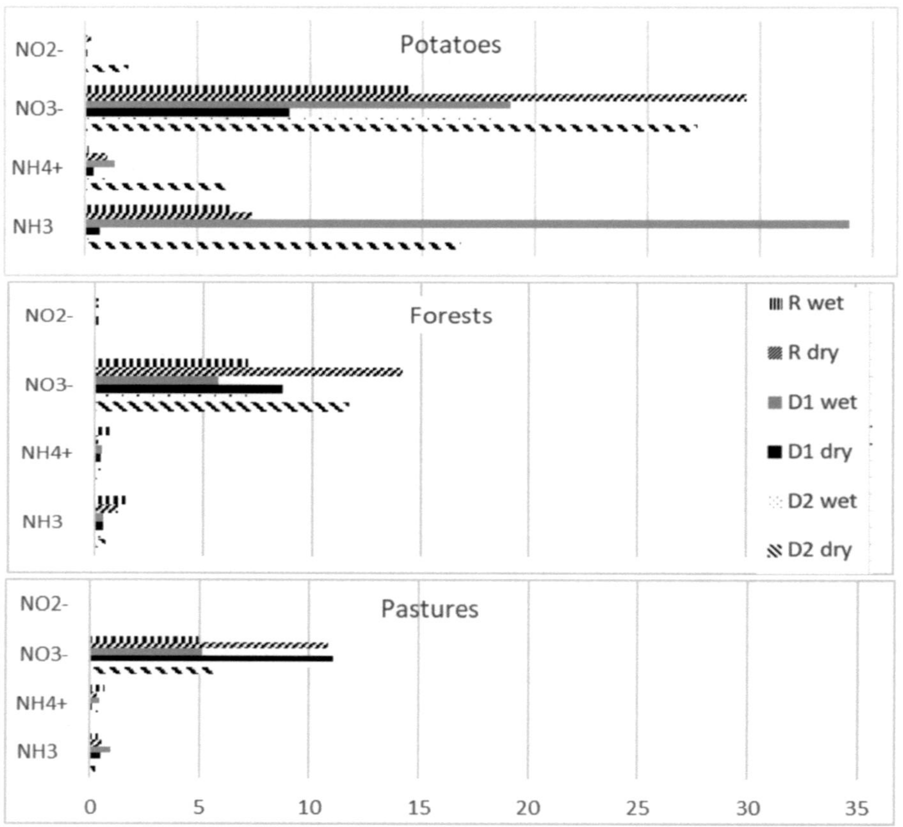

Figure 4: Nitrogenous compounds in the analyzed flows under different land uses, Las Palmas basin. (R: Runoff, D1 Depth 20 cm, D2: Depth 50cm, wet: rainy period, dry: dry period).

Figures 5 shows an example of the statistical analysis realized on the hydrological flow of chemical compounds' data. The highest nitrogenous concentrations are found in runoff and infiltration water (measured by means of the installed lysimeters, located at 0.2 m depth). These values can be derived from nitrification process, which are dominated by the pH 5.5 to 8, typical conditions on oxide-reduction conditions in water and soils. High redox values and positive potential as well as the presence of dissolved oxygen with values higher than 7 mg/L and pH between 5.5 and 6.6 is common in oxide-reduction environments [23]. Runoff shows high values in the wet period, when the concentrations of nitrogenous compounds like ammonium and ammonia are increased. The same tendency is observed for dissolved oxygen and redox potential. During dry period, the prevalence of high oxidation conditions is common that have a high correlation with nitrate and nitrite. In the last rainy period (final phase of this study), these correlations are maintained. On forest ecosystems, runoff water tends to register low presence of ammonium and ammonia.

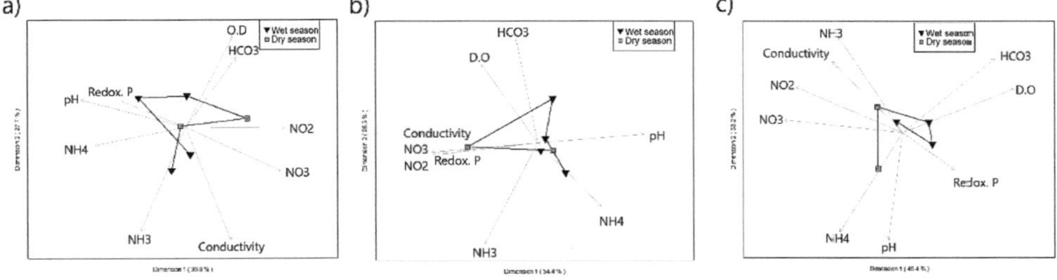

Figure 5: Forest Biplot station; relationship between nitrogen, pH, electric conductivity, bicarbonate, redox potential and dissolved oxygen, in three depths. (a) runoff, (b) depth 0.2 m, and (c) depth 0.5 m.

While chemical analyses of the rain in all land uses, it was observed that the dominant compound was ammonia. Water flow at depths of 20 cm and 50 cm in rainy seasons are more related to high levels of NH_3 and NH_4^+. Drought periods are characterized by having average rainfall values of 74 mm/month with NO_3^- concentrations of 6.1 ± 2.3 mg / L.

Forests
Forests had low moisture retention capacity in the surface soil (20 cm) below the field capacity showing shorter recharge times in rainy periods. During the drought period, soil recharge needs more than two days. There is a soil moisture loss higher than 5% after the transition from rainy to dry period due to interception by the canopy (10-21% of the total precipitation) that decreases the volume of forest internal rainfall (rain that crosses the canopy). On the other hand, the interaction of low effective rainfall and leaf litter in surface soil generates low infiltration rates (78-90%) and runoff (0.37%); the erosive capacity of the rain decreases in forestry systems. Similar conditions in tropical forest areas are described by [24-26]. Water flows had chemical characteristics closely related to soil chemistry: forest soils have acidic conditions (pH 4.6 – 6.2), high redox potential and low electric conductivity. Nevertheless, highest NO_3^- content was reported in runoff flows during wet seasons (see Figure 4). Infiltration and lateral water flows at 0.2 m were 6.2 ± 4.5 mg/L (wet season), and 4.9 ± 3.4 mg NO_3^-/L (dry season), respectively. Percolation D2 flows (0.5 m depth) were 3.9 ± 3 mg/L (wet season) and 11.7 mg/L (dry season), respectively. Some studies [6] reported that forested areas are influencers of nitrate concentrations in surface water. Similar contents of ammonium and nitrate are reported in interception water by [27, 28]. When runoff water had contact with leaf litter layer, drag ions could be formed by the breakdown organic matter and raise its concentrations. The same behavior would be expected in D1 (0.2 m depth) due to changes in soil moisture in this layer and high soil organic matter content associated with different biological processes as nitrification (especially in soils, derived from volcanic ashes with high OM content (21.8 % OM)). High infiltration (1,495 to 1,666 mm/yr), and high lateral flows (315 to 352 mm/yr) were found. High water volume in percolation and infiltration favored decreases in the presence of nitrogenous concentrations in the subsurface flows; this area of the

soil profile has greater movement of both water and chemical loads. Some authors [6] established that forests areas could generate significant impacts on nitrate contamination in surface waters.

Pastures

This land use had good water retention, and the recharge time is smaller than 1 d in wet and in dry season, where the recharge time rises. In pastures, there are gradual losses of soil water content (more than 30%) in most cases without reaching values below field capacity. Soil moisture variability (Figure 3b) in dry seasons conditions decreases and reaches mean values between 120 and 80 mm. Results show how the transitions between wet and dry season can generate changes in soil water capacity from lower to higher volumes until the soil stabilizes his moisture content. These soils can lose > 35% of soil moisture maintaining values higher than field capacity nearly the whole year. This behavior is due to lack of interception, rain erosive action, high infiltration and low runoff.

When the hydrological regime is very variable, andisols does not significantly alter its soil moisture. Depending on environmental chemical characteristics, in case of the existent acidic conditions, water has acidic conditions (pH 5.7 ± 0.7), high redox potential, and low electric conductivity. Highest NO_3^- content in pasture is related to runoff water. In wet seasons, average values of NO_3^- were 2.8 ± 1.6 mg/L, and during dry seasons 5.6 mg/L. Infiltration and lateral flows had NO_3^- values at 0.2 m deep of 3 ± 1.6 mg/L (wet season) and 11.1-8.5 mg/L (dry season), respectively. Waters at 0.5 m depths have NO_3^- values of 2.4 ± 1.1 mg/L (wet season) and 5.8-3.4 mg/L (dry season), respectively. Some researches [7] reported largest nitrate contributions from animal waste in pasture areas due to the processes of deposition of organic matter. In the study site "Las Palmas basin", grazing areas do not have a large volume of livestock (20 ha, 40 head of cattle/ha), which does not result in a significant source of increased nitrate levels. Runoff volume is controlled by rainfall dynamics (volume and intensity). Nitrate concentration has very similar behavior during wet and dry seasons. The same response is observed in subsurface flows obtained at depth 0.2 m. Nitrification process can be associated with soil moisture changes (losses less than 18%) and large amounts of organic matter (more than 20%).

Pasture was the land use with the highest infiltration rates (1,835 to 1,842 mm/yr) and subsurface flow rates (2,464 to 2,580 mm/yr) in the study area. During rainy periods, the high subsurface water volume helped to dissipate ions concentration by means of nutrient loss by percolation and recharge flows, associated with transport processes in underground water. Soils under pastures have soil moisture stability; therefore, it led to maintain a greater ions concentration. This high volume of water and its fast movement diminished the time of base flow response, affecting the catchment recharge phenomena.

Potatoes

This land use has high water retention values with low recharge times from hours to 2 d in some cases in rainy periods, rise to 4 d in dry periods. The soil has moist values near to soil capacity permanently. These results can be seen on the annual water balances (Figure 3), where it is observed a marked variability during the dry period referring to wet periods. Additionally, soil moisture trend is stable during all periods; however, the moist steep increases slightly during wet periods, reflecting few declines on April. The behavior of soil moisture in April is because of this time soils were being prepared for crops (plantation process). Soil surface a teration favored infiltration (near to 99% of total precipitation in some cases) and low runoff (0.4 to 0.6%). This response is due to high porosity, good structure and the hydrological properties of volcanic ash soils [29]. These soils have a large capacity of water storage and maintain high soil moisture values in both periods (dry and wet). Soil water has acidic conditions (pH 5.7 ± 0.7), high redox potential and low electric conductivity associated with soft water. Soils under agricultural use (potato) registered the highest content of NO_3^- in runoff flows (see Figure 3). At average runoff, nitrate amount on potatoes land use during wet periods is 3.5 ± 2.0 mg/L and in dry period 8.3 mg/L, respectively. NO_3^- in infiltration and lateral flows D1 are 5.4 ± 5.1 mg/L during wet periods and 11.1 ± 9.5 mg/L during dry period. At 0.5 m depth (D2), NO_3^- is 3.8 ± 2.8 mg/L in the wet period. In the dry period, NO_3^- was not measured, because there were not enough flow volumes for chemical analyses.

On potatoes land use, there were evidences of use of nitrogen-based products as fertilizers such as urea ($CO(NH_2)_2$), which have this type of ions in high concentration per kg. It was observed that subsurface flows at 0.2 m depth in the soil (D1 Figure 3) have the greatest concentrations, followed by runoff (R Figure 3), and it was found at low concentration at 0.5 m depth (D2). This could be inferred because of biological processes of organic matter transformation and changes in soil porosity, which can increase infiltration associated with porous system, preferential flows and good structure in soils, favoring oxidation processes at the first soil depth. Another direct influence is on the lateral flows (267 to 893 mm/yr), bring with them nutrient trawling. Nevertheless, in this area, the side refills do not have high values of water volume (580 mm/yr) compared with pasture, but is higher than values reported for forest in this study. Some researches [7, 8] reported that crop area systems have considerable nitrate contributions to water bodies.

4 Conclusions

Traditionally in tropical regions, forests are systems considered as directly influenced by the water regimes in watersheds. Soil moisture behavior and variability are subjected to rainfall intensity and volumes. The same trend can be observed for soil moisture in cultivation areas and pastures. Soil properties in volcanic ash soils are a key factor in the soil hydrological behavior; high levels of organic matter, high porosity, high infiltration rates and high soil moisture retention favors water storage even in drought periods. However in wet periods, subsurface flows and infiltration are very efficient in detriment of surface runoff.

Nitrate concentration is controlled by soil characteristics and rainfall behavior. High recharge flows have a marked influence in nitrate levels, which decreases in rainy periods in comparison with drought periods. On pastures, nitrate concentration is affected by low rates of nitrogen fertilization, low input technology, extensive grazing and high trampling.

In some cases, nitrates have high concentrations related to direct fertilizer discharges on potatoes. In all land uses, nitrate concentration is low due to the acidic environment and high water flow conditions, both characteristics are derived from the soil type, Andisols. These conditions favor natural attenuation of pollution, although change to favorable states for these compounds. Under potatoes, this attenuator effect is not enough to generate a significant impact.

In order to understand the dynamics of nitrogenous compounds under different land uses, it is necessary to address water fluxes from multiple scales. One of the main factors is the soil moisture change analysis by time and his relation with the other hydrological variables, such as runoff, precipitation and interception.

In the study site conditions, due to high soil moisture conditions, soil use and type (andisols), ions concentration is low because of the scale. On the study plot scale, the results are useful to implement theoretical models (validation and calibration), and at catchment scale they do not represent values to be considered hazardous for human health. The concept of scale in hydrology and watershed studies is discussed by [17]. The efficient recharge flows, high soil moisture conditions in combination with OM content inside soil favored transport systems of these ions and ions concentration diminished through basin.

5 Acknowledgements

We would like to thank the EXCEED Swindon project and DAAD (German Academic Exchange Service) for their support to participate at the Regional workshop on "Water in Agricultural Practices: Training of trainers" Rio de Janeiro, Brazil, held from 15 to 21 September 2019, likewise thank the Secretary of Environment and Agricultural Development of the municipality of Envigado, Antioquia (Colombia) for all the collaborations and efforts in the development of this research as well as to the Universidad Nacional de Colombia, Facultad de Minas, "Hydrolab" Research Group in Mountain Hydrology and Agrosavia advise.

6 References

[1]   FAO Food and Agriculture Organization of the United Nations. World fertilizer trends and outlook to 2018, Rome, 2015

[2]   Evangelou, V.P.: Environmental soil and water chemistry: principles and applications. John Wiley & Sons, Inc., New York, USA (1998)

[3]   Cuello, C., Correa, P., Haenszel, W., Gordillo, G., Brown, C., Archer, M., Tannenbaum, S.: Gastric cancer in Colombia. I. Cancer risk and suspect environmental agents. Journal of the National Cancer Institute, 1976, 57(5), 1015-1020.

[4]   Weiner, E.R.: Applications of environmental chemistry: a practical guide for environmental professionals. CRC press, Boca Raton, FL, USA (2010).

[5]   Akbariyeh, S., Pena, C.A.G., Wang, T., Mohebbi, A., Bartelt-Hunt, S., Zhang, J., Li, Y.: Prediction of nitrate accumulation and leaching beneath groundwater irrigated corn fields in the Upper Platte basin under a future climate scenario. Science of The Total Environment, 2019, 685, 514-526.

[6]   Sugimoto, R., Tsuboi, T., Fujita, M.S.: Comprehensive and quantitative assessment of nitrate dynamics in two contrasting forested basins along the Sea of Japan using dual isotopes of nitrate. Science of The Total Environment, 2019, 687, 667-678.

[7]   Mayo, A.L., Ritter, D.J., Bruthans, J., Tingey, D.: Contributions of commercial fertilizer, mineralized soil nitrate, and animal and human waste to the nitrate load in the Upper Elbe River Basin, Czech Republic. HydroResearch, 2019, 1, 25-35.

[8]   He, B., He, J., Wang, L., Zhang, X., Bi, E.: Effect of hydrogeological conditions and surface loads on shallow groundwater nitrate pollution in the Shaying River Basin: Based on least squares surface fitting model. Water Research, 2019, 163, 114880.

[9]   Allen, R.G., Pereira, L.S., Raes, D., Smith, M.: Crop evapotranspiration - Guidelines for computing crop water requirements. FAO Irrigation and drainage paper 56. Fao, Rome, 1998, 56.

[10]   Jaramillo, D.F.: Caracterización de la materia orgánica del horizonte superficial de un Andisol hidromórfico del Oriente Antioqueño (Colombia). Revista de la Academia Colombiana de Ciencias Exactas, Físicas y Naturales, 2011, 35(134), 23-34. *(Spanish)*

[11]   Maeda, M., Zhao, B., Ozaki, Y., Yoneyama, T.: Nitrate leaching in an Andisol treated with different types of fertilizers. Environmental Pollution, 2003, 121(3), 477-487.

[12]   Igac, U.: Atlas de la distribución de la propiedad rural en Colombia. Instituto Geográfico Agustín Codazzi, Universidad de los Andes, Bogotá. (2012). *(Spanish)*

[13]   Contraloría Municipal de Envigado: Informe del Estado de los Recursos Naturales y el Ambiente Municipio de Envigado 2015. (2016) *(Spanish)*

[14]   Botero, A.M., Vélez, J.P.: Caracterización Hidrogeológica del Municipio de Envigado. (Tesis pregrado). Universidad Nacional de Colombia, Facultad de Minas, Medellín, Colombia. (2005) *(Spanish)*

[15]   Loaiza-Usuga, J.C., Valentijn, R.N.: Desarrollo de modelos hidrológicos y modelación de procesos superficiales. Caso de estudio para vertientes de alta montaña. Gestión y Ambiente, 2011, 14(3), 23-31 *(Spanish)*

[16]   Zhang, Y., Li, Y., Walker, J.P., Pauwels, V.R., Shahrban, M.: Towards operational hydrological model calibration using streamflow and soil moisture measurements. In 21st International Congress on Modelling and Simulation. Gold Coast, Australia, 2015, 2089-2095.

[17] Sidle, R.C., Gomi, T., Loaiza-Usuga, J.C.L., Jarihani, B.: Hydrogeomorphic processes and scaling issues in the continuum from soil pedons to catchments. Earth-Science Reviews, 2017, 175, 75-96.

[18] Loaiza-Usuga, J.C., Monsalve, G., Pertuz, A., Arce, L., Sanín, M., Ramírez, L.F., Sidle, R.: Desentrañar la dinámica de una pendiente creciente en el noroeste de Colombia: variables hidrológicas y firmas geoeléctricas y sísmicas. Agua, 2018, 10, 1-17 *(Spanish)*

[19] Han, D., Zhou, T.: Soil water movement in the unsaturated zone of an inland arid region: Mulched drip irrigation experiment. Journal of Hydrology, 2018, 559, 13-29.

[20] Matiatos, I., Paraskevopoulou, V., Lazogiannis, K., Botsou, F., Dassenakis, M., Ghionis, G., Poulos, S.E.: Surface–ground water interactions and hydrogeochemical evolution in a fluvio-deltaic setting: The case study of the Pinios River delta. Journal of Hydrology, 2018, 561, 236-249.

[21] Orozco, M., Poch, R.M., Batalla, R.J., Balasch, J.C.: Hydrochemical budget of a Mediterranean mountain basin in relation to land use (The Ribera Salada, Catalan Pre-Pyrenees, NE Spain). Zeitschrift für Geomorphologie, 2006, 50(1), 77-94.

[22] WHO: Guías para la calidad del agua potable. Primer Apéndice a la Tercera Edición. World Health Organization. (2006). *1. (Spanish).*

[23] Jenkins, D., Snoeyink, V.L.: *Química del agua*. Limusa, (2004).

[24] Loaiza-Usuga, J.C., Poch, R.: Evapotranspiration in Pinus sylvestris and Pinus uncinata Forest Plots from Water Balance Component Measurements in a Mediterranean Mountain Catchment. Nova Publishers, 2011, pp. 81-100

[25] Flores Ayala, E., Guerra De la Cruz, V., Terrazas González, G.H., Carrillo Anzures, F., Islas Gutiérrez, F., Acosta Mireles, M., Buendía Rodríguez, E.: Intercepción de lluvia en bosques de montaña en la cuenca del río Texcoco, México. Revista mexicana de ciencias forestales, 2016, 7(37), 65-76 *(Spanish)*

[26] Silva, I.C., Rodríguez, H.G.: Interception loss, throughfall and stemflow chemistry in pine and oak forests in northeastern Mexico. Tree Physiology, 2001, 21(12-13), 1009-1013.

[27] Ashagrie, Y., Zech, W.: Water and nutrient inputs in rainfall into natural and managed forest ecosystems in south-eastern highlands of Ethiopia. Ecohydrology & Hydrobiology, 2010, 10(2-4), 169-181.

[28] Wilcke, W., Yasin, S., Valarezo, C., Zech, W.: Change in water quality during the passage through a tropical montane rain forest in Ecuador. Biogeochemistry, 2001, 55(1), 45-72.

[29] Nanzyo, M., Shoji, S., Dahlgren, R.: Volcanic Ash Soils: Physical characteristics of volcanic ash soils. Developments in Soil Science 21. Elsevier Science Publishers. Netherlands. 1993, 21, 189-200

MONTHLY AND SEASONALLY CROP WATER STRESSES DISTRIBUTIONS OVER THE NILE DELTA USING REMOTE SENSING TECHNIQUES

Ayat Elnmer

Irrigation and Hydraulic Engineering Department, Faculty of Engineering, Tanta University, 31734, Tanta, Egypt; Ayat_elnemr@f-eng.tanta.edu.eg

Egypt-Japan University of Science and Technology (E-JUST), Environmental Engineering Department, P.O. Box 179, New Borg Al-Arab City, Postal Code 21934, Alexandria, Egypt; ayat.abdelwahab@ejust.edu.eg

Keywords: Crop water stress, Nile delta, remote sensing, water deficit index, water stress index

Abstract

Determination of water stress and oversupply locations at a regional scale could significantly influence the water management. Recently, remote sensing techniques become an efficient tool to provide water and land surface data at a regional scale. This research aims to evaluate the monthly and seasonally water stress index (WSI) for the center portion of the Nile Delta, Egypt using remote sensing techniques. These techniques were the Simplified Surface Energy Balance algorithm (SSEB algorithm) with Landsat 8 images to estimate the actual evapotranspiration. Additionally, the Split Windows algorithm with Landsat 8 images was used to estimate the spatial distribution of the land surface temperature (Ts). The WSI algorithm was applied to the center of the Nile Delta during the summer season 2016 to assess the crop water stress and the oversupply locations using the Ts spatial distribution. Moreover, it was validated using ground data about the actual crop water stress based on the actual and wet evapotranspiration. The monthly WSI during the crops growing season was poor, where about 40% of the total irrigated area suffered from crop water stresses. However, during the beginning and harvesting seasons 2016, about 18% of the total irrigated area suffered from extreme crop water stresses. The seasonal WSI for the center of the Nile Delta was poor, about half of the total cultivated area suffered from extreme crop water stresses in summer 2016. The crop water stress at a temporal scale indicates the urgent need to efficiently distribute the water supplies in order to ensure the water security for the center of the Nile Delta.

1 Introduction

Efficient water management becomes an urgent need to meet the rapid escalation of water demands in arid and semi-arid regions. Egypt has very limited water resources followed by a water scarcity, which is considered the main challenge that faces the decision makers to manage the water resources. The Nile River is the largest water supply in Egypt, which participates 55 billion cubic meters per year (BCM/yr) [1]. Nevertheless, about 85% of total water supplies in Egypt are consumed by the agriculture sector, and the crops in the center of the Nile Delta consume about 45% of total water in the agriculture sector [2]. Therefore, the efficient management of water resources is depending on understanding the crop water consumption and the stresses that face the irrigation process.

Crop water consumption is represented by the actual evapotranspiration (*ETc*) [3]. Hence, the *ETc* should be accurately estimated to detect the crop water stresses with high accuracy. This is a main factor in the estimating the water balance in the Nile Delta [3]. There are various methods to estimate the *ETc*; i.e., empirical approaches, energy balance methods and water balance methods such as pan-evaporimeter and lysimeter [4]. The empirical approaches are used to estimate the *ETc* at point scale, hence they are very limited to estimate the *ETc* in large areas [5]. The most common empirical approach in the *ETc* estimates is the formula of FAO Penman-Monteith [6]. Energy balance methods can estimate the *ETc* at regional scale without the knowledge about water supply sources. Nevertheless, the water supply sources must be estimated in the water balance methods [7].

The Simplified Surface Energy Balance System (SSEBS) is one of the energy balance methods in estimating the *ETc*. It depends on the residual concept [8]. Senay et al. 2007 developed the SSEBS to estimate the *ETc* over large areas using the land surface temperature as a scalar, which depends on the principle of "Hot and Cold Pixels" [7]. They used moderate resolution images (MODIS sensor) to run the SSEBS to estimate the *ETc* over the agricultural lands in Afghanistan.

Crop water stresses can adversely affect the crop production, hence detecting these stresses in right time is a main challenge that faces decision makers and farmer [9]. The crop water stresses could be efficiently assessed by the water stress indexes [10]. The approaches of the water stress indexes estimates could be categories into indexes derived from optical bands and thermal bands in remote sensing techniques [11]. The used optical bands in water stress indexes estimates are near infrared band (NIR) and short wave infrared band (SWRI) [12]. This could be attributed to the ability of NIR and SWRI to detect the changes in the water content in the leaves, hence the water stress and drought could be efficiently assessed [13]. However, the SWRI has higher ability than NIR to detect the water stresses in leaves due to its sensitivity to the changes in the water thickness in the leaves. Land surface wetness index (LSWI) and Normalized difference water index (NDWI) are the most common water stress indexes derived from optical bands. The LSWI and NDWI are similar, but LSWI is more sensitive to the liquid molecules' interactions [14].

The second category of water stress indexes are the indexes derived from thermal bands. The thermal bands are used to derive the surface temperature. After that, the surface temperature (Ts) would be combined with vegetation indexes (VI) to assess the crops' statutes [15]. The merits of these approaches are assessing the crops statutes at regional scale and considering the vegetation variations. The first attempt to relate the Normalized Difference Vegetation index (NDVI) with the Ts was done by Price, J. C. 1990, who used this relation in deriving regional maps of the latent heat fluxes [16]. Maron et al. (1994) used the relation between the Ts, the air temperature (Ta) and the vegetation index to assess the crop water deficit [17]. The proposed water deficit index (WDI) was derived, based on the Et_c and the potential evapotranspiration (Et_o).

After that, the WDI was modified by the Priestley and Taylor equation to replace the *ETc*, and the wet evapotranspiration was used instead of the Et_o [18]. The new index was called the water stress index (WSI), which depends on the relation between the VI and the Ts.

The performance of the daily WSI was assessed in the center in the Nile Delta for the winter season 2015-2016 [9]. Hence, this research aims to evaluate the monthly and seasonally water stress index (WSI) for the center portion of the Nile Delta, Egypt using remote sensing techniques during the summer season 2016.

2 Material and Methods

2.1 Study area

The study area is an irrigated area in the center of the Nile Delta, which was chosen to implement the presented methodology in this paper. The locations of the study area are latitude of 30°43'43.75" - 30°57'40" N and longitudes of 30°54'42.15" - 31°1'37.35" E. The climate is an arid Mediterranean type, where the rainfall is very rare (less than 200 mm/year) and occurs only in the winter season [19]. The average temperatures in winter and summer seasons are 14 and 30 °C, respectively. The main water supply in the study area is surface water supply, whereby the abstraction of groundwater is very small and neglected [1]. The surface water supply is from a main canal: the Al-Qased Canal with its branch and distributary canals are shown in Figure 1. Moreover, the common irrigation practise in the study area is flooding irrigation. Agricultural summer season extends from May to October, while the winter season extends from October to May. The summer and winter seasons are divided into sub-seasons: beginning, growing and harvesting seasons. The main cultivated crops in the summer season are rice and maize, while wheat and clover are the main crops in the winter season [19].

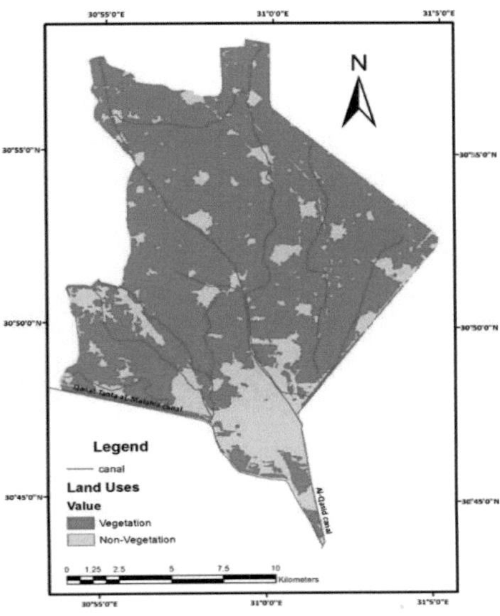

Figure 1: Study area map and the land uses

2.2 Data set

Various data sets were used in this methodology; Landsat 8 images and metrological data from a weather station in the study area (i.e., temperature, wind speed, and relative humidity *RH*). The Landsat 8 images were three cloud-free (< 10% clouds) images covering the beginning, growing and harvesting summer season.

2.3 water stress index (WSI)

Water stress index (WSI) depends on the ratio between the actual evapotranspiration and reference evapotranspiration [18]. However, the *ETc* was modified according to the Jiang and Islam method, and the ET_o was replaced by the wet evapotranspiration (ET_w), which could be referenced by the equation of Priestley-Taylor. Hence, the formula of the WSI could be expressed as follow (Eq. 1):

$$WSI = (T_i - T_{min}) / (T_{max} - T_{min}) \qquad \text{Eq. 1}$$

where; T_i expresses the pixel temperature, and T_{min} is the minimum vegetation temperature and could be expressed by the average temperature of water pixels, while the T_{max} expresses the maximum vegetation temperature.

The approach to estimate the maximum vegetation temperature is using the Ts-NDVI scatter plot, where the Ts is the land surface temperature and the NDVI the Normalized Difference Vegetation Index. The NDVI could be derived by the following equation (Eq. 2) [20]:

$$NDVI = \frac{\rho_{NIR} - \rho_{Red}}{\rho_{NIR} + \rho_{Red}} \qquad \text{Eq. 2}$$

where; the ρ_{NIR} and ρ_{Red} are optical bands of Landsat 8 images. ρ_{NIR} is the near infra-red band and has a $0.63 - 0.67$ µm width, while ρ_{Red} is the red band and its width is $0.67 - 0.90$ µm [20].

The land surface temperature could be efficiently estimated using thermal bands of Landsat 8 images through applying the split window algorithm (Eq. 3) [21]:

$$Ts = C_1 (TB_{10} - TB_{11}) + TB_{10} + C_2 (TB_{10} - TB_{11})^2 + (C_3 + C_4 W)(1-m) + C_0 + (C_5 + C_6 W)\Delta m \qquad \text{Eq. 3}$$

where; *m* expresses the mean of emissivity of the land surface, while Δm expresses the difference of the emissivity of the land surface. The atmospheric water vapor content is presented as W. The brightness temperature of thermal bands are presented as TB10-TB11, and ($C_1 - C_6$) express the split-window coefficients [21].

The Ts-NDVI scatter plot was derived using PYTHON language to efficiently determine the minimum and maximum temperatures.

2.4 Water deficit index (WDI)

Water deficit index can be defined as the ratio between actual evapotranspiration *ETc* and reference evapotranspiration *ETp*, as shown in Eq. 4 [17]:

$$WDI = \frac{ET_C}{ET_O}$$
Eq. 4

Actual evapotranspiration over the center of the Nile Delta during the summer season 2016 was previously derived from the Simplified Surface Energy Balance System (SSEBS) model with Landsat 8 images by the author. Nevertheless, the reference evapotranspiration was estimated using the formula of FAO - Penman Monteith, which depends on metrological data such as surface temperature, humidity, wind speed and precipitation [6].

2.5 Validation of the water stress index

The daily WSI results are validated using the results of the daily WDI over the center of the Nile Delta. The WSI results were compared with the WDI results using the Root Mean Square Error (RMSE) as a performance criteria measure.

3 Results and Discussion

3.1 Water stress index

The WSI estimates depend on the NDVI and Ts over the study area during the summer season 2016. Hence, the NDVI was estimated using the optical bands of Landsat 8 images over the study area during the beginning, growing and harvesting seasons of summer 2016, as shown in Figure 2. The NVDI ranged from 0.07 to 0.35 during the beginning and harvesting seasons. This could be attributed to the less vegetation cover during these seasons. Nevertheless, the NDVI was high and ranged from 0.08 to 0.522 in August 2016. This could be attributed to the ful vegetation cover during the growing season in the summer crop pattern.

The Ts was ranged from 307 to 314 K with mean and stranded deviation (SD) 310.3 K and 3.29, respectively, over the center of the Nile delta in June 2016, as shown in Figure 3. Then it started to increase to range from 303 to 316 K in August, with 312 K mean and 4.05 SD. However, it decreased in October 2016 to range from 302 to 314.05 K, with 309.46 K mean and 3.75 SD.

The relation between the NDVI and Ts was derived using python programming language to estimate the maximum temperature with NDVI equal 0.0 over the study area during the summer season 2016. The T_{mxa} was 316, 311 and 310 K, while the T_{min} were 306, 304 and 303 K in June, August and October 2016, respectively.

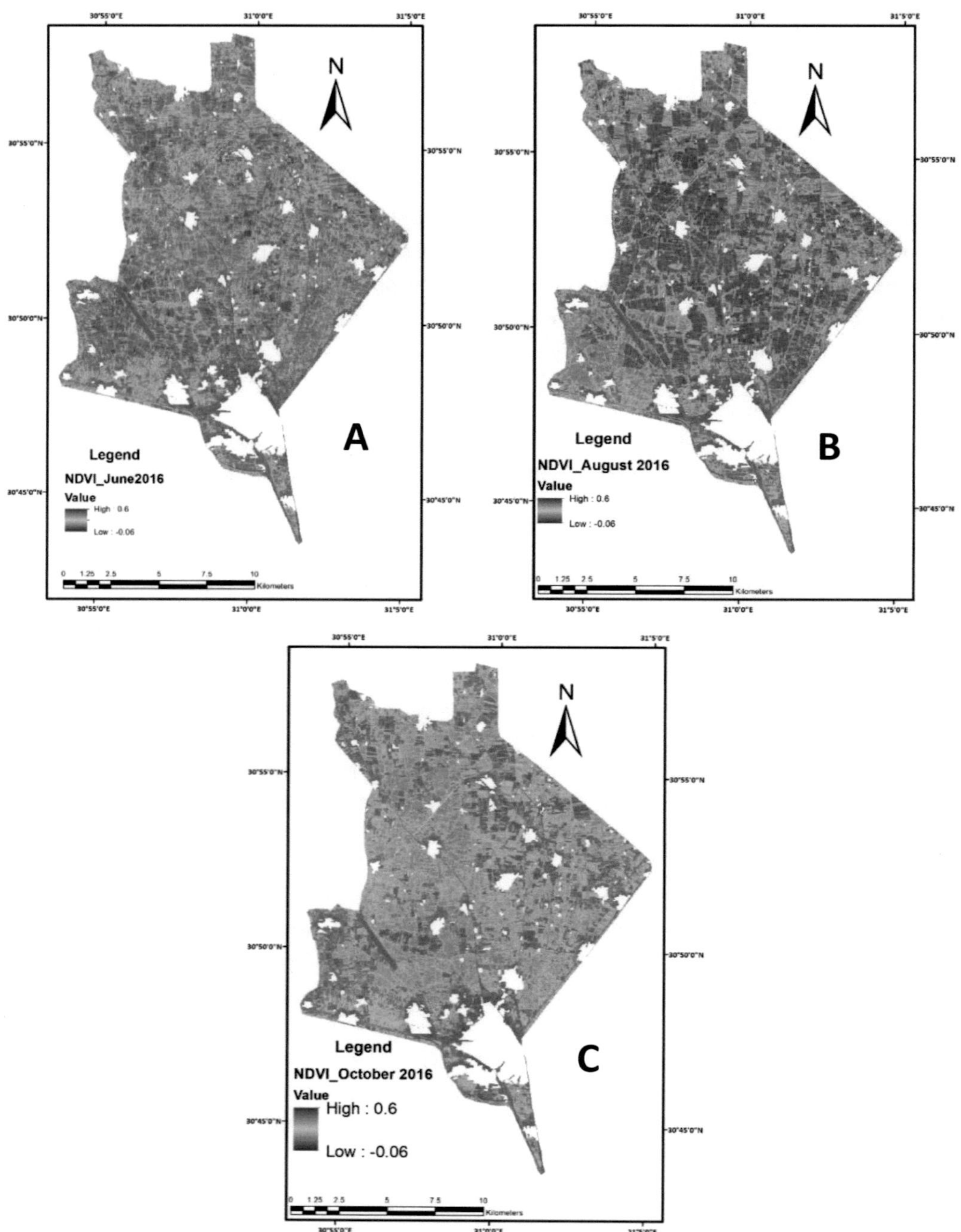

Figure 2: The Normalized Difference Vegetation Index over the study area
(A) June 2016, (B) August 2016 and (C) October 2016.

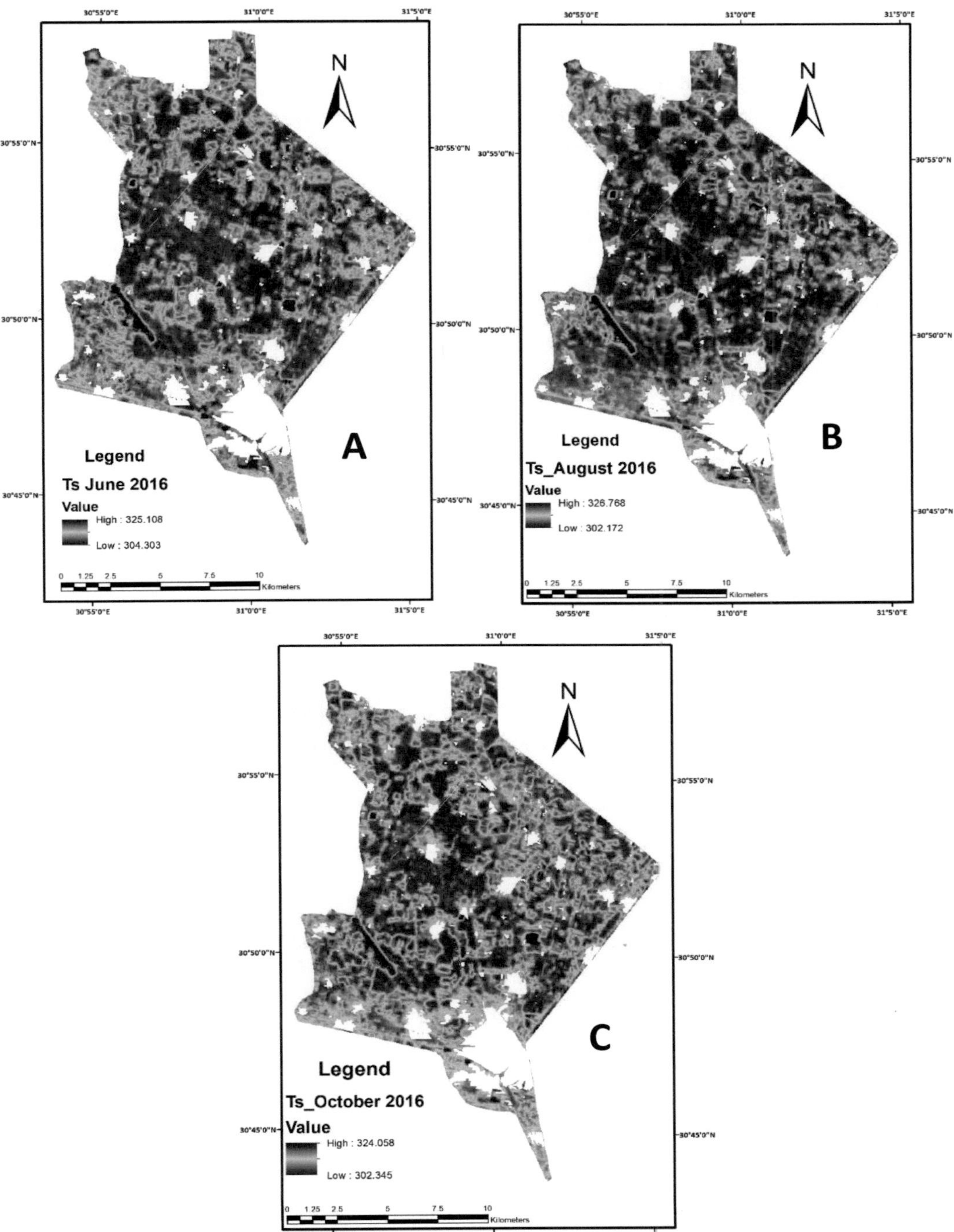

Figure 3: The surface temperature over the study area
(A) June 2016, (B) August 2016 and (C) October 2016

The WSI was derived using the T_{max} and T_{min} over the study area during the summer seasons, as shown in Figure 4 *(see after the Chapter 5 "References")*. The WSI in June was ranged from 0.05 to 0.65, with mean 0.52 and 0.328 SD, while the mean WSI in October was 0.32. This could be attributed to the low crop water requirements during the begging and harvesting seasons. Adversely, the growing season faced higher water stresses, where the WSI ranged from 0.44 to 0.8, with 0.67 mean in August 2016 due to the high crop water requirements. About 40% of total fields in the study area faced extreme crop water stress (higher than 0.5) in the growing season (August 2016), as shown in Figure 4. Nevertheless, about 18 % the irrigated area suffered from extreme crop water stresses during the beginning and harvesting seasons (June and October 2016). Hence, about half of the irrigated area in the center of the Nile Delta suffered from extreme water stresses during the summer season 2016.

3.2 Water deficit index

The ET_o was assessed using the temperature, humidity, and wind speed in the center of the Nile Delta during the summer 2016. The ET_o was 6.8, 8 and 6.2 mm/d in June, August and October 2016, respectively. Nevertheless, the *ETc* was previously assessed over the center of the Nile Delta during the summer season 2016. Hence, the WDI was estimated using the ET_o and *ETc*, as shown in Figure 5 *(see after the Chapter 5 "References")*. The mean WDI in the beginning and harvesting seasons were 0.23 and 0.34, respectively. However, it was 0.84 in August 2016 (growing season).

3.3 Validation of the WSI

The validation of the WSI was done using a comparison between the results of WSI and WDI. Group of 16 fields of rice and maize crops was used to perform this comparison. The R^2 was 73% over the study area during the summer season of 2016, as shown in Figure 6.

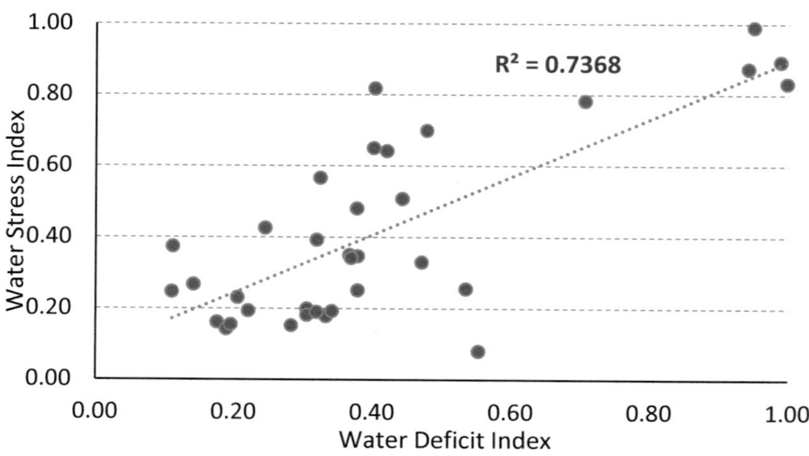

Figure 6: Relation between water stress index and water deficit index over the study area (summer season 2016)

4 Conclusions

Water stresses effect the crop health and growth, which could reduce the crop production. The water stresses over the center of the Nile Delta were assessed in this study using the water stress index (WSI) during the beginning, growing and harvesting season of summer 2016. The WSI was low in the beginning and harvesting season due to the low crop water requirements during these seasons. However, it increased in the growing season due to the increased crop water requirements in the growing stage. The averages of WSI were 0.52, 0.32 and 0.67 in June, August and October 2016, respectively. The WSI higher than 0.5, which represents an extreme water stress, was found in the study area during summer 2016. The fields with extreme water stress represented 18% of total irrigated area in the beginning and harvesting seasons, which represented 40% in the growing season. Hence, about half of the irrigated area in the center of the Nile Delta suffered from extreme water stress during the summer season 2016.

The WSI was validated with the water deficit index (WDI), which represents the old approach of water stress indexes using the Root Mean Square Error (RMSE) as a performance criteria measure. The results showed that the R^2 was 73%, which indicates the WSI could be a useful and economic tool in assessing the crop water stress location at a regional scale in the center of Nile Delta.

5 Acknowledgements

The author would like to thank EXCEED Swindon project and DAAD (German Academic Exchange Service) for support to participate at the Expert Workshop on "Water on Agricultural Practices: Training the Trainers" in Rio de Janeiro, Brazil on 15 Sep -21 Sep, 2019. The author would like to express his appreciation to Prof. Andreas Haarstrick, EXCEED's project scientific coordinator for his support.

6 References

[1]   M. Khadr, B. A. Zeidan, A. Elnmer: On-farm water management in the Nile Delta. In: The Nile Delta, ed: Springer, 2016, pp. 325-344.

[2]   A. Elnmer, M. Khadr, B. A. Zeidan: Optimal Water Productivity of Crop Pattern in Central Nile Delta, Egypt. In: Proceedings of the Eighteenth International Water Technology Conference (IWTC18), Sharm ElSheikh, Egypt, 2015, pp. 12-14.

[3]   A. Elnmer, M. Khadr, S. Kanae, A. Tawfik: Mapping daily and seasonally evapotranspiration using remote sensing techniques over the Nile delta. Agricultural Water Management, 2019, 213, 682-692.

[4]   H. Nouri, S. Beecham, F. Kazemi, A.M. Hassanli: A review of ET measurement techniques for estimating the water requirements of urban landscape vegetation. Urban Water Journal, 2013, 10, 247-259.

[5] M. Elhag, A. Psilovikos, I. Manakos, K. Perakis: Application of the SEBS water balance model in estimating daily evapotranspiration and evaporative fraction from remote sensing data over the Nile Delta. Water Resources Management, 2011, 25, 2731-2742.

[6] R. Allen, L. Pereira, D. Raes, M. Smith: Crop Evapotranspiration-Guidelines for Computing Crop Water Requirements-FAO Irrigation and Drainage Paper 56. FAO: Rome, Italy, 1998, 300, 1998.

[7] G. Senay, M. Budde, J. Verdin, A. Melesse: A coupled remote sensing and simplified surface energy balance approach to estimate actual evapotranspiration from irrigated fields. Sensors, 2007, 7, 979-1000.

[8] G. Calcagno, G. Mendicino, G. Monacelli, A. Senatore, P. Versace: Distributed estimation of actual evapotranspiration through remote sensing techniques. In: Methods and tools for drought analysis and management, ed: Springer, 2007, pp. 125-147.

[9] A. Elnmer, M. Khadr, A. Tawfik: Using Remote Sensing Techniques for Estimating Water Stress Index for Central of Nile Delta. In: IOP Conference Series: Earth and Environmental Science, 2018, 151, 012026.

[10] L. Song, S. Liu, W. P. Kustas, J. Zhou, Z. Xu, T. Xia, M. Li: Application of remote sensing-based two-source energy balance model for mapping field surface fluxes with composite and component surface temperatures. In: Agricultural and Forest Meteorology, 2016, 230, 8-19.

[11] L. López, R. Arteaga, P. Vázquez, C. López, C. Sánchez. Water stress index as an indicator of irrigation timing in agricultural crops. Agricultura Técnica en México, 2009, 35, 97-111.

[12] H. Nouri, E.P. Glenn, S. Beecham, S.C. Boroujeni, P. Sutton, S. Alaghmand, B. Noori, P. Nagler: Comparing three approaches of evapotranspiration estimation in mixed urban vegetation: Field-based, remote sensing-based and observational-based methods. Remote Sensing, 2016, 8, 492.

[13] M.W.A. Halmy, P.E. Gessler, J.A. Hicke, B.B. Salem: Land use/land cover change detection and prediction in the north-western coastal desert of Egypt using Markov-CA. Applied Geography, 2015, 63, 101-112.

[14] Z. Du, W. Li, D. Zhou, L. Tian, F. Ling, H. Wang, Y. Gui, B. Sun: Analysis of Landsat-8 OLI imagery for land surface water mapping. Remote Sensing Letters, 2014, 5, 672-681.

[15] L.O. Serbina, H.M. Miller: Landsat and water: Case studies of the uses and benefits of Landsat imagery in water resources: US Department of the Interior, US Geological Survey, 2014.

[16] J.C. Price: Using spatial context in satellite data to infer regional scale evapotranspiration. IEEETtransactions on Geoscience and Remote Sensing, 1990, 28, 940-948.

[17] M. Moran, T. Clarke, Y. Inoue, A. Vidal: Estimating crop water deficit using the relation between surface-air temperature and spectral vegetation index. Remote Sensing of Environment, 1994, 49, 246-263.

[18] Z. Su, A. Yacob, J. Wen, G. Roerink, Y. He, B. Gao, H. Boogaard, C. Diepen: Assessing relative soil moisture with remote sensing data: theory, experimental validation, and application to drought monitoring over the North China Plain. In: Physics and Chemistry of the Earth, Parts A/B/C, 2003, 28, 89-101.

[19] A. Elnmer, M. Khadr, A. Allam, S. Kanae, A. Tawfik: Assessment of irrigation water performance in the nile delta using remotely sensed data. Water. 2018, 10, 1375.

[20] E. Glenn, A. Huete, P. Nagler, S. Nelson: Relationship between remotely-sensed vegetation indices, canopy attributes and plant physiological processes: What vegetation indices can and cannot tell us about the landscape. Sensors, 2008, 8, 2136-2160.

[21] O. Rozenstein, Z. Qin, Y. Derimian, A. Karnieli: Derivation of land surface temperature for Landsat-8 TIRS using a split window algorithm,. Sensors, 2014, 14, 5768-5780.

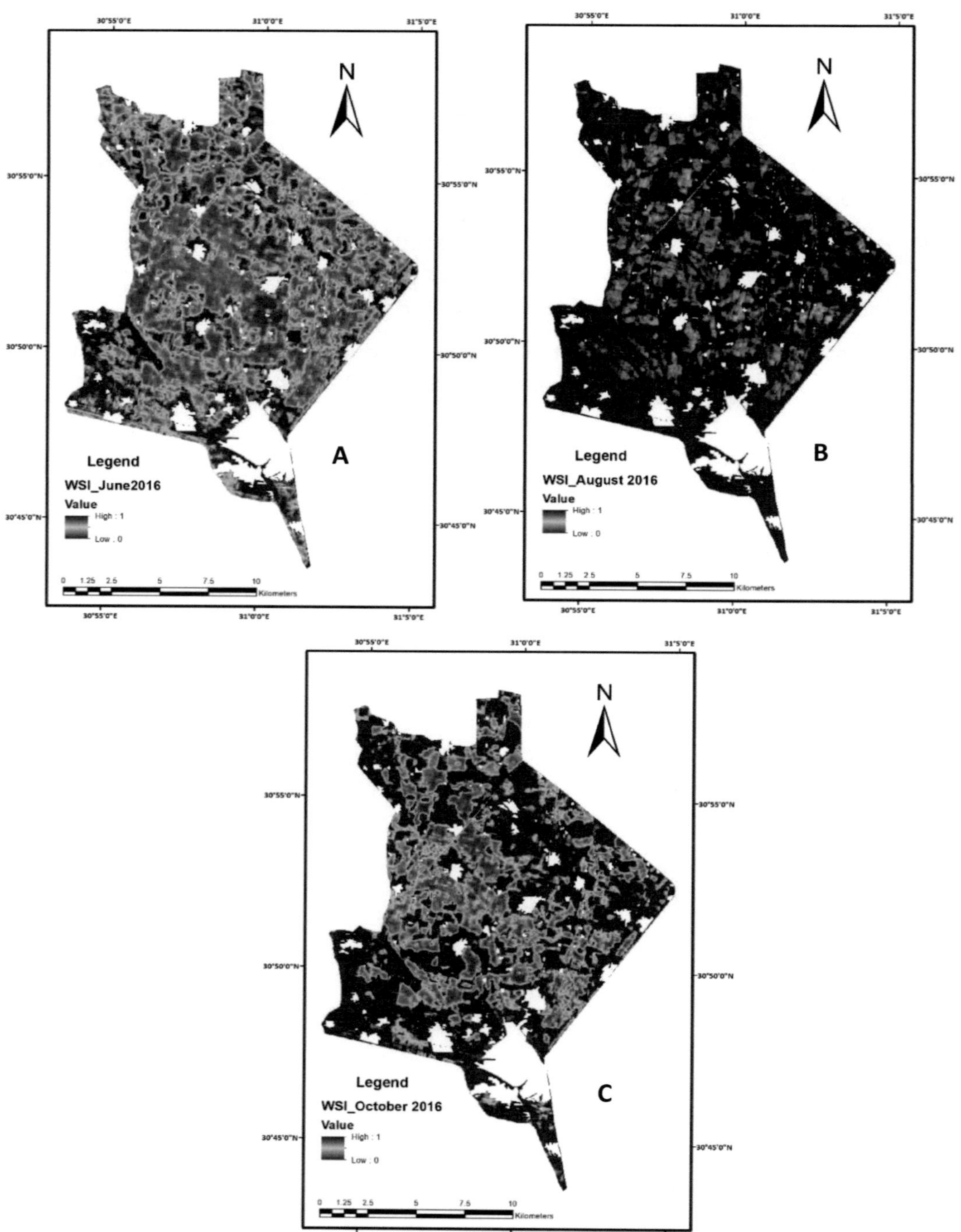

Figure 4: Water stress index over the study area
(A) June 2016, (B) August 2016 and (C) October 2016.

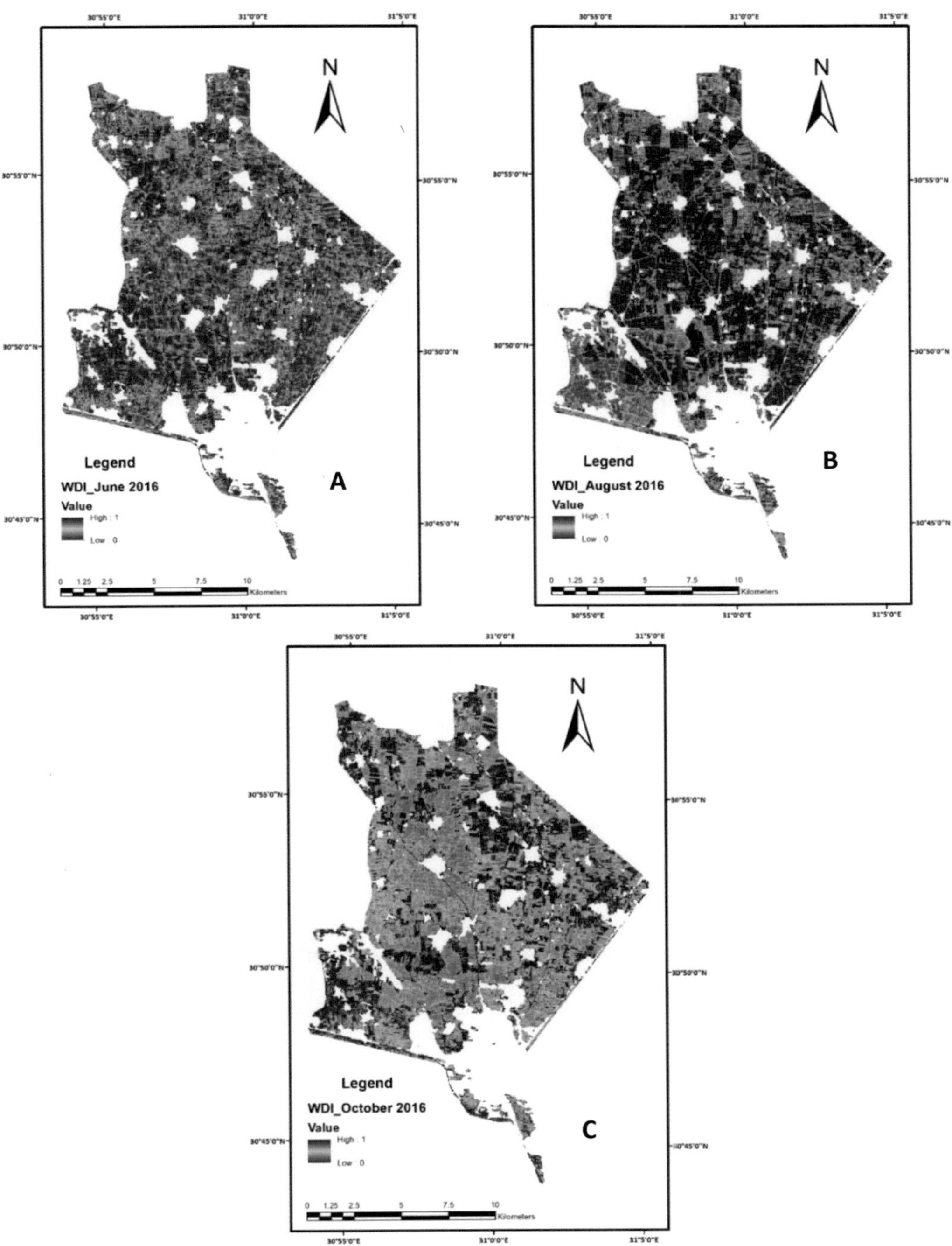

Figure 5: Water deficit index over the study area
(A) June 2016, (B) August 2016 and (C) October 2016

DPSIR FRAMEWORK ASSESSMENT OF LAND DEGRADATION BY WATER EROSION IN KENYA: A REVIEW

Kipkirui Edwin

UNEP- Tongji University, Institute of Environment for Sustainable Development, College of Environmental Science, 1239, Siping Road, Shanghai, 200092, P.R. China; lagatedu@gmail.com

Keywords: Land degradation, water erosion, DPSIR, sustainable land management

Abstract

Water erosion is a leading environmental threat to the sustainability and productive capacity of agriculture in many tropical and sub-tropical regions of the world. Soil erosion by water is by greater margin the most common type of soil degradation, affects approximately 1,100 million ha worldwide. The aim of this paper is to highlight the causes and state as well as effects of land degradation in Kenya through the DPSIR model framework for land degradation analysis. The general pressure and driving factors attributed to the land degradation are varied depending on the climate, human activities and vulnerability of the soil to these effects. Rainfall erosion potential and soil erodibility in most regions in Kenya present an environment, which is very susceptible to soil erosion. The dynamic change towards intensive agriculture has removed much of the protective land cover vegetation cover and led to the need for the introduction of soil and water conservation measures in order to reduce soil erosion to acceptable rates.

1 Introduction

Water erosion is a growing global concern in the past couple of years. Water erosion is by far the most common type of soil degradation, impacting approximately 1,100 million ha worldwide.[1] Particularly as a repercussion of its negative impacts on declining agricultural productivity, the general environment and the general quality of life are suffered from. The economic negative impacts of land degradation have been very drastic in most parts of Africa [2], where productivity has reduced by 50% [3].

In response to negative impacts of global climate change, the 16 Sustainable Development Goals (SDGs) was adopted by the UN General Assembly on 25th September 2015. SDG goal 15 includes a target (15.3) to: *"combat desertification, restoration of the degraded soils as well as areas affected by desertification, floods and work towards achieving land degraded-neutral world"* by 2030 [4, 5]. Currently, the issues of sustainable agriculture, food security, and an end to hunger are receiving increasing attention by the policy makers and general public especially small holder farmers [4].

'Land Degradation' refers to the reduction of utility of land and the failure of soil to potentially provide biological and economic productivity as a result of processes of soil erosion by either wind or water, compaction of the soils and loss of nutrients in the soil leading to reduction of production of pasture, rangelands, cropland, forests, woodland, habitat loss and destruction of biodiversity [6]. Globally, land degradation has a lot of consequences on the environment and general quality of lively hood. Such potential impacts include:

Climate change: Land degradation greatly contributes to global climate change by leading to greenhouse gas emission (GHGs) and reduction of carbon sinks due to the reduction of terrestrial ecosystems.
Biodiversity: Land degradation can result in loss of biodiversity, both in areas already degraded and by inducing additional clearing of natural habitats.
International waters: Land degradation can result in damages to shared international water bodies.

It is worth noting that almost 80% of the terrain globally has been affected by water erosion in one way or another [7].

2 Materials and Methods

The methodological approach for this paper includes searching, identifying, review and analysis of peer review journals related to the topic along with the secondary grey literature, including government reports, non-governmental stakeholders, institutional releases and all locally published information, using the keywords *sustainable land management, DSPIR, land degradation Kenya, conservation agriculture* on mainly Google and Google Scholar, Baidu and other web search engines. The relevant research papers, manuscript and periodical releases were obtained and reviewed accordingly.

3 DPSIR Framework

The Driver-pressure-state-impact-response (DPSIR) Framework is an Environmental analysis framework developed by European Environmental Agency in the year 1999 (Figure 1). This model has since be adopted and applied in solving environmental challenges [8]. The DPSIR framework is an analysis framework that focusses on causal-effect relationships amongst the intertwined components of social, economic and biophysical factors [9]. The DPSIR framework has been used in many environmental resource applications.

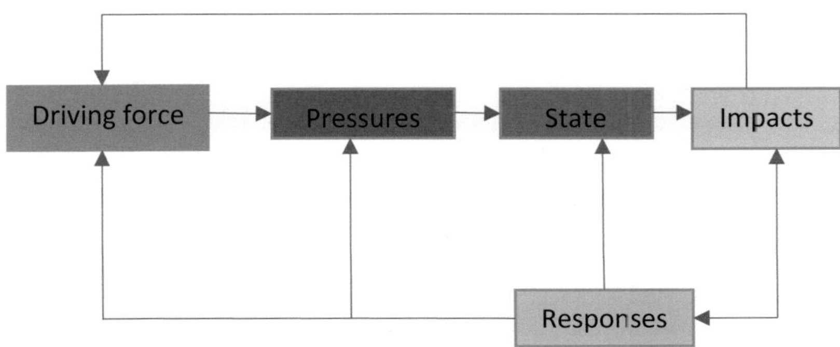

Figure 1: DPSIR framework and relationship between the factors [9]

The general driving forces attributed to the land degradation are varied depending on the climate, human activities and the vulnerability of the soil to these effects. The typical factors include unsustainable agricultural practices, deforestation; underlying factors include soil erodibility, slope of the land, land use and rainfall intensity [10]. The aim of this paper is to highlight the state of land degradation by water erosion using the driver-pressure-state-impact-responses analysis framework.

4 DPSIR analysis of land degradation by water erosion

Driving forces of land degradation

Driving factors are economic, social or ecological factors that cause changes in the system; they are often referred as driving forces and underlying factors. These are day-to-day community activities that make soil more vulnerable to land degradation by water erosion. Proximate causes are also called unsustainable practices of land management that cause excessive land conversion. Proximate causes can be grouped to two: biophysical factors and anthropogenic land use practices.

Table 1: Driving forces of land degradation

Type of cause		Driving Forces
Proximate causes	Biophysical	Steep slopes, flood prone regions, droughts, soil erodibility, climate change threats
	Anthropogenic	Encroachment of water catchment areas, deforestation, unsustainable exploitation of biodiversity, unsustainable agronomic practices, poor rangeland management, coastline ecosystem degradation
Underlying causes		Lack of available environmental knowledge, lack of information about appropriate alternative technologies, unplanned land use change, unplanned urban growth, land use pressure, livestock pressure, population pressure, poverty, breakdown of the indigenous (local) institutions

Pressures of land degradation

Pressures for the land degradation are anthropogenic activities that directly affect the system and are solely caused by the driver causes. It is found out that agriculture is the leading cause of environmental change thus causing the alteration of water circles, droughts, reduced land productivity and increased built up of greenhouse gases, a recipe for climate change [11]. Consequently, the land resources in the past decades has come under increasing demands from competing usages agriculture, forests, and pasture, energy source as well as unplanned extraction of raw materials, over-exploitation of resources, urbanization, etc. [11].

Unsustainable land-use practices have potential impacts on land and is attributed to major land degradation, and reduce the ecological and social resilience of landscapes [12]. These areas also continue to experience increased *fragmentation* and *deforestation* due to increasing pressure for new cultivation and grazing lands as well as for settlement. Land scarcity has caused increased migration of people to the fragile arid and semi-arid regions in search of land for cultivation and settlement.

State of land degradation

State is the most current condition of the system at a specific time and is represented by a set of descriptors of system attributes that are affected by pressures and the type, degree and rate of land degradation. Due to pressures, the 'state' of the environment is affected. Many studies on soil erosion in Kenya have shown that land degradation is on the rise in many areas of the country in both magnitude and extent with over 20% of areas under cultivation, 30% of forests, and 10% of grasslands being subject to severe unhealthy land practices. Unfortunately, some of the areas, which experience the highest degradation risks, coincide with the most productive areas in the country. The land cover changes are summarized in the Table 2 as spatial analysis of land use changes in Kenya (1990-2010) [13].

Table 2: The land cover changes: Spatial analysis of land use changes in Kenya (1990-2010[13]

Land use / Land cover changes	Changed Area (km^2)	Percent Change
Forest 1990-2000	5,199	-0.8
Forest 2000-2010	983	-0.2
Total loss of Forest (1990-2010)	6,182	-1.0
Agriculture 1990-2000	25,159	+3.9
Agriculture 2000-2010	222,237	+3.4
Total Increase in agricultural land (1990-2010)	47,397	+7.3
Range Land 1990-2000	46,399	+7.1
Range Land 2000-2010	12,153	+1.9
Total increase I Range Lands 1990-2010	58,552	+9.0
Bare Lands 1990-2000	26,457	+4.1
Bare Land 2000-2010	9,508	-1.5
Total change in Bare Lands (1990-2000)	16,948	+2.6

Land degradation was determined through RS/GIS mapping by combining various factors including Aland use/land cover, rainfall amounts, soil erodibility, and topography after delineating protected areas obtained. The results are shown in Figure 2. Spatial analysis obtained that only about 2.2% of the country's land area has minimal risk of degradation through soil erosion [13]. The rest experiences moderate to very acute degradation with the largest proportion (61.4%) being at risk of severe land degradation (Table 3). Moreover, high erosion risk extends also onto protected areas such as national parks due to encroachment of human activities and natural propensity for erosion, e.g., high soil erodibility, drought and flood risks as well as declining vegetation cover. This is in agreement with land cover changes in other regions of the world as a consequence of global climate change [14].

Although soil erosion is a natural geomorphic process, economic and social activities such as cultivation, overgrazing and deforestation accelerate the process beyond the acceptable levels. Excessive erosion is associated with diverse negative on- and off-site impacts, including loss of soil nutrients, leading to a reduction in crop yields, decreasing stream competence and capacity because of sedimentation, and siltation of reservoirs.

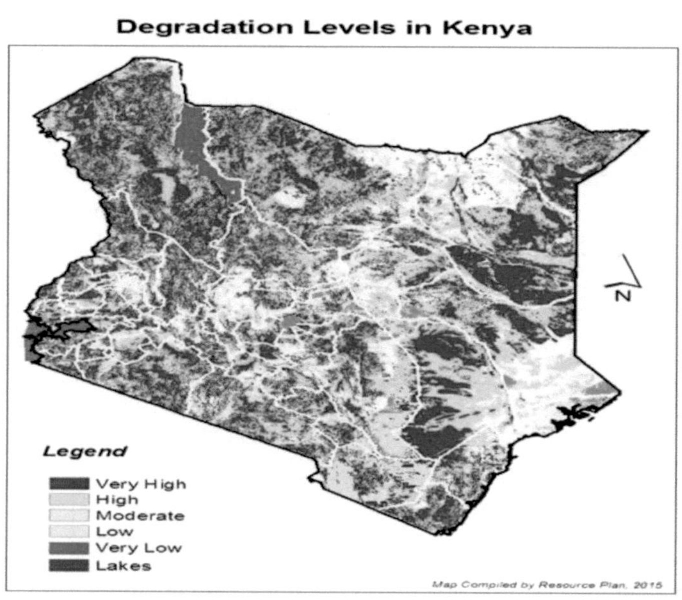

Figure 2: Map showing various zones vulnerable to soil erosion in Kenya [13]

Table 3: Spatial analysis of land use changes in Kenya (1990-2010)[13]

Degradation class	Area (km^2)	Percentage (%)
Very Low	4,548	0.8
Low	8,225	1.4
Moderate	52,383	9.2
High	350,992	61.4
Very High	155,264	27.2
Total area (Km^2)	571,416	100

The results of land use mapping by LADA [13] show that Savannah grasslands is a combination of various vegetation types with and without shrubs, covers approximately 69.7% of the land area of Kenya and is thus the major land use type. Agriculture at 17.7% follows next, comprising of both rain fed and irrigated. The LADA mapping obtained a forest cover of 8.0% inclusive of both closed and open forests as well as mangroves.

The relatively larger proportion of forest cover calculated from this study is agreeable with recent government reports, which have indicated that recent re-afforestation efforts have seen resurgence of forest cover, attaining 6.99% national forest coverage by 2014 [15]. More so, about 2.2% of the land area of Kenya was classified to be bare lands. The implications for this are that some 71.9% of the land is at the risk of degradation in Kenya due to the poorer natural and cover in these zones.

Impacts

Thus, soil erosion is reducing the productivity of land, requiring farmers to increase the inputs in terms of fertilizers and chemicals so as to help check declining productivity. The resultant excess use of these fertilizers and some chemicals contributes to soil degradation and water pollution. Cultivation of such arid areas characterized by unreliable rainfall results in frequent crop failures and leaves most of the land bare for long periods, hence increasing the vulnerab lity of the land to soil erosion.

Responses to land degradation

Responses are the efforts made by individuals, people and society as result of the changes manifested in the impacts. As directed actions, responses typically take the form of program activities, campaigns and combined efforts to control with attempt to reverse soil erosion as a positive result of more awareness.

Adaptations to changing climate

Conservation agriculture

Based unprecedented climate change, introduction of water-conserving tillage practices in places, where rainfall is expected to decrease, and to put more emphasis on erosion control practices in areas, where rainfall is expected to increase are necessary, so as to ensure less degradation [16]. There are varieties of tillage practices, which should be adopted especially by small holder farmers as an adaptation mechanism to climate change that in long-run will help to reverse the land degradation.

No tillage & minimum tillage

The current researchers have found out that minimum tillage of land can significantly lower the nutrient losses and soil erosion without compromising the production yields [17]. Thus, there is an obligation for farmers in Kenya to modify conventional tillage practices especially in areas with high erodibility.

Agroforestry and afforestation

Agroforestry should also be widely adopted as a measure to reverse land degradation. Agroforestry, reforestation and other conservation practices absorb tons of other dangerous greenhouse gases such as methane (CH_4) and nitrogen(I)oxide (N_2O). Hence, the adoption of conservational farming systems serves other ecological functions. They enhance the crop yields, conserve environment and offer alternative adaptation mechanisms to climate change [18].

Improving farming practices by small holder farmers
Stone lines used for terracing, agroforestry – growing crops with trees, Stubble mulch tillage in Laikipia, Retention ditches Flood diversion for banana crop in Mbeere (Figure 3).

Figure 3: Stone lined terracing and retention ditches

5 Conclusions

Rainfall erosion potential and soil erodibility present in most regions in Kenya an environment, which is very susceptible to soil erosion. The dynamic change towards intensive agriculture has removed much of the protective land cover vegetation cover and led to need for the introduction of soil and water conservation measures in order to reduce soil erosion to acceptable rates. There are important agricultural and other benefits to be derived from the embracement of sustainable agricultural practices, and with high erosion rates of 10 mm/y, the soil has a life of only 100 years in some areas [19].

There is a need for more monitoring and assessment on state of land cover and soil erosion. This study only applied DPSIR model and thus is limited to socioeconomic indictors as part of encouraging use of simplified tools for monitoring the menace of water erosion in Kenya. Regular monitoring using advanced erosion models along with their field work should be often carried out in order to improve the accuracy of soil erosion trends for formulating sustainable land management policies and practices in the future that is thus suggested.

6 Acknowledgements

The author would like to thank EXCEED Swindon project and DAAD (German Academic Exchange Service) for support to participate at the *Regional Workshop on Water on agricultural practices* held in Rio de Janeiro, Brazil on September 15-21, 2019.

7 References

[1]   Fidele, K., Chi Z., Felix N., Hua S., Alphonse K., Xia F., Lamek, N., Enan, M., Guangjin, T.: Extent of cropland and related soil erosion risk in Rwanda. Sustainability, 2016, 8(7), p. 609.

[2]   Ligonja, P., Shrestha, R.: Soil erosion assessment in kondoa eroded area in Tanzania using universal soil loss equation, geographic information systems and socioeconomic approach. Land Degradation & Development, 2015, 26(4), 367-379.

[3]   Barungi, M., Ng'ong'ola, H., Edriss, A., Mugisha J., Tukahirwa J.: Factors Influencing the Adoption of Soil Erosion Control Technologies by Farmers along the Slopes of Mt. Elgon in Eastern Uganda. Journal of Sustainable Development, Canadian Centre of Science and Eduction, 2013, 6(2), 9 doi: 10.5539/jsd.v6n2p9

[4]   Terlau, W., Hirsch, D., Blanke, M.: Smallholder farmers as a backbone for the implementation of the Sustainable Development Goals. Sustainable Development, 2018, doi:10.1002/sd.1907

[5]   Nations General United Asembly: A/69/l.85 2015. https://sustainabledevelopment.un.org/post2015/summit

[6]   Pagiola, S.: The Global Environmental Benefits of Land Degradation Control on Agricultural Land. World bank, 1999, doi: 10.1596/0-8213-4421-8

[7]   Fidele, K., Chi Z., Felix N., Hua S., Alphonse K., Xia F.: Extent of Cropland and Related Soil Erosion Risk in Rwanda. Sustainability, 2016, 8, 609.

[8]   Carr, E.R., Philiop, M., Sarah C., Mary C., Natalie J.: Applying DPSIR to sustainable development. International Journal of Sustainable Development & Word Ecology, 2007, 14(6), 543-555.

[9]   Bradley, P., Yee, S.: Using the DPSIR framework to develop a conceptual model: technical support document. US Environmental Protection Agency, Office cf Research and Development , EPA/600/R-15/154, 2015. doi: 10.13140/RG.2.11870.7608

[10]   Gessesew, W.: Application of DPSIR Framework for Assessment of Land Degradation: A Review. Approaches in Poultry, Dairy and Veternary Sciences, 2017, 1(5), doi: 10.3031/APDV.2017.01.000522

[11]   Stavi, I., Lal, R.: Achieving zero net land degradation: challenges and opportunities. Journal of Arid Environments, 2015, 112, 44-51.

[12]   Mubiru, D., Rockstro J., Kaumbutho P., Mwalley, J.: Conservation Farming and Changing Climate: More Beneficial than Conventional Methods for Degraded Ugandan Soils. Sustainability, 2017, 9(7), 1084-1099.

[13]   Land Degradation Assessment (LADA) in Kenya. Ministry of environment and natural resourses report, 2016, http:/landportal.org/library/resources/landdegradationassesment-kenya

[14]   Latocha, A., Mariusz, S., Justyna j., Monica R., Magdalena S.: Effects of land abandonment and climate change on soil erosion — An example from depopulated agricultural lands in the Sudetes Mts., SW Poland. Catena, 2016, 145, 128-141.

[15]   National Forest Policy: Ministry of Environment, Ministry of Water and Natural Resources. Government of Kenya report release, 2014. www.environment.go.ke/wp-content/uploads/2016/08/national-environment-policy-2014.pdf

[16]   Olesen, J.E., Trnka, M., Kersebaum, C., Skjelvåg, A.: Impacts and adaptation of European crop production systems to climate change. European Journal of Agronomy, 2011, 34(2), 96-112.

[17]   Atreya, K., Sharma S., Bajracharya R.,Rajbhandari, N.: Developing a sustainable agro-system for central Nepal using reduced tillage and straw mulching. Journal of Environmental Management, 2008, (3), 547-555.

[18]   Mkonda, M., He, X.: Sustainable Environmental Conservation in East Africa through Agroforestry Systems: A Case of the Eastern Arc Mountains of Tanzania. International Journal of Sustainable and Green Energy 2017, 6, 49-56.

[19]   Moore, T.R.: Land Use and Erosion in the Machakos Hills. Annals of the Association of American Geographers, 1979, 69(3), 419-431.

MORPHODYNAMICAL CHANGES IN MAGDALENA RIVER DELTA (COLOMBIA) AND ITS INFLUENCE IN RIVERINE POPULATIONS

F.J. Gomez[1], R.R. Gutierrez[2], J. Biswell[3], R. Doria[4], G. Rivillas-Ospina[5]

[1,2,4]*Universidad Del Norte, Civil and Environmental Engineering Department, Institute of Hydraulic and Environmental Studies (IDEHA), km 5 Puerto Colombia, Barranquilla, Atlántico, Colombia; gomezjf@uninorte.edu.co, rgutierrezll@uninorte.edu.co, doriar@uninorte.edu.co*

[3]*Universidad Del Norte, College of Humanities and Social Sciences; jbiswell@uninorte.edu.co*

[5]*Universidad Del Norte, Civil and Environmental Engineering Department, PIANC COLOMBIA; grivillas@uninorte.edu.co*

Keywords: River dynamics, Affectation, Erosion, Sedimentation, Social Fabric

Abstract

The Magdalena River is the largest river in Colombia. It plays an important role in the circulation of goods and services for most of the country. And in the region, where its mouth is located (department of Atlántico and Magdalena), it is vital for sustaining agricultural activities for formal landowners and precarious ones, namely those, who settle on riparian strips and represent the most vulnerable socio-economic portion of the population. Thus, fluvial erosion and sedimentation processes induced by both natural and anthropogenic controls severely impact on the latter group. This research focuses on describing the interrelationship between the river dynamics and the precarious landowners for the lowest 34 km stretch of the Magdalena River. To this end, the river migration was quantified, based on freely accessible satellite observations between 1986 and 2019 and field hydrological, economic and social information provided by Colombian public agencies. The results show that in the study stretch, the river exhibits sedimentation rates ranging 12-20 m/yr and erosion rates up to 65 m/yr with total losses of 630 ha land dedicated to agricultural activities. The fluvial dynamics directly affect the inhabitants that live on the banks, reducing the areas of cultivation and the economic income. This generates an irruption in the social fabric and migration of farmers, who must look for other economic activities for their subsistence. It is an institutional need on the part of the government to improve regional policies for the protection of farmers in the area and to ensure economic productivity, taking into account the risk generated by the river's dynamic processes.

1 Introduction

The natural river dynamics are described by natural channel migration, flood plain erosion, and accretion processes. However, human activities related to the extractive industry, infrastructure supply, mining, and riverine urbanization trigger changes in river hydrodynamics and sediment supply, which oftentimes intensifies river erosion and accretion processes. This modifies the bank land use and redefines riverine deposits affecting the most socioeconomically sensitive populations, which depend on agriculture and fishing for their subsistence [1, 2].

The Magdalena River is the main river of Colombia, flowing about 1,528 km across the country; a considerable amount of the national population sits on its banks and from it, multiple economic activities are carried out. About 80% of the Colombian population is located in the basin, and it accounts for 84% of the national GDP [3]. The upper and middle sections of the Magdalena River have been modified by intensive deforestation and agriculture. Specifically, deforestation has been registered in many areas of the catchment, which significantly decreased water retention capacity and increased riverbank erosion, and thus affecting the efficiency of sediment-water transfer and magnifying flooding events [2].

There is a close relationship between climate variation, land use, vegetation cover density, river morphological changes and sediment yield, which have negative repercussions. That could be exacerbated by climate change in the lower basins. Climate change will likely modify water balance, which subsequently might alter land cover density and erosion rates [4]. In recent years, there has been a notable increase in the frequency and magnitude of processes related to the effects of rainfall in the Magdalena River basin, especially in human and material losses. For instance, during the 2010-2011 ENSO event (cold stage La Niña), heavy precipitation along with land-use changes negatively affected the economy and the development of basin.

Many riverine settlers depend on the Magdalena River for their subsistence. This area attracts low-income families that rely on subsistence agricultural and fishery. The erosive processes that occur both naturally and induced have significant consequences on the quality of life and safety of the inhabitants, and eventually affect the economic dynamics generated by the municipalities. Figure 1 show an eroded road section near the municipality of Malambo, which isolated 40 families from the urban center and made it impossible for them to bring their agricultural products for sale [5].

Figure 1: Magdalena river erosion in the Municipally of Malambo, Colombia [5]

Since Colombia is now a member of OECD (Organization for Economic, Cooperation and Development), there is a governmental need to improve the institutional response to events that are rare but could cause significant damage to many farmers and more, if they are low-income people [6]. Thereby, there is a need for identifying the current socio-economic conditions of the low-income riverine population of Magdalena River and for quantifying the river dynamics. This also requires improving the understanding of how these dynamics have evolved spatially and temporarily in the lower basin.

The main objective of this study is to quantify the morphological changes of the Magdalena River and its impact on economic activities (e.g., agriculture) using remote sensing observations. In particular, it is aimed to answer the following research questions: (1) What metrics could describe the Magdalena River dynamics by using freely accessible satellite observations? (2) How does Magdalena River dynamics trigger socio-economic impacts on the riverine population? and (3) What are the future steps to improve the understanding of the role of the Magdalena River dynamics on riverine population socio-economic sustainability? To that end, Landsat satellite images for years 1986 to 2019 and socio-economic data from public agencies were collected and subsequently statistical analysis on them performed.

2 Materials and Methods

2.1 Study area

The Magdalena River discharges into the Caribbean Sea at Bocas de Ceniza located at 11° 06' 22.78" N 74° 51' 08.19" W near the city of Barranquilla. The river flows through the central region of Colombia between the East and Central Andes, and its basin occupies circa 24% of the country. The lower river basin has a drainage area of 43,360 km^2 and is hydro-sedimentologically characterized by annual mean precipitation of 1,632 mm/yr, a mean annual water discharge of 7,500 m^3/s, a sediment discharge of 160 Mt/yr and sediment yield of 560 t/km^2.yr [7, 8].

The delta of the Magdalena River has an extension of approximately 1,690 km^2. In this area inhabit about two million people in the municipalities of Barranquilla, Soledad, Malambo and Sitio Nuevo [9] (Figure 2). The delta is morphologically characterized by plains of unfinished alluvial deposits of sedimentary origin conformed by sands, some lagoon complexes and coastal cords in the eastern zone. The Western area of the delta is highly urbanized; some areas have high resistance rock formations [10].

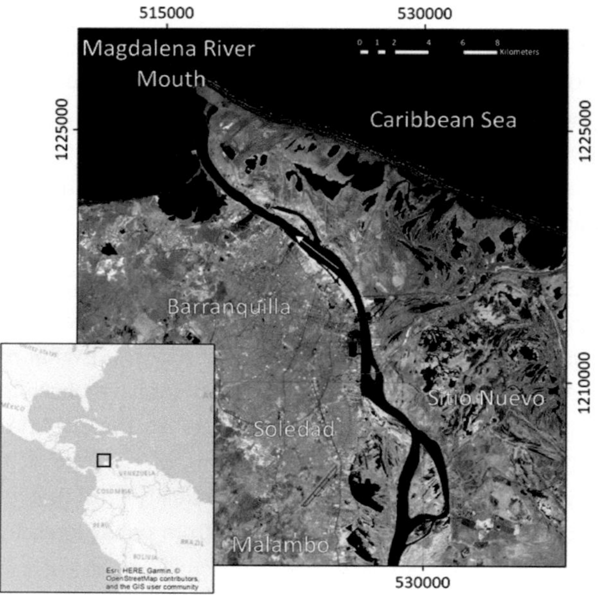

Figure 2: Satellite image of the study area

The spatial location of the municipalities shows that considerable portions of these are within the area known as "main channel"; this area is characterized by being the flood area in periods of increasing, and it is said that the river can flow in this area over time. Being an area, in which river sediments are deposited, the soil is characterized by having a high concentration of minerals that help crop growth, which is why it is an effective area for agricultural and livestock activities.

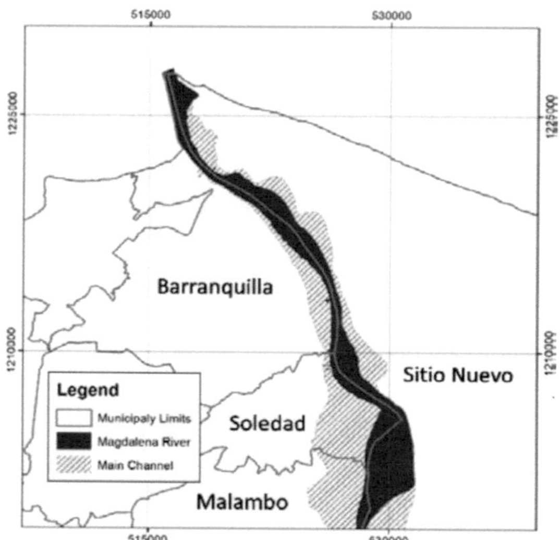

Figure 3: Main and active channel

In recent years, the processes of stiffening and construction of infrastructure have been carried out, such as contraction dikes in the area near the mouth to stabilize the river and to allow safe navigation in the river. Together with the intervention processes of human origin and climatic changes along the Magdalena river basin affecting the behavior of the upstream river, this is reflected in the increase of sediment transport and accelerating the processes of river dynamics in the area.

2.2 Data

The methodology of this study is based on recent erosion-accretion processes' analysis and land use evolution using the software ArcGIS 10.4 and tools for historical river bank positions created through the digitalization of Landsat images from 1986 to 2019 for the study area. The analysis of the satellite images was performed by combining different bands in the RGB spectrum to achieve greater contrast between water, terrain and areas with vegetation. These combinations differ between satellites given the technology of each [11]. Table 1 summarizes the characteristics of the satellite information gathered.

Table 1: Landsat satellites information

Satellite	Period	Resolution	Used bands (RGB)	Location (Path-Row)
Landsat 8	2014-2019	30 m	6-5-2	9-52
Landsat 7	2000-2013	30 m	5-4-2	9-52
Landsat 5	1986-1999	30 m	5-4-2	9-52

For the development of this research, hydrological information was provided by the Institute of Hydrology and Meteorology and Environmental Studies (IDEAM), which is the entity in charge of the management of scientific, hydrological, meteorological information and everything related to the environment in Colombia. Information of water levels, liquid and solid flows of the river were provided by the Magdalena River Observatory.

Socio-economic information was collected from data and statistics generated by the National Administrative Department of Statistics (DANE), responsible for the planning, compilation, analysis, and dissemination of the official statistics of Colombia such as population, agricultural production, and economic reports. Specific information on annual agricultural production by municipality is reported in conjunction with the Colombian Ministry of Agriculture. Geospatial information was also obtained, such as maps and shapefiles of land and land uses from the Geographic Institute Agustín Codazzi (IGAC). This institute is responsible for producing the official maps and basic cartography of Colombia and managing the national cadastral infrastructure and the national soil survey.

In many cases, this information is restricted, outdated or there are no reports of certain variables in several municipalities or meteorological gauging stations, so there is no total access to public information through portals on the web of the country's public entities. Limitations are thus generated in the development of academic research, since these data are essential to model or to characterize a condition, and it ends up excluding important variables due to insufficient information [12, 13].

2.3 Analysis of the Magdalena River dynamics

Erosion and sedimentation rates are evaluated with the Digital Shoreline Analysis System (DSAS v5); a tool developed by de United States Geological Survey (USGS) that calculates the change rates using statistical methods. This tool is intended for use in coastal zones, but it is useful in other scenarios to compute rates of change for any boundary-change problems that incorporates a clearly identified feature position at discrete times, such as riverbanks or land use/cover boundaries [14]. This tool allows quantifying river dynamics in the sector by characterizing the areas, where sedimentation has occurred, and in which erosion processes are experienced through the estimation of the riverbank rate of change. Several studies are conducted using DSAS in rivers, such as Matane, De La Roche and Yamaska in Canada [15], the Matanuska River in the southcentral Alaska [16], the Kilim River in Malaysia [17] and the Mekong River in Vietnam [18], in which the spatial-temporal variation of the riverbanks are analyzed. DSAS was used also for spatial analysis of deltas and estuaries worldwide, such the Krishna-Godavari delta [19], some river mouths in the region of Andalusia in Spain [20], in the Nile River delta in Egypt [21] and in the coastal area of the municipally of Puerto Colombia very close to the Magdalena river delta [22].

In order to estimate the erosion potential, the riverbanks of each available Landsat satellite image were digitalized. The Baseline for the DSAS model is the most recent bank corresponding to the month of June 2019. The temporal analysis was carried out for each riverbank separately and taking into account the geological composition and resistance of the river banks characterized by geotechnical studies conducted in the study area.

3 Results and Discussion

3.1 Riverbanks migration analysis and hydrologic variability

The evolution of the banks of the river using the DSAS tool from 1986 to 2019 generated several statistical adjustments, based on the spatio-temporal variation of each digitized shore as Net Shoreline Movements (NSM), Shoreline Change Envelopes (SCE), End Point Rate (EPR), Linear Regression Rate (LRR), and Weighted Linear Regression Rate (WLR), depending on the quantity and quality of the shoreline digitization due to the frequency and uncertainty of the information source. The analysis of the spatio-temporal evolution of the erosion and sedimentation processes in the study was performed using the LRR statistic, which fits a least-squares regression line to all historical points for each transect. This method uses all information collected unlike other methods such as the EPR. The linear regression rate is the slope of the line calculated for each transect, and the result is the evolution in m/yr, the results can be displayed as a colorized output of transects based on the rate of change.

The delta of the Magdalena River is a particular case. The anthropogenic interventions that have been developed in this as contraction dikes between the mouth and km 14 have largely stabilized the river, restricting certain processes of river dynamics in order to ensure the port activities that take place in the city of Barranquilla. The channeling of the river in this sector implies that in order to maintain the natural equilibrium condition of the river flow and sediment transport, the upstream natural processes are altered, accelerating certain processes of erosion and sedimentation fluvial dynamics. The analysis shows certain critical sectors that have presented high dynamics in the last 33 years, mainly the sector between the km 18 to 32 in the jurisdiction of the municipalities of Soledad, Malambo and Sitio Nuevo.

On the left bank of the river, the most notorious processes have been presented. At km 28, an erosive rate of about 65 m/yr is quantified with the DSAS tool. Several kilometers downstream, there is considerable sedimentation between 12 to 20 m/yr, which has generated the deposition of alluvial material. The right bank of the river has presented total erosion rates of about 16 m/yr (km 26). This area is the outer shore to the curve, on which the road connecting the municipalities of Sitio Nuevo is located, and Barranquilla, this route has the functionality of being also a flood protection dike, but erosion has seriously affected the bank. On several occasions, the dike has been broken by compromising the residents of the area and isolating the populations (Figure 4). Once the erosion and sedimentation rates have been determined, the total area lost and gained due to dynamic processes can be determined, a total erosion of 631 ha and a gain of 485 ha were estimated. This area is characterized by having a low to medium-low resistance to erosion due to its geological conformation of unconsolidated alluvial material. Therefore, the areas, in which material has been deposited recently, are not suitable for the development of urbanization or economic production (agriculture or animal husbandry). Figure 5 shows the average total variation of the riverbank for the period between 1989 and 2019 and soil resistance for the area.

Figure 4: Imminent erosion of the road that connects the municipalities of Sitio Nuevo and Barranquilla *(Source: Google Street View)*

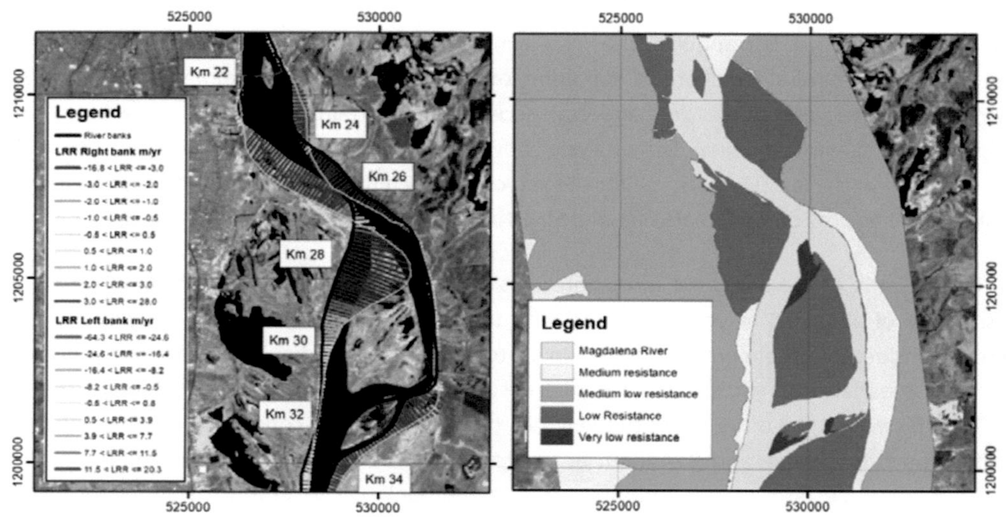

Figure 5: Erosion and sedimentation zones in contrast with soil resistance

It was found that none of the statistical tools present in DSAS specifically determines, which are the forces influencing the river morphodynamics. It is only a tool that allows the calculation of erosion and sedimentation rates that have been presented [23]. It was identified that there are certain time intervals, in which the channel has presented a greater variability in the progression of erosion-sedimentation processes, since the channel is morphologically conditioned to the river's hydrological variability. This variability is strongly influenced by the occurrence of the ENSO. During La Niña cold phase events, the flow of the Magdalena River can reach up to 12,000 m^3/s, and in El Niño warm phase, when rainfall is very low, the flow rates of the Magdalena River are 1,500 to 2.000 m^3/s [24, 8]. Figure 6 shows the climatic variation of the ENSO through the Niño Ocean Index (ONI) and the hydrological variability of the Magdalena River represented with the flow rate reported in the study area in the last 35 years. In the warm phase phenomena, the flow decreases given the reduction in precipitation along the basin and the sedimentation processes intensify, contrary to what happens in the cold phase.

Magdalena River in this sector has several curves that have suffered from lateral migration processes typical of rivers of meandering nature such as the Magdalena River in its lower basin. This migration is the result of low resistance erosion processes of external banks combined with sedimentation on the inner banks. This is a dynamic process that is reflected along the river, since morphological changes are not local, and the magnitude of these is closely linked to the occurrence of low-frequency hydrological events as explained above. Alluvial deposits located on the inner bank of the river are a consequence of low speeds and shear stresses that allow sediment to deposit, when this sediment do not get submerged (situation between 2014 and 2016), vegetation grows and stabilizes the sediment to form islands and to reduce the active channel width [25].

Similarly, floods are triggers of accelerated erosion processes due to the high hydraulic capacity of sediment transport, modifying the morphology of the river in short periods of time [26, 27].

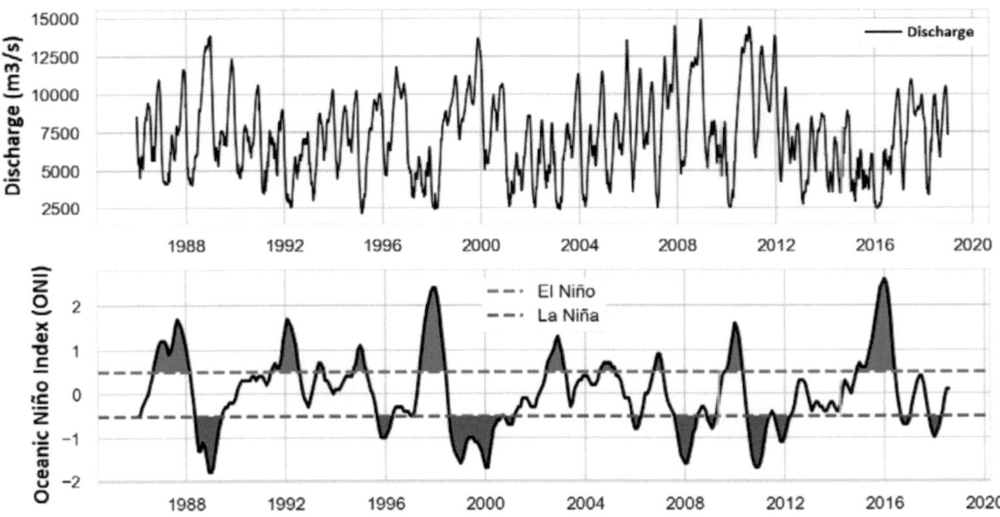

Figure 6: Averaged monthly discharge of Magdalena River and ONI anomalies

Lateral migration of the rivers is linked to changes in the river islands in the middle of the channel. There are two important river islands, which are Rondón Island and Cabica Island, both are located in the area of greatest mobility of the river analyzed above, and in these, it was identified that several families inhabit there for years have also used the land to cultivate. The processes of river dynamics have constantly modified the size of these islands to the point that Rondón Island has lost about 87 ha, equivalent to 80% of its surface, just like Cabica Island, which had a migration concordant to the movements that have suffered the bank of the river. Morphological changes generated by the 2010-2011 flood event caused that the dominant flow at Cabica Island to migrate from the right arm to the left, intensifying erosion-sedimentation processes downstream. Figure 7 shows the reduction in the surface of these islands through the riverbank digitization.

Understanding the morphological changes in fluvial systems is very important for river erosion mitigation, and it is a critical input for the generation of cartography and tools to spatially locate the historical sectors that have been flooded. It is necessary to generate in developing countries policies that seek to improve the risk management of communities in the event of changes in their environment and to allow identifying, which areas are unsuitable for developing human settlements [28].

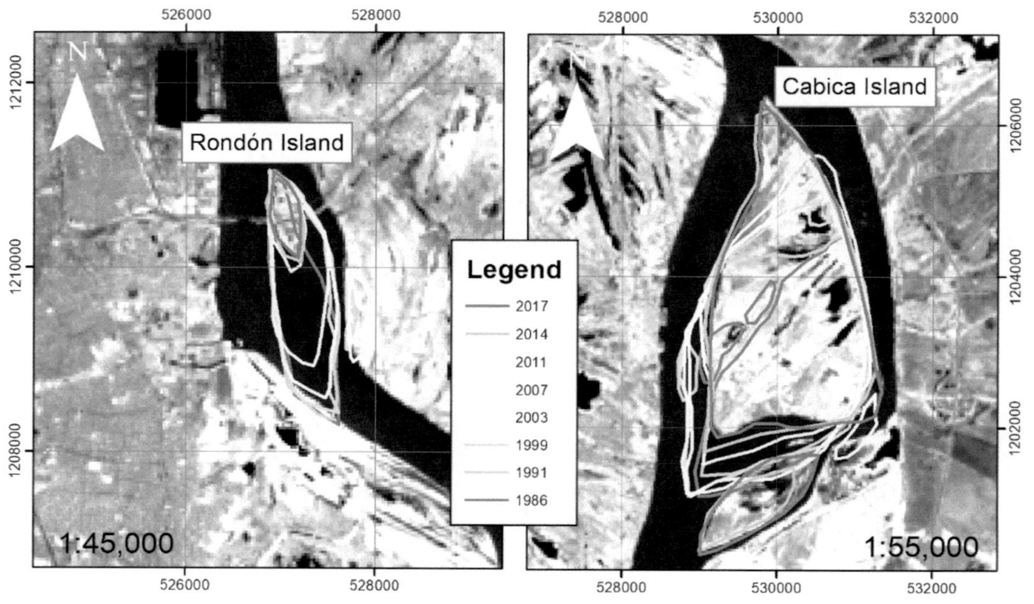

Figure 7: Morphological evolution of Rondón Island and Cabica Island

3.2 Impact on economic and social aspects

Magdalena River dynamics have eventually brought problems to the families in the area. The socioeconomic impacts originate around the decrease in the income of the inhabitants due to erosion in farming areas or livestock activities and the loss of these lands. In the area, agriculture is concentrated in cassava, plantain, lemon, mangoes and oil palm crops, which have a good profit if they are extensive crops [29]. Figure 8 shows land use in the study area. The reduction of the ecosystem services of the riverbank due to erosion has increased the Index of Unsatisfied Basic Needs in these populations. This situation without any compensation strategies raises the issue of problems regarding the coverage of basic needs, difficulties in family dynamics and social problems that affect the well-being and quality of life of individuals. The search for different livelihoods and the possibility of seeing their territory affected cause riparian communities to transform their social dynamics by being forced to move to other territories in search of improving living conditions or transforming their economic activities. These transformations sometimes involve the disarticulation of the social fabric, loss of cultural identity or poverty. In the case of the migrations that are generated by the loss of territory due to erosion, the individual culturally and emotionally detaches himself from his place of origin, facing new adaptation processes and different social systems, which can be overcome, depending in turn of the resources that the receiving place can provide.

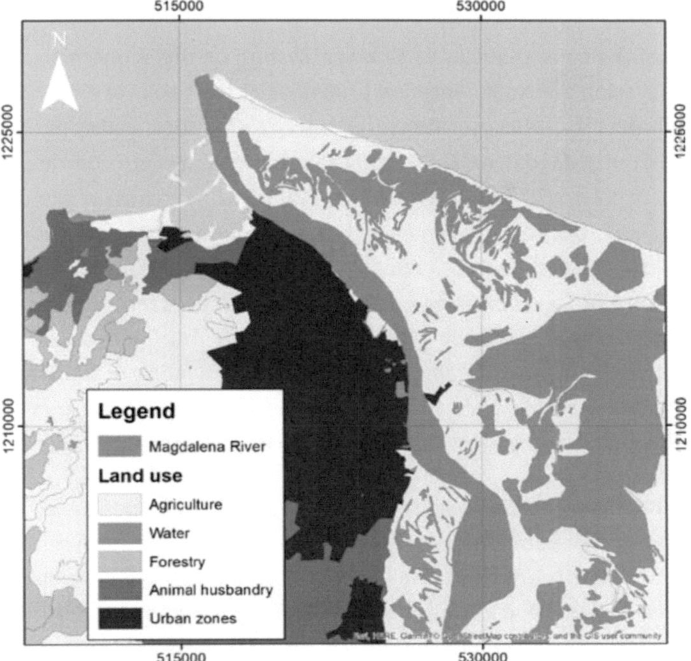

Figure 8: Land use reported in study area [30]

To this panorama, the disarticulation of the social fabric is added, which so far has shaped them, the displacement of economic activities or the search for other new activit es generates the breakdown of support networks. Which throughout life have built the people in their communities, facing new scenarios, in which their networks must be reconfigured, establishing new relationships as the main mechanism of adaptation to the new place. This same social fabric has allowed and safeguarded their customs, beliefs, and ways of life. Therefore, they fear that there will be a loss of cultural identity due to possible migrations and changes in occupations, especially amphibious culture, "in which the fisherman establishes a complex, dynamic and adaptive relationship with the cycles of the dynamics of the river. Which has been possible thanks to the awareness of the existence of common goods, such as marshes, beaches and banks" [31].

Of course, something expressed by the leaders of these communities is that due to the lack of occupation in the locals, mixed with the incipient need to get food for their families, they are exposed to other social problems, specifically to the increase in crime and child exploitation. It is required the development of economic and social policies that allow the integration of the inhabitants within the processes of river dynamics, considering the exposure and risk of living near the banks of the river and how to improve the economic returns for residents, reducing the number of victims and improving the living conditions of the residents.

4 Conclusions

The understanding of the consequences of a natural system on the economic activities was carried out in an area. The erosion and sedimentation processes of the banks of a river are key in the daily dynamics of the inhabitants, who use the water resource for their activities. The river dynamics that has developed in the Magdalena River study sector presented erosion rates of about 65 m/yr and sedimentation of 12 to 20 m/yr, which are quantified in losses of about 631 ha and the accretion of 485 ha land, but that present a high danger due to their low consolidation. This has generated trauma to the communities, since it has caused the loss of many hectares of land dedicated to agriculture or livestock or have left dozens of families incommunicado due to erosion of the roads, which were erroneously built a few meters from the shore. All morphological changes, which the river can generate an irruption in the social fabric of the area, change habits in agriculture activities many times, since the economic performance of the areas in production has been reduced and the inhabitants must seek other sources of income for its support.

In developing countries like Colombia, there is an institutional deficiency in the collection, updating, and publication of data. The generation of models and statistical results requires a large amount of information that is often not available, or access is often a problem. The methodology used in this paper allowed the description of these erosion-sedimentation phenomena from a statistical quantification of erosion and sedimentation rates, based on open satellite information and open-access tools, which can be related to basic hydrological parameters of stations of measurement in the river. It is a necessity to generate tools that allow the identification of vulnerable areas, improve decision-making in order to contribute to the welfare and economic production of the villagers, and raise awareness among the population before the risks, to which they are exposed to be located in danger zones.

5 Acknowledgments

This work was carried out by the Institute of Hydraulic and Environmental Studies of Universidad del Norte. The authors are also grateful to the Federal Ministry for Economic Cooperation and Development, Germany (BMZ), German Academic Exchange Service (DAAD), Excellence Center for Development Cooperation, Sustainable Water Management (EXCEED/ SWINDON), and Technical University of Braunschweig (TUBS) for their financial and technical support, which enabled participation at the International Expert Workshop on Water on Agricultural Practices: Training the Trainers, September 2019 in Río de Janeiro, Brazil.

6 References

[1]   J. Best: Anthropogenic stresses on the world's big rivers. *Nature Geoscience,* 2019, 12, 7-21,.

[2]   J. D. Restrepo: El impacto de la deforestación en la erosión de la cuenca del río Magdalena (1980-2010). *Revista Académica Colombiana de Ciencias Exactas, Fisicas y Naturales,* 2015, 39, 250-267.

[3] Banco de la Republica: El Río Magdalena: Escenario primordial de la patria. Biblioteca Cultura, June, 2013.

[4] J.D. Restrepo, J.P.M. Syvitski: Assessing the Effect of Natural Controls and Land Use Change on Sediment Yield in a Major Andean River: The Magdalena Drainage Basin, Colombia. Ambio, 2006, 35(2), 65-74.

[5] L. Jiménez: El río Magdalena se está llevando nuestras tierras: habitantes de Caimital, Barranquilla, El Heraldo, 29.03.2019.

[6] OECD: Agriculture and fisheries, 2019.

[7] IDEAM: River Database of the Magdalena drainage basin. Bogotá, 2003.

[8] J.D. Restrepo: B. Kjerfve: Magdalena river: Interannual variability (1975-1995) and revised water discharge and sediment load estimates. Journal of Hydrology, *2000, 2*35, 37-149.

[9] DANE: Proyecciones nacional y departamental de Población, 2006 – 2020. 2007. https://www.dane.gov.co/index.php/estadisticas-por-tema/demografia-y-poblacion/proyecciones-de-poblacion

[10] G. Vargas: Estudio geológico, geomorfológico, dinámica fluvial y alternativas de conexión al mar caribe. 2014.

[11] C.R. Leslie, L.O. Serbina, H.M. Miller: Landsat and agriculture—Case studies on the uses and benefits of Landsat imagery in agricultural monitoring and production. Open-File Report, p. 27, 2017. https://pubs.er.usgs.gov/publication/ofr20171034

[12] L.M. Garcia Corrales, H. Avila Rangel, R. Gutierrez Llantoy: Land-use and socioeconomic changes related to armed conflicts: A Colombian regional case study. Environmental Science and Policy, 2019, 97, 116-124,.

[13] R. Gutierrez Llantoy: Vías navegables de Sudamérica : retos y oportunidades. Barranquilla, 2018.

[14] D. Henningson, M. Berggren: Fluid Dynamics : Theory and Computation. Stockholm, August 2005. Lecture Notes at the Dept. of Mechanics and the Dept. of Numerical Analysis and Computer Science (NADA) at KTH.

[15] P.M. Biron, T. Buffin-Bélanger, M. Larocque, G. Choné, C.A. Cloutier, M.A. Ouellet, S. Demers, T. Olsen, C. Desjarlais, J. Eyquem: Freedom Space for Rivers: A Sustainable Management Approach to Enhance River Resilience. Environmental Management, 2014, 54, 1056-1073,.

[16] J.H. Curran, L. McTeague: Geomorphology and Bank Erosion of the Matanuska River, South-central Alaska. Scientific Investigations Report 2011 – 5214. p. 64, 2011. https://pubs.usgs.gov/sir/2011/5214/

[17] H.M. Suhaimi, M.H. Jamal, A. Ahmad: Assessment of river bank erosion at Kilim River, Langkawi using geospatial technique. IOP Conference Series: Earth and Environmental Science, vol. 169, 2018. https://iopscience.iop.org/article/10.1088/1755-1315/169/1/012012

[18] L.D. Nguyen, N.T. Minh, P.T.M. Thy, H.P. Phung, H.V. Huan: Analysis of changes in the riverbanks of Mekong River, Vietnam by using multi-temporal remote sensing data. International Archives of the Photogrammetry, Remote Sensing and Spatial Information Science, 2010, 38, 287-292,.

[19] A. Kallepalli, N. Rao, D.B. James, M.A. Richardson: DSAS-based change detection along the highly eroding Krishna-Godavari delta front. pdf document online, pp. 1-22.

[20] A. Prieto, J. Ojeda, S. Rodríguez, F.J. Gracia, L.D. Río: Procesos erosivos (tasas de erosión) en los deltas mediterráneos andaluces: herramientas de análisis espacial (DSAS) y evolución temporal (servicios OGC). XV Congreso Nacional de Tecnologías de la Información Geográfica, Madrid, AGE-CSIC, 2012, pp. 19-21.

[21] K. Dewidar, O. Frihy: Pre- and post-beach response to engineering hard structures using Landsat time-series at the northwestern part of the Nile delta, Egypt. Journal of Coastal Conservation, 2007, 11, 133-142.

[22] M. Bolívar, G. Rivillas-Ospina, W. Fuentes, A. Guzmán, L. Otero, G. Ruiz, R. Silva, E. Mendoza, M. Maza, L. García, Y. Berrío: Anthropic Impact Assessment of Coastal Ecosystems in the Municipality of Puerto Colombia, NE Colombia. Journal of Coastal Research, 2019, 92, 112-120.

[23] T.D.T. Oyedotun: Shoreline Geometry: DSAS as a Tool for Historical Trend Analysis. Geomorphological Techniques (online edition), 2014, 2, 1-12.

[24] J.C. Restrepo, J.C. Ortíz, J. Pierini, K. Schrottke, M. Maza, L. Otero, J. Aguirre: Freshwater discharge into the Caribbean Sea from the rivers of Northwestern South America (Colombia): Magnitude, variability and recent changes. Journal of Hydrology, 2014, 509, 266-281.

[25] J. Pierre: River Mechanics, 2nd Ed. C U Press, Cambridge, 2018, p. 499.

[26] S. Yousefi, S. Mirzaee, S. Keesstra, N. Surian, H.R. Pourghasemi, H.R. Zakizadeh, S. Tabibian: Effects of an extreme flood on river morphology (case study: Karoon River, Iran). Geomorphology, 2018, 304, 30-39.

[27] M. Gharbi, A. Soualmia, D. Dartus, L. Masbernat: Floods effects on rivers morphological changes application to the Medjerda River in Tunisia. Journal of Hydrology and Hydromechanics, 2016, 64, 56-66.

[28] M. Spaliviero, M. De Dapper, S. Maló: Flood analysis of the Limpopo River basin through past evolution reconstruction and a geomorphological approach. Natural Hazards and Earth System Sciences, 2014, 14, 2027-2039.

[29] M. d. Agricultura: Agronet - Participación Departamental en la Producción y en el Área Cosechada,» 2017. [online]. Available: https://www.agronet.gov.co/estadistica/Paginas/home.aspx?cod=2.

[30] IGAC: Datos Abiertos – Agrología. Instituto Geográfico Agustín Codazzi, 2017. [Online]. Available: https://geoportal.igac.gov.co/contenido/datos-abiertos-agrologia.

[31] R. Delvalle: El proyecto de recuperación de la navegabilidad del rio magdalena como generado de conflictos ambientales de la llanura inundable del rio magdalena. Universidad Nacional, Bogotá, 2017.

THE USE OF HUMIC SUBSTANCES TO REDUCE SOIL ERODIBILITY

<u>Gricel Portillo</u>[1,3], José Araruna[1], Eduardo Brocchi[2]

[1]*Department of Civil and Environmental Engineering, Pontifical Catholic University of Rio de Janeiro, R. Marquês de São Vicente, 225 - Gávea, Rio de Janeiro - RJ, 22451-90, Brazil;* <u>*gricelp@gmail.com*</u>

[2]*Department of Chemical and Materials Engineering, Pontifical Catholic University of Rio de Janeiro, Rio de Janeiro, Brazil;*

[3]*Materials Testing Institute, UMSA University, La Paz, Bolivia;*

Keywords: Humic substances, erosion, erodibility, restoration, soil conditioners

Abstract

The research evaluates the efficiency of incorporating HS (Humic Substances) into a material to reduce the material's erodibility. Erodibility (i.e., the susceptibility to erosion) depends on a series of parameters, including particle-size distribution, mineralogical composition, presence of organic matter, and moisture content. The HS was synthesized from the waste of mineral coal from the company "Copelmi Mineracao" in the state of Santa Catarina, and is expected to minimize erosion caused by agricultural practices. The characteristics of the soil in Santo Antônio de Padua and the HS used for erosion control were evaluated by performing laboratory tests to determine the erodibility of soil samples with added HS to reach a content of 3% HS by weight, and comparing these to soil samples with no HS added. The soil mineralogy and the cation exchange capacity (CEC) of the samples were analyzed to interpret the results. To analyze the effect of contact time of the HS with the soil, samples were stored for 2 months or 1 year to preserve the moisture content and specific weight of the samples throughout the storage time. HS addiction was shown to increase cation exchange capacity and to interact with clay minerals presented in the soil. Therefore, these characteristics contribute to the stability of aggregates and reduce soil erodibility.

1 Introduction

Chemical deterioration is part of the soil degradation in the region of Santo Antônio de Padua. The region was progressively cleared for establishment of coffee plantations which, over time, caused the loss of soil fertility [1]. This chemical deterioration refers to the loss of nutrients from soil or organic matter; some of these nutrients may also have been lost due to erosion, since torrential rains can cause several nutrients to be washed away, especially in unprotected soils. Physical deterioration was also observed, identifying possible soil compaction resulting from trampling of cattle. This compaction interferes with water infiltration and causes greater surface runoff and, consequently, increases erosion due to water movement. This research is part of research projects carried out by the research and development network "Rede Carvao", which received financial support from CAPES, CNPq, and FINESP, aiming to support the sustainability of coal production in

Brazil. The contribution of this research is to provide an alternative use for the HS obtained from a coal extraction waste. In addition, the research was also part of the project entitled "Obtaining Humic Substances and their Applications in Sustainable Development and Environmental Sanitation: Waste Treatment and Recovery of Degraded Areas supported of FAPERJ".

The coal mining wastes have a significantly high level of contaminants [2, 3], and the environmental impacts from coal extraction can include the contamination of groundwater, alteration of water bodies, and soil contamination [4-7]. The management of mining waste and minerals processing is, therefore, a fundamental issue in sustainable development of these activities [8].

In Brazil, coal is an important source of energy, but its use generates serious environmental impacts in the State of Santa Catarina. Although there has been a positive impact on economic development of the region, environmental damage has been sustained [9, 10]. Therefore, some researchers have been investigating the use of waste from this mining activity to obtain HS [11, 12]. The HS's present in the organic matter of the soil are universally recognized as essential components in productive soils [13].   Different organic amendments (for example, sewage, sludge, green waste, or compost) generally increase the organic carbon content of the soil, plant growth, microbial activity, and improve the physical proprieties of the soil by stimulating formation of the particle aggregates and reducing soil erosion [14].

Humic substances influence soil parameters, imparting characteristics of moisture retention and high cation exchange capacity (CEC), which favor physicochemical interactions. These are related to soil fertility, improving water penetration and infiltration capacity, heavy metal retention, and soil aggregation. Other research analyzing the behavior of the HS on the proprieties of a silty-clay-loamy soil concluded that these substances increase the porosity and stabilize the aggregates of the soil [15].

The risk of erosion depends as much on the natural conditions of the soil as on the use of the soil. Natural conditions refer to climatic factors, topography, vegetation cover, and the characteristics of the vegetation. With respect to soil use, any human activity that requires the removal of vegetation cover may promote erosion, as well as geological processes and improper exploitation of natural resources in different activities, which contribute to the emergence of degraded areas by modifying the original features of soil [16]. Chemical deterioration can occur in soil, which refers to the loss of nutrients or organic matter, and some of these nutrients are also lost by erosion during heavy rains, especially in unprotected soils [17].

Erosion is a process that may have geological or anthropic origin. Geological erosion is a product of geological activity (water, wind, and ice) on the earth's surface, and is a natural process. Anthropic erosion refers to that, which is the product of man's interference with the environment, intensifying the action of rainwater and wind on the ground [18, 19].

The mechanisms of erosion are considered to be three fundamental processes: detachment, transport, and deposition of soil particles.

The detachment occurs due to the impact of raindrops on soil surface without vegetation cover. The vegetation coverage may have been removed to use the land for mining activities, livestock, or to grow some crops that degrade the soil, all of which can cause soil sealing, reduction in water infiltration, and surface runoff. Subsequently, the transport of the particles occurs due to the superficial runoff of water that does not infiltrate the soil. Depending on the intensity of runoff, surface erosion may occur (laminar erosion), or erosion may occur in shallow or medium-deep channels opened by the force of water runoff (erosion in gullies), or through large furrows, which concentrate large amounts of water, be it superficial or underground (erosion in ravines). The deposition process consists of storing the eroded soil in rivers, lakes, dams, terraces, which causes various environmental impacts.

The soil structure is composed of the arrangement of particles and the system of forces between them, which reflect all facets of soil composition, past and present, and the environmental state [20].

Soil texture may be indicative of soil erodibility; silt and clay particles are light, and since the strength of water overcomes the cohesion attributed to their chemical composition, they are displaced. Larger particles such as coarse sand, gravel, and boulders tend to accumulate on the surface, because their mass is responsible for the frictional force that resists erosion. As for texture, aconsiderable silt content in soil generally contributes to high erodibility [21]. Clays as isolated particles are organized into flocculated structures, oriented structures, or dispersed structures [22]. Flocculation results from face-to-edge interactions that combine negative and positive charges, which cause the formation of a cavity in the floc aggregates and thereby provides cohesion and increased erosion resistance. Soil aggregation is the process, by which aggregates of different sizes are joined and held together by different organic and inorganic materials [23]. Erosion resistance depends on the nature of the stabilizing agents formed between the aggregates. Particle adhesion is caused by substances with properties that help bind the particles together. Within this set of substances are organic matter and mineral substances such as iron oxides and clay minerals that act on the soil giving it greater aggregation and greater structural stability, making it less erodible [24].

One of the main agents involved in erosion is the organic matter content. Iron oxides, aluminum oxides (amorphous), and organic matter exert a marked influence on the formation of microaggregates (mainly in soils rich in these oxides). Some authors emphasize the importance of soil organic matter content to its erodibility and aggregate stability [23]. Additionally, microbiological processes in organic matter that occur in the soil matrix can produce substances that aid in the aggregation of particles in the soil [26].

Another important parameter regarding soil erodibility is the soil's moisture content. In areas of the Federal District affected by gulling processes, it was observed that suction stresses have an important effect on unsaturated soil resistance, and saturation fronts generate significant resistance reduction in soils [27]. In order to analyze this effect, a qualitative evaluation of erodibility was also conducted by analyzing the behavior of undisturbed samples against capillary rise, and the results of the study showed that the presence of organic matter, roots, and iron oxide tend to stabilize aggregates [25]. The authors also concluded that the moisture content of both soil and sedimentary rock studied is decisive for the behavior of the material against the erosive action of water. During dry periods, the exposed layers lose moisture to the atmosphere and are easily eroded in a torrential rain event. In the study of the evolutionary process of some gullies, it was observed that moisture content exerts a marked influence on the erosion process, based both on the evaluation of erosive processes in the field during periods of rain and drought and on soil tests conducted in the laboratory under saturated and unsaturated conditions [28].

Humic substances and clay minerals form compounds that contribute to particle aggregation and thereby reduce soil erodibility [18]. The structure of humic substances contains voids of different sizes, where other organic or inorganic compounds may be housed. In this sense, HS's can contribute in different ways to prevent soil erosion, through interaction with metals, retention of other organic substances, and forming stable humic-clay complexes that favor the aggregation of particles. Soil aggregation is influenced by the content of soil moisture [29]. This aggregation is defined as the result of the union of the primary particles: sand, silt and clay, and other soil components such as organic matter and calcium and magnesium carbonates, yielding distinct masses with stable aggregates. Researchers also presented a work that analyzed the behavior of humic substances in the properties of a silty clay soil and concluded that these substances increased the soil porosity and stabilized the soil aggregates [15]. It is important to mention that adsorption of humic substances in soil is limited and states that there is a limit to the amount of carbon and nitrogen, that can be associated with clay and silt particles [30].

A dynamic exists in particle aggregation in the face of climatic conditions, cultivation practices, and decomposition of organic matter. It is known that conceptual models exist that describe the effect of different organic products added to the soil. According to the proposed model, aggregate stability varies at different scales after incorporation of these products; they may vary between weeks, months, or years after their incorporation into the soil. Monnier's model suggests that fresh organic matter has an intense effect after one month of incorporation. On the other hand, decomposed organic matter has a very low effect over this period but its effectiveness increases over time [31].

The aim of this study was to test a humic substance from coal mining waste for reducing the soil erodibility in Santo Antonio de Padua. The evaluation was carried out on specimens with the addition of 3% of HS in weight and two different storage times, and to observe the behavior of the soil with this HS over time.

2 Material and Methods

The soil characterization tests were performed in the facilities of the Geotechnical and Environmental Laboratory of the Civil Engineering Department of Pontifical Catholic University of Rio de Janeiro, Brazil (PUC-Rio). The determination of moisture content (ω) was performed according to NBR 6457. The value of grain density (Gs) was obtained according to the methodology established in NBR 6508. Particle size analysis was performed following the procedures established in NBR 7181. The test to determine the Liquid Limit (ωL) was performed according to NBR 6459 and the Plasticity Limit (ωP) followed the procedures established in NBR 7180, all referred in [32].

In order to determine the content of soil organic matter, organic carbon determination test was performed according ASTM 5291 [33] at the Marine and Environmental Studies Laboratory of the PUC-Rio, Department of Chemistry, using the SHIMADZU TOC 5000A equipment. According to the procedures established in the EMBRAPA Handbook [34], the organic matter (OM) content can be determined as a function of the total organic carbon content (TOC): OM (%) = $TOC \times 1.724$.

The identification of the minerals present in the soil was conducted with X-ray diffraction using the Bragg method [35]. The analyses were performed at the facilities of the Diffractometry Laboratory of the Department of Chemical Materials and Process Engineering of PUC-Rio (Siemens diffractometer, model D5000). For the X-ray fluorescence analysis, a Bruker model S2 Ranger was used to perform a semi-quantitative analysis with the Espectra EDX program.

The HS from mineral coal waste was tested, consisting of a liquid part and a precipitated part. The supernatant contained mainly humic acid and the precipitated fulvic acid, humine, and other inert substances. The density of the substance was 1.1 kg/L and pH 8. The elemental composition of the humic substance was found to be 11.6% carbon, 1.8% Nitrogen and 1.3% hydrogen.

For the erodibility evaluation, the disintegration test was performed. The disintegration test evolved from the crumb test, which are tests used to identify dispersive soils by immersing soil aggregates in water. However, the main objective of this test was to verify the particular breakdown of a cubic or cylindrical soil sample, regardless of material dispersion.

The samples tested were made using dry soil, enough water necessary to reach the moisture content equal to that present in the field, and HS at 3% by weight. A control sample without any HS was also created for comparison. After obtaining a homogenous mixture, the sample was placed into a metal cylindrical mold measuring 10 cm in diameter and 6 cm in height. The mold was then placed into a mini-CBR manual press, where it was compacted until the sample obtained a specific weight equal to that it would have in the field.

The cylindrical specimens were wrapped in PVC film and stored in plastaform boxes with the aim of preserving the moisture over a specific storage period. Two storage periods were tested: two months and one year. Once the desired storage period was achieved, the cylindrical specimens were removed and cubic samples measuring 5 cm on each side were cut from the center of the cylindrical specimens. The cubic samples were then partially immersed in deionized water in a metal tray, where the level of the deionized water was at half the height of the tray.

3 Results and Discussion

In the unified soil classification system, this soil is classified as silty sand and the minerals identified in the X-ray diffraction are kaolin, Illite, feldspar, and quartz. The characteristics of the soil that was tested are shown in Table 1.

Table 1: Characteristics of the soil used in the tests

Granulometric Analysis	Gravel 13.6%
	Sand 48.0%
	Silt 28.5%
	Clay 9.9%
Atterberg limits	ωL 32%
	non plastic
Gs	2.68
OM	≤ 0.3%
Semi-quantitative analysis using fluorescent X-rays	47.3% O
	24.6% Si
	16.3% Al
	5.84% Fe
	2.56% K
	1.09 Mg
	0.64% Ti
	0.5% Ca
	0.37% Na
	0.13% Ba

The specimens were constructed and stored for two months and one year, respectively, before they were evaluated. The natural specific weight of the specimens was on average 16.75 kN/m^3 and the moisture content 22.28%.

After the specimens were stored and wrapped with film paper, the presence of fungi was observed on the surface of the cylindrical specimens in contact with the film paper. Cubes measuring 5 cm on each side were cut from the center of these cylinders, but no fungi were observed on the cubes.

In the disintegration test, it was observed that increasing the HS content in the soil decreased the soil erodibility. And one can say that the improvement in soil erodibility due to mixture with the HS is also a function of the storage time of the sample, as noted from specimens with the same characteristics and one year of storage.

Around the world, products with high organic carbon content are being used to control soil erosion [36, 37]. In this way, the waste of coal mining used in the tests contains 11.6% carbon, and this content contributes to the decrease of the erodibility of the soil.

The cation exchange capacity (CEC) of the soil is a characteristic directly linked to its erodibility, as the CEC refers to the sum of the negative charges present in the clay fraction and the organic matter that holds cations. The tendency of increased cation exchange from the application of humic compounds was also reported in other research [38, 39].

EC values in soil without added HS also vary as a function of time, as observed in the results of this research, and range from 1.4 to 1.8 cmol/kg in soils stored for two months and one year, respectively. Those specimens with 3% of added HS varied their CEC between 1.6 and 2 cmol/kg in soils stored over the same periods. This may have happened due to possible microbial activity in the presence of fungi that grew in the soil during the storage time. The presence of these microorganisms was evidenced mainly in the soil, which was in contact with the film paper, since this was observed to conserve the moisture content. Darkness and humidity are a favorable environment for the development of these microorganisms. In addition, the average temperature around 22 °C may also have contributed to the microorganism growth. The fungi present in the soil may have an impact on soil erodibility, because these microorganisms may be partly responsible for changing the cation exchange capacity of the analyzed soils.

4 Conclusions

The soil studied is a silty sand that has a high erodibility, and, therefore, the addition of humic substances is a measure that can be used to improve areas degraded by erosion in the region of Santo Antônio de Padua. The evaluation of HS from the waste produced from coal mining proved to be efficient for preventing soil erodibility, increasing cation exchange capacity, and interacting with soil minerals. For this reason, this HS can be used to mitigate erosion problems in agricultural practices.

With regards to the physicochemical interaction, it can be noted that feldspars containing potassium that are present in the soil of Santo Antônio de Padua can attach to the HS compounds due to the retention power of van der Waals forces and electrostatic connections resulting from cation or anion exchange. The erodibility of soil with added HS decreases, because the kaolinite and mica in the soil also interact with the HS, which has a structure containing voids, capable of accommodating inorganic compounds present in the soils studied.

5 Acknowledgements

The authors would like to thank the following entities that provided the financial support to make this research possible CAPES, CNPq, FINEP and FAPERJ. The corresponding author would also like to thank to DAAD and EXCEED-SWINDON project for support to participate at the International Expert Workshop on "Water on Agricultural Practices: Training the Trainers" held in Rio de Janeiro, Brazil, September 15 – 21, 2019.

6 References

[1] Calderano, S.B.: Geo-environmental assessment of the Municipality of Santo Antonio de Pádua, RJ Land Use Potential and Limitations. MSc Thesis, Federal University of Rio de Janeiro, Geoscience Institute, URFJ, 2005. *(In Portuguese)*.

[2] Giam, X., Olden, J., Simberloff, D.: Impact of coal mining on stream biodiversity in the US and its regulatory implications. Nature Sustainability, 2018, 1, 176–183

[3] Hussain, R., Luo, K., Chao, Z., Xiaofeng Z.: Trace elements concentration and distributions in coal and coal mining wastes and their environmental and health impacts in Shaanxi, China. Environmental Science and Pollution Research, 2018, 25, 19566–19584.

[4] Komnitsas, K., Modis, K.: Soil risk assessment of As and Zn contamination in a coal mining region using geostatistics. Science of the Total Environment, 2006, 371, 190-196.

[5] Bian, A., Inyang, I., Daniels, J., Otto, F., Struthers, S.: Environmental issues from coal mining and their solutions. Mining Science and Technology (China), 2010, 20, 215-223.

[6] Ribeiro, J., Taffarel, S.R., Sampaio, C.H., Flores, D., Silva, F.L.O.: Mineral speciation and fate of some hazardous contaminants in coal waste pile from anthracite mining in Portugal. International Journal of Coal Geology, 2013, 109–110, 15-23.

[7] Pandey, B., Agrawal, M., Singh, S.: Ecological risk assessment of soil contamination by trace elements around coal mining area. Journal of Soils and Sediments, 2016, 16, 159- 168.

[8] Franks, D.M., Boger, D.V., Cote, C.M. Mulligan, D.R.: Sustainable development principles for the disposal of mining and mineral processing wastes. Resources Policy, 2011, 36(2), 114-122.

[9] Silva, L.F.O., Oliveira, M.L.S., da Boit, K.M.: Characterization of Santa Catarina (Brazil) coal with respect to human health and environmental concerns. Environmental Geochemistry and Health, 2009, 31(4), 475- 485.

[10] Zocche, J.J., da Silva, L.A., Damiani, A.P.: Heavy-Metal Content and Oxidative Damage in Hypsiboas faber: The Impact of Coal-Mining Pollutants on Amphibians. Archives of Environmental Contamination and Toxicology, 2014, 66(1), 69-77.

[11] Pehlivan, E., Arslan, G.: Comparison of adsorption capacity of young brown coals and humic acids prepared from different coal mines in Anatolia. Journal of Hazardous Materials, 2016, 138, 401-408.

[12] Firpo, B.A., Amaral Filho, J.R., Scheneider, I.A.H.: A brief procedure to fabricate soils from coal mine wastes based on mineral processing, agricultural, and environmental concepts. Minerals Engineering, 2015, 76, 81-86.

[13] Brunetti, G., Plaza, C., Clapp, C.E., Senesi, N.: Compositional and functional features of humic acids from organic amendments and amended soil in Minnesota, UMSA. Soil Biology and Biochemestry, 2007, 39, 1355-1365

[14] Bastida, F., Hernández, T., García, C.: In: The Future of Soil Carbon - Its Conservation and Formation (Eds:. Garcia, C., Nannipieri, P., Hernandez, T.), Ch. 8 - Soil Erosion and C Losses: Strategies for Building Soil Carbon, Academic Press, Elsevier, Spain, 2018, p. 215-238.

[15] López-Cervantes, R., Gallegos Del Tejo, A., Peña-Cervantes, E., Reyes-López, A., Castro-Franco, R., Chavez-Gonzales, J.: Humic substances from different sources on some physical properties of a silty-clay-loamy soil. Terra Latinoamericana, 2006, 2006, 24(3), 303-309. (*In Spanish*)

[16] Araujo, G.H.S., Almeida, J.R., Guerra, A.J.T.: Environmental management of degraded areas. 4th edition. Rio de Janeiro, Bertrand: Brazil, 2009 *(In Portuguese)*

[17] Li, Q., Liu, G., Zhang, Z., Tuo, D., Miao, X.: Structural Stability and Erodibility of Soil in an Age Sequence of Artificial *Robinia pseudoacacia* on a Hilly Loess Plateau. Polish Journal of Environmental Studies, 2016, 25(4), 1595-1601.

[18] Guerra, A.J.T., Da Silva, A.S., Botelho, R.G.M.: Soil erosion and conservation. Concepts, themes and applications. Rio de Janeiro, Bertrand, Brasil, 2010. (*In Portuguese*)

[19] Guerra, A.J.T., Jorge, M.C.O. (Organizers): Erosive processes and recovery of degraded areas. São Paulo, Oficina de Textos, 2013. *(In Portuguese)*

[20] Mitchell, J.K., Soga, K.: Fundamentals of Soil behavior. John Wiley & Sons, New Jersey, 2005.

[21] Camapum De Carvalho, J. (Org), Cordão Net, M.P. (Org), Aguiar, L.A. (Org): Commemorative book of the 20th anniversary of the Graduate Program in Geotechnics of the University of Brasília. Editora FT, 2009. *(In Portuguese)*

[22] Santos, P.S.: Clay technology. Vol. 1: Fundamentals. São Paulo. Ed. Universidade de São Paulo, 1975. (*In Portuguese*)

[23] Amézketa, E.: Soil Aggregate Stability: A Review. Journal of Sustainable Agriculture, 1999, 14(2-3), 83-151.

[24] Lepsch, I.F.: Soil Formation and Conservation. Oficina de Textos. São Paulo, 2002. *(In Portuguese)*

[25] Avila, L.O, Nummer, A., Pinheiro, R.B.: Erosion in the Sarandi stream microbasin: Buraco Fundo ravine, Santa Maria / RS- Brazil. In: II Ibero-American Seminar on Physical Geography, University of Coimbra, 2010.

[26] Haynes, R.J., Francis, G.S.: Changes in microbial biomass C, soil carbohydrate composition and aggregate stability induced by growth of selected crop and forage species under field conditions. European Journal of Soil Science, 1993, 44(4), 665-675.

[27] Mortari, D.: Geotechnical Characterization and Analysis of the Erosion Evolutionary Process in the Federal District. MSc Thesis, Department of Civil and Environmental Engineering, Faculty of Technology, UnB, Brasilia, DF., 1994.

[28] Lima, M.C.: Physicochemical and mineralogical degradation of massifs together as gullies. PhD Thesis. Department of Civil and Environmental Engineering, Faculty of Technology, UnB, Brasília, DF, 2003. *(In Portuguese)*.

[29] Kiehl, E.J.: Organic Fertilizers. Ed. Agronômica Ceres, São Paulo, 1985. *(In Portuguese)*

[30] Hassink, J.: The capacity of soils to preserve organic C and N by their association with clay and silt particles. Plant and Soil, 1997, 191, 77–87.

[31] Abiven, S., Menasseri, S., Chenu, C.: The effects of organic inputs over time on soil aggregate stability – A literature analysis. Soil Biology and Biochemistry, 2009, 41(1), 1-12.

[32] ABNT: Brazilian Association of Technical Standards. Rio de Janeiro, 1984. (*In Portuguese*)

[33] ASTM: American Society for Testing Of Materials. ASTM D5291, Standard Test Methods for Instrumental Determination of Carbon, Hydrogen, and Nitrogen, Pennsylvaria. USA, 2010.

[34] EMBRAPA: Methods of soil analysis *(In Portuguese)*. Brazilian National Soil Research Center. Rio de Janeiro, RJ, 1997.

[35] Klein, C., Hurlbut Jr., C.S.: Manual of Mineralogy. 21st Ed., Dana J.D. (ed.), John Wiley& Sons., New Jersey, 1993.

[36] Jien, C., Wang, S.: Effects of biochar on soil properties and erosion potential in a highly weathered soil. Catena, 2013, 110, 225-233.

[37] Peng, X., Zhu, Q., Xie, Z., Darboux, F., Holden, N.: The impact of manure, straw and biochar amendments on aggregation and erosion in a hillslope Ultisol. Catena, 2016, 138, 30-37.

[38] Borgues, C.S., Ribeiro, B.T., Cabral, D.A., Wendling, B.: Humic substances in different type of soil aggregates treated with organic waste. Organic matter and environmental quality. In: X Brazilian Meeting of Humic Substances. Brasilia, DF: EMBRAPA, 2013. *(In Portuguese)*

[39] Ronquim, C.C.: Soil fertility concepts and proper management for tropical regions. EMBRAPA Satellite Monitoring. Research and Development, Campinas, 2010. *(In Portuguese)*

POTENTIALITY OF RADICAL TERRACING, A PREVALENT WATER EROSION CONTROL STRATEGY IN RWANDA

I. Jean Pierre Bavumiragira

UNEP Tongji University, Institute of Environment for Sustainable Development, Department of Environmental Engineering, Shanghai, 200092, P.R. China; jpbavumiragira@gmail.com

Keywords: Erosion control, land degradation, radical terrace, Rwanda, steep slope

Abstract

Water use in agriculture is at the core of any discussion of water and food security. It accounts for 70% of water withdrawal globally. Agriculture is the main economic activity for the people of Rwanda, providing employment to about 86% of the total population. The country's topography has been the main cause of land degradation due to severe soil erosion, which recently reflected to poverty and food insecurity. The objective of this study was to highlight the main reason of the potential applicability of radical terracing as the best soil water conservation method in Rwanda's highly steep slopes, population perception in adopting the technology and recommending a way of improvement. Recent nationwide researches conducted about radical terracing implementation and its effectiveness revealed that radical terracing is an environmentally friendly, economical, profitable and effective method, which often increases farm's productivity. Farmers perceive that it is undoubtable solution, although it is labor intensive and hard to adopt, it increases yields when properly planned, maintained and combined with other erosion control methods such as the application of manure, compost or mineral fertilizers. This would be improved, when governmental and NGOs incentives continue and strengthen community-based mobilization to work together in associations or cooperatives to fully adopt the technology, because no farmer working alone can afford its cost.

1 Introduction

In agriculture, a terrace is a levelled section of a hill cultivated area, designed as a method of soil conservation to slow or prevent the rapid surface run-off of irrigation water. Often such land is formed into multiple terraces, giving a stepped appearance. This form of land use is prevalent in Rwanda and is used for crops requiring a lot of water. Terraces are also easier for both mechanical and manual sowing and harvesting than a steep slope would be. Rwanda is often characterized as having very high rural population densities, lush vegetation due to high rainfall, and steeply sloping highlands [1]. Erosion and land degradation have long been assumed to be severe and a major reason for the poverty and food insecurity in the country [2-4]. In a nationwide survey taken in the early 1990's, it was found that soil fertility was declining in all parts of the country with only one prefecture, Gisenyi which is currently Rubavu district, showing a lower degree of impact [5]. It was found that the area of the country that was experiencing the worst poverty and social problems was also the region experiencing the worst land degradation [6, 7]. One estimate was that in 1990, erosion caused the loss of productivity equivalent to 8,000 ha/yr., enough to feed 40,000 people [8-10]. Rwandan farmers depend primarily on seasonal crops, such as maize, sorghum, beans and tubers, and hand hoe their soil at least twice a year, exposing the soil to erosion and rapid decomposition of organic matter [11]. The dominance of seasonal crops increased following the decline of international coffee prices starting in the early 1990s. As erosion control measures, a large increase in tree planting around homes and on field boundaries reduced the fuelwood shortage, but governmental and NGOs efforts to introduce agroforestry and other forms of biological erosion control have not been widely adopted, partly because of the perception that they occupy much space on the fields and compete with crops for nutrients [12]. Terracing had been relatively common on small-scale farms but now is a prevalent, effective and profitable soil water conservation method in Rwanda [13, 14]. However, it is labor intensive, time consuming, requires regular inspections and large amounts of material, which is a barrier for its adoption to farmers. This paper describes the origin of this technology, its types, technical design specification, potential advantages and socio-economic and environmental benefits, its cost-benefit analysis and finally the current people perception for its adoption and the recommended way of improvement.

2 Erosion Control Methods in Rwanda

Many researches have been conducted in Rwanda testing various techniques to reduce erosion. The effects of competition with the crop for light, water and nutrients would need to be considered before specific techniques could be promoted [15-17]. A replicated finding was biological erosion control [18]. Farmers have been already using various form of biological control to reduce erosion, including leaving crop residues on the fields, planting trees and bushes around the fields and multi-cropping to cover the soil during the rainy season. Nevertheless, controlling soil erosion and increasing soil organic matter and nitrogen was insufficient to increase productivity on the acidic, ferrallitic soils of Rwanda due to phosphorus deficiency. Applications of mineral fertilizers and dolomite, in addition to erosion control, produced a reasonable yield [19]. In the past, the government focused almost exclusively on controlling erosion to the exclusion of other soil management or improved agronomic practices. An approach that had been promoted by the government "radical terraces" (bench terraces) would be more effective in reducing erosion compared with erosion ditches and grass lines established in the 1980's. The construction of such terraces, however, would require a significant amount of labor and probably an intensive community-based program with technicians and extension agents [20].

Table 1: On-farm erosion control strategies [20]

Agronomic or Biological Measures	Soil Management Strategies	Mechanical or Physical Methods
Mulching	**Conservation Tillage**	**Terracing**
Crop Management	Minimum tillage	**Contour Bunds**
Cover Crops	Improved fallows	**Infiltration Galleries**
Improved Fallows	No -tillage	**Waterways**
Intercropping	**Contour Tillage**	**Gully Controls**
Planting Pattern/Time	**Strip farming**	Stabilization structures
Crop rotation		Stone check dam
Agroforestry		Gabion baskets
		Reno Mattresses
		Stone lining

3 Historical Origins and Introduction of Radical Terracing in Rwanda

In Rwanda, a unique method of back-slope terracing, originally introduced by missionaries growing wheat in the Northern Province precisely at Kisaro on ex Buyoga District in ex Byumba Province in the 1970s, has been widely adopted by smallholder farmers in many parts of the country. The farmers are careful to isolate the topsoil, then they re-work the subsoil to create the required reverse–slope bench, after which the topsoil is spread over the surface. The riser is planted with short runner grass for stabilization, all within the same day. Radical terracing is usually done manually with hoes and shovels, mostly by communal group-work involving a number of farmers. Where radical terraces have been constructed, the effects have been dramatic, achieving optimum water and soil conservation on slopes exceeding 50%, while adoption rates have been quite extensive. This high adoption of radical terracing is related to the existing policies and programs such as land consolidation, management, and crop intensification programs. These policies and programs boost the use of radical terraces by providing farmers more opportunities to easily access inputs such as improved seeds and manure for increasing the productivity of constructed radical terraces. Recent studies assert that radical terraces in highlands of Rwanda are only financially viable, when the opportunity cost of labor and manure are below the local market price levels and the agriculture area on these radical terraces can be substantially intensified [21, 22].

4 Types of Radical Terracing

The ultimate success of soil conservation schemes depend on how the nature of the erosion problem has been identified and on the suitability of the conservation measures selected to deal with the problem and related to the agricultural or land-use system so that farmers and others are willing to implement them. There are two main types of radical terraces: Irrigation or level radical terraces, which are used where crops, such as rice, need flood irrigation and impounding water. Upland radical terraces, which are used mostly for rain-fed crops or crops that only require irrigation during the dry season. They are generally sloped for drainage. In humid regions, reverse sloped radical terrace types are used. In arid or semi-arid regions, outward-sloped radical terraces types are used [23].

Figure 1: An aerial view of radical terraces in Gishwati area, Rubavu District, West of Rwanda

5 Severity of Erosion in Rwanda and Current Status Coverage Area by Terraces

Agriculture practiced on the slopes of hills and mountains caused land degradation and soil erosion. The increasing levels of soil erosion and reduced soil fertility in the acid-soil mountainous areas of Rwanda have resulted in ecosystem degradation and lowered agricultural yields, which severely impacted on rural livelihoods and the national economy [23]. The study in 2008 showed that over one ton of soil per hectare was swept away by erosion of rivers and lakes every month. Erosion has been responsible of soil degradation with losses of soil nutrients estimated at 945,200 t of organic materials, 41,210 t of nitrogen, 200 t of phosphorus and 3,055 t of potash annually, an estimation of 39.1% of land in the country. Land degradation continues to worsen in the country despite efforts to prevent it. Comprehensive long-term programs and policies for soil and water conservation measures are required [2].

Table 2: Erosion risk by land category in Rwanda [24, 25]

Number	Parameter	Area (%)	Hectare	Slope class
1	Very high erosion risk	17.6	358,000	Slope class over 55%
2	High erosion risk	21.5	437,000	Slopes class 25-55%
3	Average erosion risk	37.5	763,000	Slopes class 13-25%
4	Low erosion risk	16.7	340,000	Slope class 6-13%
5	Very low erosion risk	6.7	137,000	Slope class less than 6%

In fact, 39% of all cultivated land in Rwanda fall under the high erosion risk categories, 37,5% in the middle risk category and only 23% are classified under the "no or low erosion risk". GIS models can estimate the rate of sediment accumulation to evaluate average soil loss rates within catchment areas [26]. Sedimentation measurements provide information on the extent of soil erosion that was estimated using the Universal Soil Loss Equation (USLE) GIS model as presented in Figure 2.

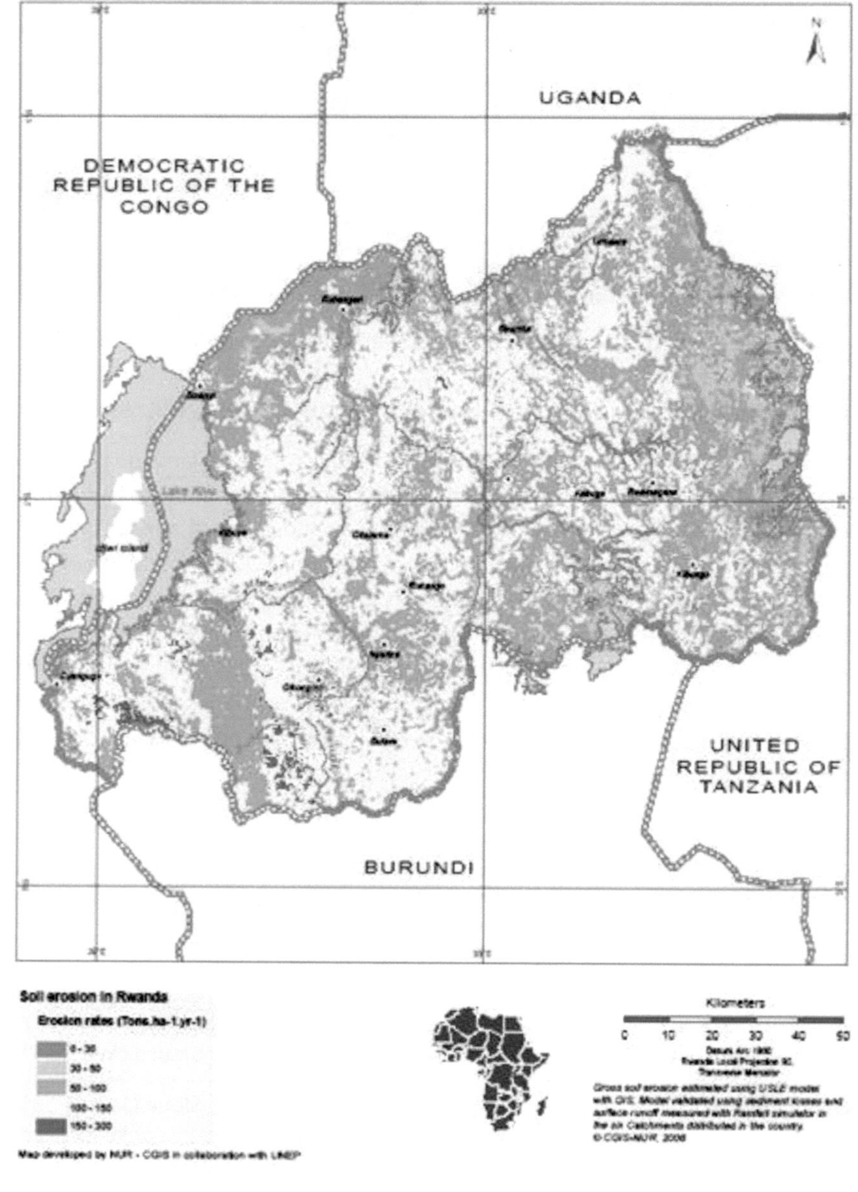

Figure 2: GIS Modelling of Soil Erosion Rates [20]

Currently, 1,013,454 ha out of 1,502,727 ha arable land are protected against soil erosion. This represents 67.4% of the required coverage. Current area covered by Progressive Terraces is 911,115 ha and another covered by radical terraces is 102,339 ha. Rwanda Agriculture Board (RAB) continues to put more efforts to increase national protected area against soil erosion through establishment of new radical and progressive terraces combined with regular maintenance of already established structures [27].

6 Design Specifications

These typical design specifications are commonly used in Rwanda.

6.1 Length

The length of a terrace is limited by the size and shape of the field, the degree of dissections, and the permeability and erodibility of the soil. The longer the terraces, the more efficient they will be. But it should be kept in mind that long terraces cause accelerated runoff and greater erosion hazards. A maximum of 100 m in one draining direction is recommended for typical conditions in a humid tropical climate. The length can be slightly increased in arid and semi-arid regions [28, 29].

6.2 Width

The width of the bench (flat part) is determined by soil depth, crop requirements, and tools to be used for cultivation, the land owner's preferences and available resources. The wider the bench, the more cut and fill is needed and hence the higher the cost. The optimum width for handmade and manual-cultivated terraces range from 2.5 to 5 m; for machine-built and tractor-cultivated terraces, the range is from 3.5 to 8 m [28, 29].

6.3 Gradients

Horizontal gradients range from 0.5% to 1% depending on the climate and soils. For example, in humid regions and on clay soils, 1% is safe for draining the run-off. In arid or semi-arid regions, the horizontal gradients should be less than 0.5%. The reverse grade for a reverse-sloped terrace is 5%, while the outward grade for an outward sloped terrace is 3% [28, 29].

6.4 Slope limit

If soil depths are adequate, hand-made terraces should be employed on 7 to 25 degree (12-47%) slopes [28]. The bench terraces are constructed in 16-40% slope categories but not in higher slope categories than 40%. This is average slope range. Their effectiveness varies in the way we space the bench terraces for each slope category. For 20% slope at 1.5 m vertical interval, the spacing will be every (100/20) x 1.5 = 7.5 m, while spacing for 39% slope would be (100/39) x 1.5 = 3.85 m [30]. If the soil depths and slopes are not adequate for bench terraces, hillside ditches or other types of rehabilitation measures should be used.

6.5 Risers and riser slopes

Riser material can be either compacted earth, protected with grass or rocks, so after cutting a terrace, its riser should be shaped and planted with grass as soon as possible. The riser slopes are calculated by the ratio of the horizontal distance to the vertical rise hand-made with earth material: 0.75:1, hand-made with rocks: 0.5:1, and machine-built with earth material: 1:1 [28]. In order to ensure easy maintenance, terrace riser height should not exceed 2 m.

6.8 Vertical interval

The vertical interval (VI) gives the height of the terrace; provides basic data for calculating the cross-section and volume of soil to be cut and filled [28].

6.7 Water ways and cut-off-drains

The water ways and cut off drains are made before starting terracing to avoid different problems caused by runoff [14].

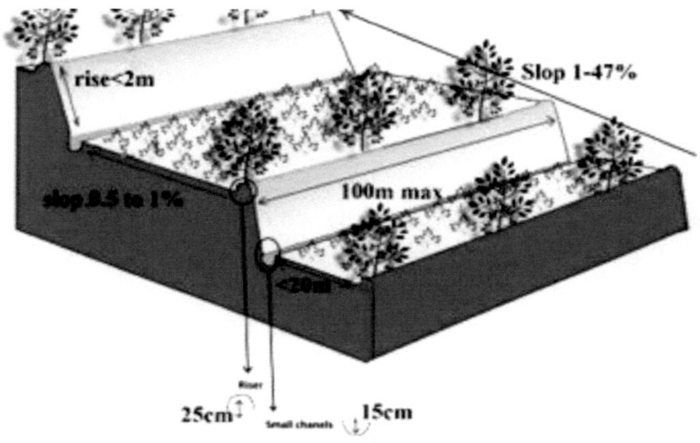

Figure 3: Technical drawing of radical terrace

7 Construction Methods

The terrace must be built, when the soil is neither too dry nor too wet. The cut and fill of the terraces should be done gradually and at an equal pace so that there is neither an excess nor a lack of soil. Bench terrace is constructed either from top to bottom or from bottom to top; both methods have their own advantage and disadvantage, the decision is left for development agents in agreement with community or land owners. In the Top-Down method starts building the terrace from the top of a hill and proceeds down slope. Although it is not labor consuming compared with the bottom-up approach, this type of construction needs a berm for foundation stability of the riser, which competes with land for cultivation. The Bottom-Up is the most preferred method; it allows the construction of support the whole height of the riser starting from the bottom with good foundation, which gives more stability to the bench. Hence, berm is not needed, allowing the whole bench width to be used for cultivation [31, 32].

8 Protections and Maintenance

New terraces should be protected at their risers and outlets, and should be carefully maintained, especially during the first two years. After cutting a terrace, its riser should be shaped and planted with grass as soon as possible. Rhizome-type grasses are better than those of the tall or bunch-type. Although tall grasses may produce considerable forage for cattle, they require frequent cutting and attention. The rhizome-type of local grass has proved very successful in protecting risers. Stones, when available, can also be used to protect and support the risers. An additional protection method is hydro-seeding. The outlet for drainage–type terraces is the point, where the run-off leaves the terrace and goes into the waterway. Its gradient is usually steep and should be protected by sods of earth. A piece of rock, a brick or a cement block is sometimes needed to check the water flow on steeper channels [22].

9 Advantages

Radical terraces significantly reduce the run-off and the loss of soil, but they also contribute to increasing the soil moisture content through improved infiltration and reducing peak discharge rates of rivers. It was stated that terracing promotes rock weathering and eventually increases soil build up and crop growth [33]. On technical aspects, the structural competence of terraces suggested that the efficacy of terraces was determined by local conditions along with their dimensions, form and stability [34]. The efficiency of a terrace system increases by applying additional conservation practices, such as appropriate land preparation (contour ploughing and sowing), appropriate cultivation (e.g., strip cropping) and maintenance of a permanent soil cover [19]. Terraces are often promoted as effective soil and water conservation (SWC) measures on sloping land. It was specified that the advantage of bench terraces is to reduce the slope and erosion on one hand, and to facilitate the work on soil on the other hand [19]. In fact, radical terraces play an important role; they suppress completely the slope and totally the runoff, increase available water for the plant, capitalizes the acquired fertility through organic and mineral manure and allow establishment of small irrigation schemes on the hills and mountains [35].

9.1 Environmental benefits

Often, land is formed by multiple terraces, giving a stepped appearance. Radical terraces are common in Rwanda's rain fed agriculture. They have the benefit that they are easier for both mechanical and manual sowing and harvesting and irrigation than a steep slope would be. Reduce soil losses through enhanced retention and infiltration of runoff promotes permanent agriculture on steep slopes, and intensive land use and consolidation.

9.2 Socio-economic benefits

Furthermore, it is often found that cultivation on terraces is so intensive that a quarter of a hectare can generate full-time employment for one person. The construction of the terraces can be divided over several years in order to create a quarter of a hectare of cultivable land; the upland farmer may work one month per year over four years during periods of low agricultural activity. Permanent structures of these kinds are effective soil conservation technologies as excessive soil loss and silting up of the fields are reduced.

9.3 Cost-benefit analysis on bench terracing

A study conducted in Rwanda on a first analysis of costs and benefits, based on farmers' estimates and market prices showed that gross margins on terraced plots are not much higher than those on non-terraced plots and that bench terracing is hardly profitable. However, since the use of labor and manure were found to be the main determinants of profitability, and these are mostly available on farm, the cost-benefit analysis was subsequently also undertaken with opportunity cost for labor and manure (both at 50 % of market prices) [36]. This plot level cost benefit analysis, using both farmers' estimates and official standard figures, showed that bench terraces in that case were profitable [35]. Increased yields on bench terraced plots are found to be a key stimulus for further adoption of bench terraces. Other variables like soil properties, farm management, and crop and rainfall patterns determine the magnitude of this potential and actual increase of yields [36].

10 Population Perceptions and the Way of Improvement

10.1 Soil erosion

Understanding farmers 'perception of soil erosion and its impact is important in promoting soil and water conservation technologies. The survey results done in Ethiopia by [37] showed that higher proportions (82.7%) of the sampled households were aware of the problem of soil erosion, and majority of these households (54.5%) perceived erosion on their land as severe. The responses of sampled households about the rate of soil erosion in their area for the last ten years based on their knowledge showed that 37.1% were of the opinion that erosion was happening very rapidly, 11.9% moderately and 51% slowly. They were also asked when erosion becomes severe in their area. Accordingly, 19.6 % reported that severe erosion started 20 years ago and before, 24.4% as 15-20 years, 29.3% as 6-14 years and the rest 25.4% as the last 5 years, 1.3% reported that there is no erosion at all. The analysis of responses of farm households on the severity of fertility decline on their farm shows, 28.1% perceived less severe, 57.9% severe and 13.9 % very severe problem in fertility decline. Concerning the perception of Rwandan farmers on the causes of soil fertility decline on their farms in research done in Nyaruguru District, most of respondents ranked soil erosion and lack of manure and mineral fertilizers the first reasons for the decline of soil fertility [38]. According to the research conducted in Nyaruguru District of Rwanda, a great majority of respondents/farmers ranked bench terraces at the first position in soil erosion control methods and affirmed that this method improves soil fertility, and few of them had a bad experience from the bench terraces done in the last years, which were badly done in terms of technique and soil treatment, and farmers cultivated other resisting crops such as cassava, sweet potatoes and trees on these terraces.

10.2 Radical terraces

The argument most often heard against radical terracing is its cost, a conclusion often reached by multiplying the amount of soil to be cut and filled and the resulting work required per hectare by the official daily wage. The results usually show a cost per hectare, which no farmer can afford it. However, the fact that farmers in Rwanda have constructed terraces for centuries shows clearly that when population density and intensity of cultivation reach certain thresholds, radical terraces are a workable solution. Incentives to farmers may be necessary to accelerate the development of terraces. However, high labor intensity, time-consuming regular inspections, high consumption of scarce farmland, and the large amounts of construction material required are factors that stop farmers from installing or maintaining terraces.

11 Discussions

In Rwanda, radical terraces are principally designed to reduce soil losses through enhanced retention and infiltration of runoff, to promote permanent agriculture on steep slopes, and promoting land consolidation and intensive land use. The history of bench terraces in Rwanda is linked to state policies and regulations and to interventions by NGOs. The approach used to promote these terraces has shifted over time from top down to somewhat participatory. Various development policies promoted by the current government, such as the 'performance contracts' (known as Imihigo), collective community work (Umuganda) and Agasozi Ndatwa (literally meaning a 'model hill'), entail certain aspects of community-based development, promotion of farmers' associations and co-operatives, and a self-reliance mentality towards rural development. In the case of soil and water conservation, these policies are geared primarily towards collective awareness and soil erosion control. At the same time, farmers operate in small-scale associations and co-operatives, from which different forms of social capital originate [35].

12 Conclusions

Radical terraces have the potential of improving farmers' livelihoods and increasing the resilience of a degraded environment. Generally, the most recently researches confirmed that radical terracing was found to be the most effective soil erosion control measure on both two consecutive rainy seasons. The topography of the country is dominated by high mountains, which is a serious challenge for environmental protection. The relief of Rwanda is one of the causes of soil erosion, a big barrier to the improvement of farm produce. In addition, most of Rwandese population lives in the rural area and 83.4% depends on agricultural production, therefore, radical terraces have been seen as solution for soil water conservation (SWC) in highly steep slopes areas, and the government and NGos intervened with new policies and community mobilization in the extension of the technology as well as by providing technical and financial support.

13 Recommendations

In order to increase farm productivity, environmental protection and sustainable terracing, a way of improvement was recommended:

- During terracing activity, farmers should carefully isolate the topsoil, then they re-work the subsoil to create the required reverse–slope bench, after which the top soil is spread over the surface. The riser should be planted with short grass for stabilization, all within the same period.

- Erosion control measures are not sufficient to reduce land degradation and increase yields. A combination of erosion controls such as increasing organic matter, and adding mineral fertilizers is required, especially in areas, where it is already severe, also the use of improved agronomic and management practices by considering agro-ecological zones conditions.

- Installation of radical terraces can increase the risks of landslides and the leaching of nutrients, if these are not well constructed and maintained. They require regular care and maintenance. If small break is neglected, large-scale damage will result. Radical terraces are generally accepted as the ultimate intensity in physical management of soil runoff and water retention management.

- A continual community-based mobilization and intensive training of technicians and engineers for well-equipped personnel is highly recommended.

- As the technology is a labor intensive one, farmers should develop and strengthen the spirit of working together, either in associations or cooperatives to ease and fasten the implementation of the technology in their area.

14 Acknowledgements

The author would like to thank EXCEED/SWINDON project and DAAD (Germany Academic Exchange Service) for support, to participate at the international Expert Workshop on "Water on Agricultural Practices; Training the trainers ", held in Rio de Janeiro, Brazil 16-20 September 2019.

15 References

[1]   Havugimana, E.: State policies and livelihoods-Rwandan Human Settlement Policy. Case Study of Ngera and Nyagahuru Villages. 2009.

[2]   PMI Country Profile, Rwanda: US Census Bureau, International data Base / PMI Malaria Operational Plan FY09; 2009, 1-2.

[3]   Giblin, J.D., Fuller, D.Q.: First and second millennium AD agriculture in Rwanda: archaeobotanical findsand radiocarbon dates from seven sites. Veg. Hist. Archaeobot. 2011, 20, 253.

[4]   [4] Bizoza AR (2014). Population Growth and Land scarcity in Rwanda: the other side of the coin. Conference on Land Policy in Africa, Addis Ababa, Ethiopia 11-15th November 2014.

[5]   König, D.: Degradation et Erosion des Sols au Rwanda. Cahiers d'Outre-Mer 1994, 47(185), 35- 48.

[6]   Olson, J.M.: Farmer Responses to Land Degradation in Gikongoro, Rwanda. Ph.D., Geography Department, Michigan State University, East Lansing. 1994b.

[7]   Nsengimana, C., Gascon, J.-F.: Resultats de l'Enquete sur les Exploitations les Plus Touchées per la Disette. Gikongoro, Rwanda: Préfecture Gikongoro. 1991.

[8]   Gasana, J.: Remember Rwanda? Washington, D.C.: World Watch Institute. 2002.

[9]   Habimfura, V., Fabiola, H.: Statut Nutritionnel et Sécurité Alimentaire au Rwanda. Kigali, Rwanda: Division des Statistiques Agricoles, MINAGRI. 1993.

[10] National Institute of Statistics (NISR): Rwanda 4th Population and Housing Census report, Kigali/Rwanda. 2012.

[11] Kaberuka,.: Programme d'appui à la stratégie de la réduction de la pauvreté. Phase II. Kigali/Rwanda, 2002, 20-56

[12] Egli, A.: La conservation des sols à l'aide des méthodes agrofrorestières: le cas du Rwanda. In: Colloque international "Devéloppement agricole et conservation du patrimoine naturel duns les pays du Tiers Monde" 9-11 octobre 1985. Gembloux, 1985, 1-31.

[13] Mupenzi J, Jiwen, G., Habiyaremye, G., Mkukakayumba. U.: Impact of radical terraces on environment: a case of Kaniga Sector in Gicumbi District/Rwanda. Int. J. Sustain. Sci. Stud. 2009, 67-72.

[14] Zuazo, V.H.D., Ruiz, J.A., Raya, A.M., Tarifa, D.F.: Impact of erosion in taluses of subtropical orchard terraces. Agriculture, Ecosystems and Environment, 2005, 107(2), 199- 210.

[15] Fialho, J.S., De Aguiar, M.I., Dos Santos, M., Magalhães, R.B., De Araújo, F.D.C.S., Matoso, M.: Soil quality, resistance and resilience in traditional agricultural and agroforestry ecosystems in Brazil's semiarid region. Afr. J. Agric. Res. 2013, 8(40), 5020-5031

[16] Ministry of Lands, Environment, Forests, Water and Mines: National Land Policy, Kigali, Rwanda. 2004.

[17] Ministry of Natural Resources: Mining Policy in Rwanda. Kigali, Rwanda. 2010.

[18] Moeyersons, J.: Les Essais Récents de Lutte Anti-Erosive au Rwanda. Cahiers d'Outre-Mer 1994, 47(185), 65-78.

[19] Roose, E. Barthès, B.: Organic Matter Management for Soil Conservation and Productivity Restoration in Africa: A Contribution from Francophone Research. Nutrient Cycling in Agroecosystems 2001, 61, 159-170.

[20] REMA: Rwanda State of Environment and Outlook Report, Rwanda Environment Management Authority, P.O. Box 7436 Kigali, Rwanda, 2009. http://www.rema.gov.rw/soe/

[21] Bizoza, A.R., de Graff, J.: Financial cost-benefit analysis of bench terraces in Rwanda, Land Degradation & Development, 2012, 23(2), 103-115.

[22] Kagabo, D.M. L., Stroosnijder, L., Visser, S.M., Moore, D.: Soil erosion, soil fertility, and crop yield on slow –forming terraces in the highlands of Buberuka, Rwanda. Soil and Tillage Research, 2013, 128(7), 23-29.

[23] Gascon, J.-F.: Pauvreté á Gikongoro. Gikongoro, Rwanda: MINAGRI and FAO. 1992.

[24] MINITERE: National strategy and action plan for the conservation of Biodiversity in Rwanda. Kigali, 2003, 67.

[25] Ministry of Disaster Management and Refugees Affairs: Impacts of floods and landslides on socio-economic development profile. Kigali, Rwanda. 2012.

[26] Amore, E., Modica, C., Nearing, M.A., Santoro, V.C.: Scale effect in USLE and WEPP application for soil erosion computation from three Sicilian basins. J. Hydrol., 2004, 293(21), 100–114.

[27] Rwanda Development Board: Agriculture sector report, Rwanda skills survey, Kigali, Rwanda. 2012.

[28] FAO: Method of erosion control. Food and Agricultural Organization, Rome. 1985.

[29] [Sheng, C.T.: Bench Terrace Design Made Simple, 12th ISCO Conference, Beijing, China, 2002.

[30] Morgan, G., Powell, A., Mc Vay, K.A.: Terrace Maintenance, Kansas State University, 2004. Retrieved: July 1, 2014 from: https://www.bookstore.

[31] Liu, B.Y., Nearing, M.A., Risse, L.M.: Slope gradient effects on soil loss for steep slopes. Rome, Italy. 1994. *(This paper was peer-reviewed for scientific content.)*

[32] Dabney, S.M., Liu, Z., Lane, M., Douglas, J., Zhu, J., Flanagan, D.C.: Landscape benching from tillage erosion between grass hedges. Soil Tillage Res., 1999, 51(2), 219-231

[33] Beach, T., Dunning, N.P.: Ancient Maya terracing and modern conservation in the Peten rain forest of Guatemala. J. Soil Water Conserv., 1995, 50(2), 138-145.

[34] Rufino, R.L.: Terraceamento. In: Manual Técnico do Subprograma de Manejoe Conservação do Solo, Curitiba. Secretaria da Agricultura e do Abastecimento, Paraná, 1989, 2(1), 218-235.

[35] Inbar, M., Llerena, C.A.: Erosion processes in high mountain agricultural terraces in Peru. MT Res. Dev., 2000, 20(1), 72-79.

[36] Bizoza, A.R., de Graff, J.: Financial cost-benefit analysis of bench terraces in Rwanda. Land Degrad. Dev., 2012, 23(2), 103-115.

[37] Kassa, Y., Beyene, F., Haji, J., Legesse, B.: Farmers's perception of the impact of Land Degradation and Soil and Water. Conservation Measures in West Harerghe Zone of Oromia National Regional State, Addis Ababa, Ethiopia. 2013.

[38] Karemangingo, C., Bugenimana, E.D., Bimyebebe, M.: Development of catchment management plan for Akanyaru sub-catchment, Nyaruguru district, Rwanda, Kigali, Rwanda. 2014.

SOIL AND WATER LOSSES IN AN ACRISOL UNDER NATURAL RAINFALL IN DIFFERENT MANAGEMENT SYSTEMS AT SOUTH OF BRAZIL

Luciana da Silva Corrêa Lima, Cláudia Alessandra Peixoto de Barros, Elemar Antonino Cassol, Liana Dambros

[1]*Federal University of Rio Grande do Sul, Valley Campus, Faculty of Agronomy, Department of Soils, 7712 Bento Gonçalves, 91540-000, Porto Alegre, Rio Grande do Sul, Brazil;*
limaluciana@outlook.com

Keywords: Soil loss, water loss, soil and water conservation, bounded plots, field measurements

Abstract

Among the distinctive types of soil degradation processes, water erosion is the most worrisome one, especially in humid climate regions, as it causes the soil to detach and to be transported more easily. Field studies in erosion are scarce in Brazil, mainly due to the costs and time of experimentation involved to gain representative data. About 50% of the experimental studies have 2 years or less of monitored data. It is important to prolong the monitoring period of the experimental sites in Brazil in order to reduce the variability of the observed data, to allow the development of models and to support the decision makers. Data from 1978 to 1983 from Santa Maria, at southern Brazil, were analyzed, thus, a 5 years series. This study aims to evaluate soil and water losses in an acrisol under natural rainfall in different management systems at south of Brazil. Soil and water losses were higher in conventional tillage than in non-tillage management systems for this 5 years series. Also, for the non-tillage treatments, lupine and corn were more efficient in controlling soil and water losses than oat and soybean crop succession. Water losses are better explained than soil losses, when correlated with EI_{30} and rainfall during summer and winter periods.

1 Introduction

Among the distinctive types of soil degradation processes, water erosion is the most worrisome one, especially in humid climate regions, as it causes the soil to detach and to be transported more easily [1]. In Brazil, the rate of soil loss has been estimated between 600 and 800 Mg/yr [2–4], but the contributions from different agricultural activities that add up to this large amount are still not well understood.

Also, field studies in erosion are scarce in Brazil, mainly due to the costs and time of experimentation involved to gain representative data. In Brazil, the first experimental study on plot scale was carried out in the 1940s. However, since the 2000s, where a peak of experimental studies on soil erosion occurred, there was a decrease of approximately 86% of these field studies [5]. In addition, 50% of the experimental studies have 2 years or less of monitored data [5]. It is important to prolong the monitoring period of the experimental sites in Brazil to reduce the variability of the observed data and to allow the development of models and support the decision makers [4, 5].

Furthermore, it is agreed in the area of soil and water management and conservation that in conventional till systems soil loss is greater than in non-tillage systems [6, 7]. The positive effect of higher residues produced by non-tillage on the soil surface in southern Brazil is undeniable [8]. However, when the water dynamics is analyzed in these systems, it is not possible to observe a standard behavior, in some cases there is a greater loss of water in non-tillage systems [9, 10], which is generally not expected. Deuschle et al. [11] discussed that the behavior of greater or less losses of soil and water in non-tillage systems is dependent on the magnitude and intensity of the rainfall event, and when there are events of greater magnitude and intensity, the area is more vulnerable to runoff and soil erosion.

The database that is being discussed here was part of an experiment that belonged to a monitoring network built in the late 70's. There were other two studies like this one placed at Central Depression [12, 13] and Plateau [14] regions of the Rio Grande do Sul state. At the present time, the data from this study in the municipality of Santa Maria at Rio Grande do Sul state in southern Brazil, which were not processed and analyzed before, therefore, unpublished, are still being processed. Its series is expected to be composed of 10 years (1978-1989), but until this moment, the data from 1978 to 1983 were analyzed, thus, a 5 years series. In this context, this study aims to evaluate soil and water losses in an acrisol under natural rainfall in different management systems at south of Brazil.

2 Materials and Methods

The data of this study was collected in the late 1970s and early 1980s in one of the research centers of the current Department of Agricultural Diagnosis and Research (DDPA), located in the district of Boca do Monte, in the municipality of Santa Maria (Figure 1). The region's climate, according to Köppen's classification, is Cfa type, humid subtropical, without a drier period, the average annual temperatures being between 10 °C and 22 °C and average annual rainfall 1769 mm [15]. The soil of the study area is classified as acrisol [16]. The experimental design was performed with 5 treatments and 5 replicates, consisting of 10 bounded plots (Figure 2).

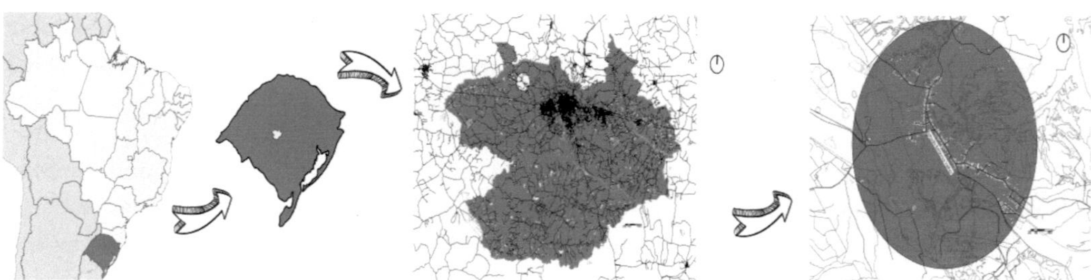

Figure 1: Study site in Santa Maria, Rio Grande do Sul, Brazil

Figure 2: Experimental design

NOTE: Because there were no any picture of this study, the pictures used here are from an experimental area that belonged to the monitoring network mentioned in this text before, in the Central Depression region, in the municipality of Eldorado do Sul in Rio Grande do Sul state in Brazil [12, 13]. All the studies from this monitored network were installed the same way and using equal materials. The highlighted difference here is that in the present study the slope of the area was 9%, while the slope of the area in the picture was 12%.

The plots had dimensions of 22.0 × 3.5 m, with an area of 77 m^2 and 9% slope. The plots were delimited by galvanized plates fixed in 10 cm soil depth on the sides and at the upper end, with a drainage system at the lower end, consisting of a trough and two reservoirs. The first reservoir collected the coarse and the second one the fine sediments (Figure 2). Treatments were denominated by BS – uncovered soil under conventional tillage; LCCT – lupine and corn succession under conventional tillage; OSCT – oat and soybean succession under conventional tillage; LCNT – lupine and corn succession under non-tillage; OSNT – oat and soybean succession under non-tillage. There was one replicate for each treatment; therefore, occurring a total of 10 bounded plots.

At each rainfall event, where surface runoff was generated, field measurements were performed and samples of sediment and water were collected. To determine the runoff volume, the water level in the reservoir was measured and related to its area. The samples were transported to the laboratory, weighted and placed at the laboratory oven under 55–60 °C temperature until reaching constant mass. After laboratory procedures, soil losses (SL) and water losses (WL) were evaluated by the equations (1) and (2), respectively.

$$SL = \frac{(DS_{runoff} \times 10)}{A_{plot}} \qquad (1)$$

$$WL = \left(\frac{V_{runoff}}{V_{H_2O\,plot}}\right) \times 100 \qquad (2)$$

DS_{runoff} means total dry sediment (*coarse + fine*) (g); A_{plot} represents the plot area that is 77 m^2; WL is in %; V_{runoff} is the runoff volume (cm^3), and $V_{H_2O\,plot}$ is the water volume that was caused by rain in the plot (cm^3).

Final SL data were extrapolated to Mg/ha and WL units converted to mm. Soil and water losses were estimated for each treatment and its replicate, and then summed and averaged. The EI_{30} index was also calculated, where E is the total kinetic energy of rainfall in MJ/ha and I is the maximum intensity in 30 continuous minutes of rainfall, given in mm/h. Additionally, the accumulated rainfall was estimated. Then, correlations were made between the EI_{30} index (MJ.mm/ha.h) and soil and water losses as well as between rainfall (mm) and soil and water losses. The study aimed to estimate the total losses by seasons (winter and summer) and each year as well as the total of the 5 years of experimentation. Besides, in order to compare the conditions of each study region of the monitoring network mentioned here, the K factor (soil erodibility) for the Santa Maria's acrisol was estimated. The K factor was calculated according to [17].

3 Results

Soil losses were higher in the BS treatment during the five years of experimentation. Also, soil losses decreased in the lupine and corn crop succession treatment from conventional tillage (LCCT) to non-tillage (LCNT) (Table 1). The same occurred for the oats and soybean crop succession, where losses were higher from conventional tillage (OSCT) to non-tillage (OSNT), too (Table 1).

When comparing losses between the same management systems, for conventional tillage, the LCCT treatment had higher losses than the OSCT treatment (Table 1). On the other hand, for the non-tillage system, the OSNT treatment had higher soil losses than the LCNT (Table 1).

Table 1: Soil losses in treatments monitored over 5 years of experimentation in Santa Maria

Crop year	Duration (days)		Soil losses by treatment (Mg/h)				
	Winter	Summer	BS	LCCT[1]	OSCT	LCNT[1]	OSNT
1978-79	181	180	108.80	11.55	11.69	9.63	11.46
1979-80	173	216	243.78	17.74	11.24	4.14	5.81
1980-81	152	183	248.12	62.53	46.53	14.00	12.68
1981-82	175	243	409.27	50.23	49.55	12.63	22.31
1982-83	149	177	527.70	165.84	105.24	8.73	9.66
Total			1,537.67	307.89	224.24	49.12	61.92

[1] *In the first two crop years period, barley was sown during winter periods instead of lupine, but as the crop straw were not measured in any moment of this study, the losses values of these two first years were considered for the total estimations, as the management systems (conventional and non-tillage) were always the same.*

As for soil losses in Table 1, water losses were higher in the BS treatment during the five years of experimentation, too (Table 2). Likewise, water losses decreased in the lupine and corn crop succession treatment from conventional tillage (LCCT) to non-tillage (LCNT). In the oats and soybean crop succession, water losses were also higher from conventional tillage (OSCT) to non-tillage (OSNT).

In conventional tillage systems, the OSCT treatment had higher losses than the LCCT treatment (Table 2). The same happened in non-tillage systems, where OSNT treatment had higher water losses than the LCNT (Table 2).

Table 2: Water losses in treatments monitored over 5 years of experimentation in Santa Maria

Crop year	Rainfall (mm)		Water losses by treatment (mm)				
	Winter	Summer	BS	LCCT	OSCT	LCNT	OSNT
	3863.80	4132.20					
1978-79	1202.30		349.25	210.11	155.06	168.07	235.51
1979-80	1518.60		297.47	135.13	177.53	117.54	198.37
1980-81	1154.80		347.07	242.56	165.07	186.35	185.80
1981-82	1722.40		623.21	362.13	483.61	215.22	277.48
1982-83	2397.90		679.06	772.58	880.01	375.09	418.54
Total	7,996.00		2,296.05	1,722.52	1,861.28	1,062 27	1,315.70

As shown in Table 3, the K Factor value for Santa Maria's acrisol was estimated as 0.0363 Mg.ha.h/ha.MJ.mm.

Table 3: Soil losses and EI30 index for K factor determination during 5 years of experiments monitoring from 1978-79 to 1982-83 in Santa Maria.

Crop year	Soil losses by treatment $\left(\frac{Mg}{ha}\right)$					EI_{30} $\left(\frac{MJ.mm}{ha.h}\right)$	K Factor $\left(\frac{Mg.ha.h}{ha.MJ.mm}\right)$
	BS	LCCT	OSCT	LCNT	OSNT		
1978-79	108.80	11.55	11.69	9.63	11.46	4.791.0	0.0227
1979-80	243.78	17.74	11.24	4.14	5.81	7.171.4	0.0340
1980-81	248.12	62.53	46.53	14.00	12.68	5024.6	0.0494
1981-82	409.27	50.23	49.55	12.63	22.31	11178.2	0.0366
1982-83	527.70	165.84	105.24	8.73	9.66	13653.7	0.0386
Mean						8,363.78	0.0363

In the following series of figures, the outcomes are correlated for the different treatment strategies.

For the 5 years of monitoring data, a high coefficient of determination (R^2 of 0.8276) was observed between the EI_{30} index and SL during the winter period (Figure 3 a). For the WL in this same treatment, moderate values of R^2 were found, when it was correlated with the EI_{30} index and the rainfall amount (in mm) (Figures 3 c, g and h).

In the LCCT treatment, the SL were better explained by the EI_{30} index during the winter period, with a R^2 of 0.7372 (Figure 4 a). WL were better explained by the rainfall during winter, with a moderate R^2 of 0.4597 (Figure 4 g). The SL also had a moderate R^2 of 0.4301 when related with rainfall (Figure 4 e).

At OSCT treatment, SL were not well explained by any relation. The higher estimated R^2 for SL was of 0.3835, when it was related with rainfall during winter (Figure 5 e). On the other hand, WL were better explained by the analyzed factors in this treatment, where R^2 was 0.5083, when WL were related with rainfall (Figure 5 g).Also, moderate R^2 were found for WL related with EI_{30} index during winter (R^2 = 0.3514) (Figure 5 c) and summer (R^2 = 0.4479) (Figure 5 d), and with rainfall during summer (R^2 = 0.4259) (Figure 5 h).

In the LCNT treatment, SL were not well explained by any relation as well. WL were better explained with a R^2 of 0.5468, when related to rain in winter (Figure 6 g) and an R^2 of 0.421, when related to EI_{30} in summer (Figure 6 d). When related to EI_{30} in winter, it had an R^2 of 0.3553 (Figure 6 c) and, when related to summer rain, it had an R^2 of 0.265 (Figure 6 h).

Finally, at the OSNT treatment, for the 5 years of monitoring data, SL were not well explained by any relation either. WL had a better correlation with rainfall during winter (R^2 = 0.5778) (Figure 7 g). When related with EI_{30} index, it had a R^2 of 0.5291 in summer period (Figure 7 d), and a R^2 of 0.4637 in winter period (Figure 7 c). In the correlation with rainfall in summer, the R^2 had a value of 0.2415 (Figure 7 h).

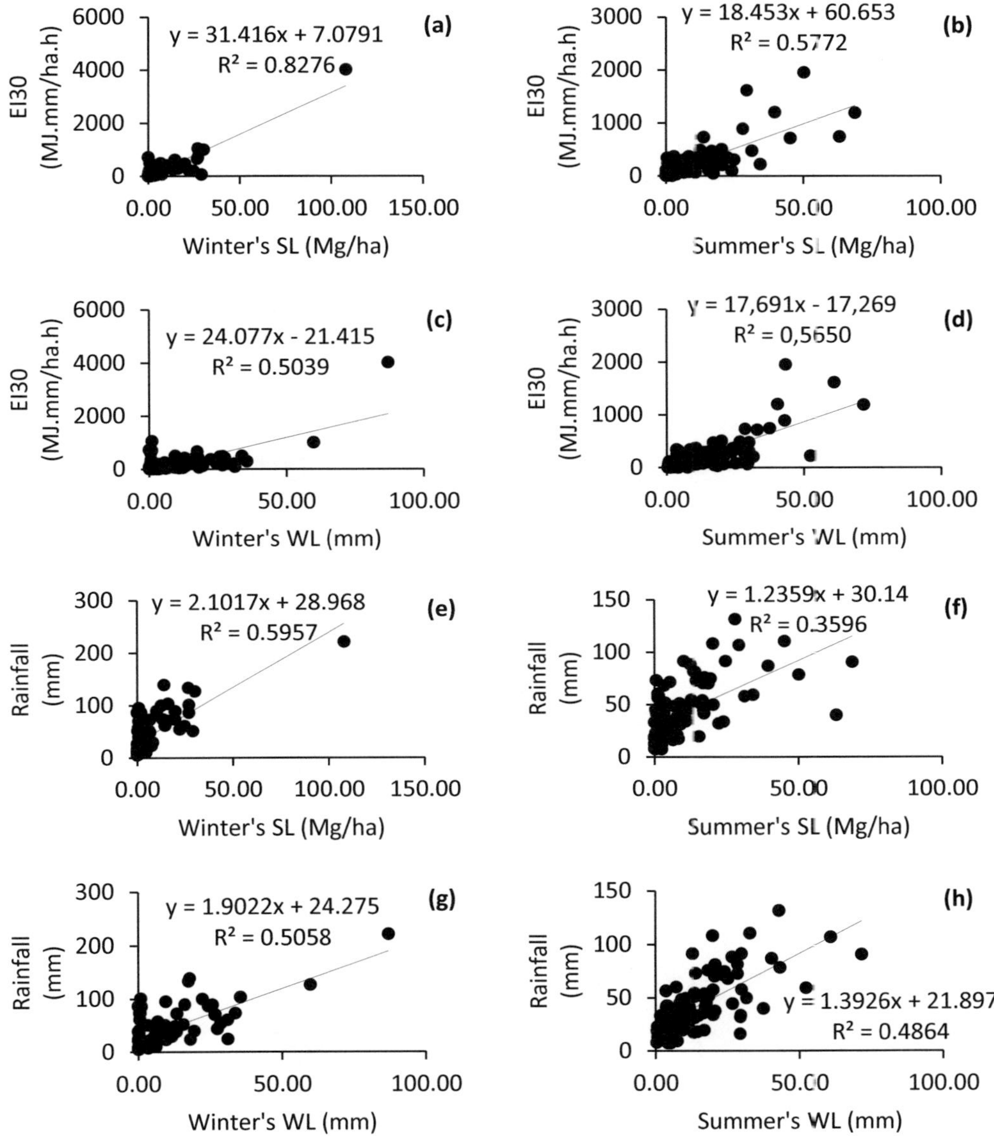

Figure 3: BS treatment correlations during winter and summer periods of 5 years of experimentation. **(a)** between EI_{30} and SL in winter; **(b)** between EI_{30} and SL in summer; **(c)** between EI_{30} and WL in winter; **(d)** between EI_{30} and WL in summer; **(e)** between rainfall and SL in winter; **(f)** between rainfall and SL in summer; **(g)** between rainfall and WL in winter; **(h)** between rainfall and WL in summer.

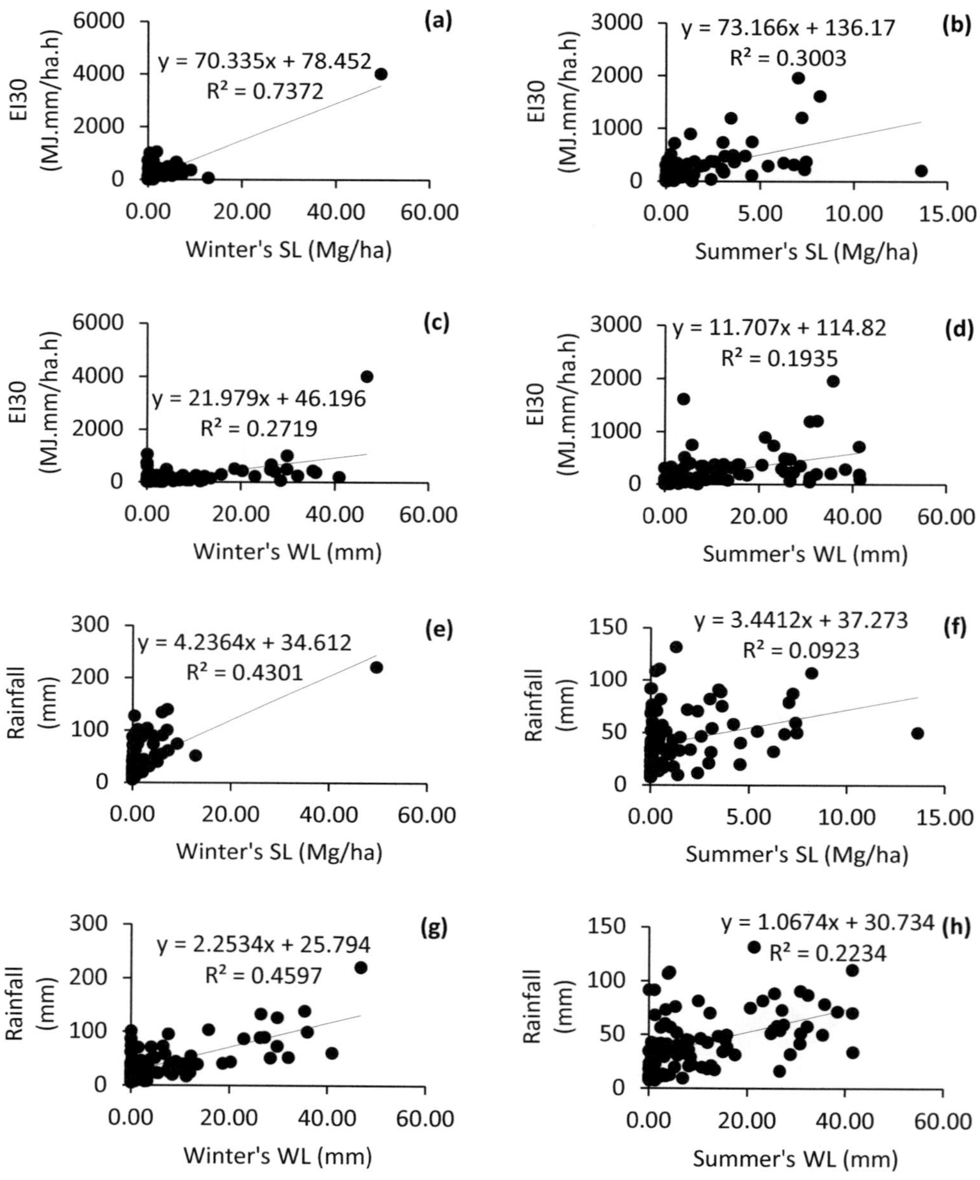

Figure 4: LCCT treatment correlations during winter and summer periods of 5 years of experimentation. **(a)** between EI_{30} and SL in winter; **(b)** between EI_{30} and SL in summer; **(c)** between EI_{30} and WL in winter; **(d)** between EI_{30} and WL in summer; **(e)** between rainfall and SL in winter; **(f)** between rainfall and SL in summer; **(g)** between rainfall and WL in winter; **(h)** between rainfall and WL in summer.

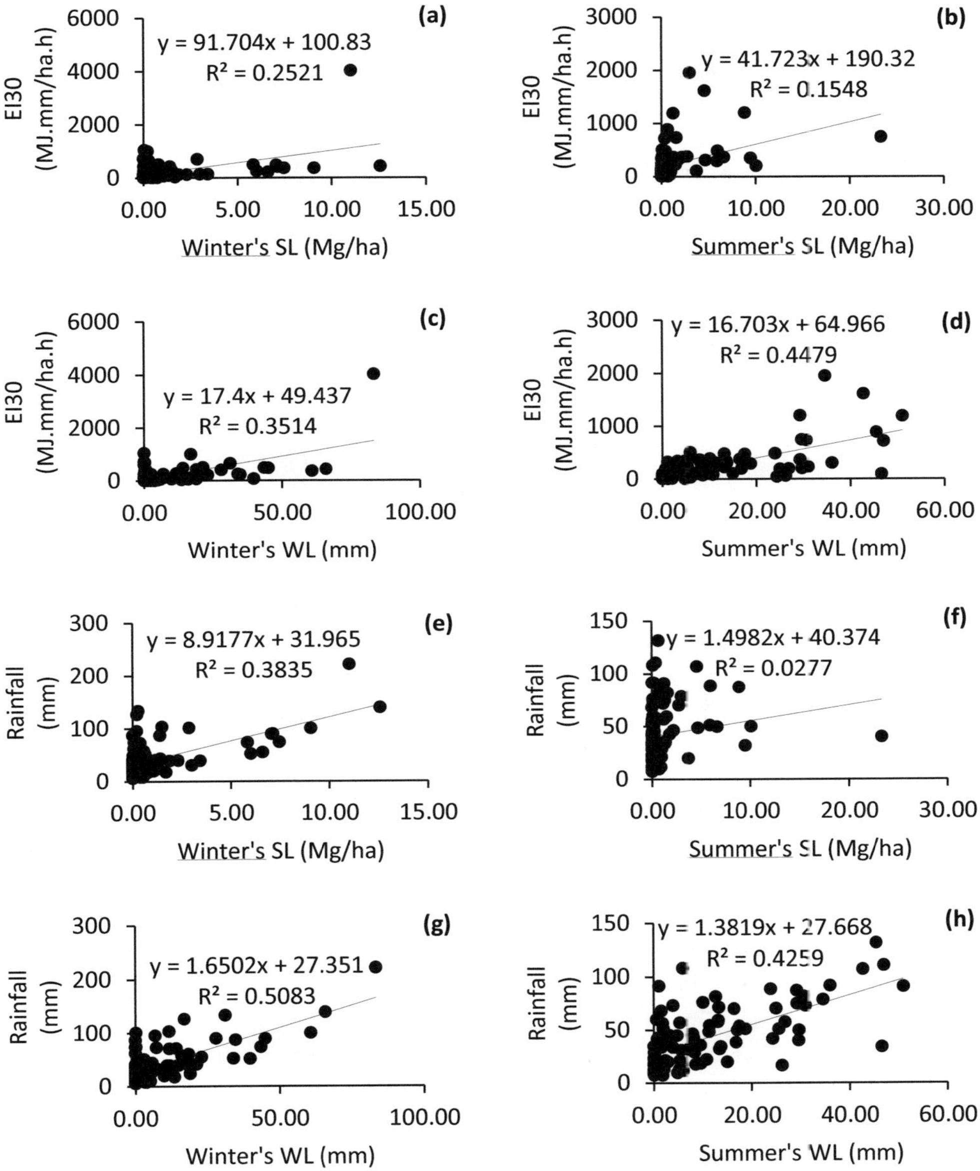

Figure 5: OSCT treatment correlations during winter and summer periods of 5 years of experimentation. **(a)** between EI_{30} and SL in winter; **(b)** between EI_{30} and SL in summer; **(c)** between EI_{30} and WL in winter; **(d)** between EI_{30} and WL in summer; **(e)** between rainfall and SL in winter; **(f)** between rainfall and SL in summer; **(g)** between rainfall and WL in winter; **(h)** between rainfall and WL in summer.

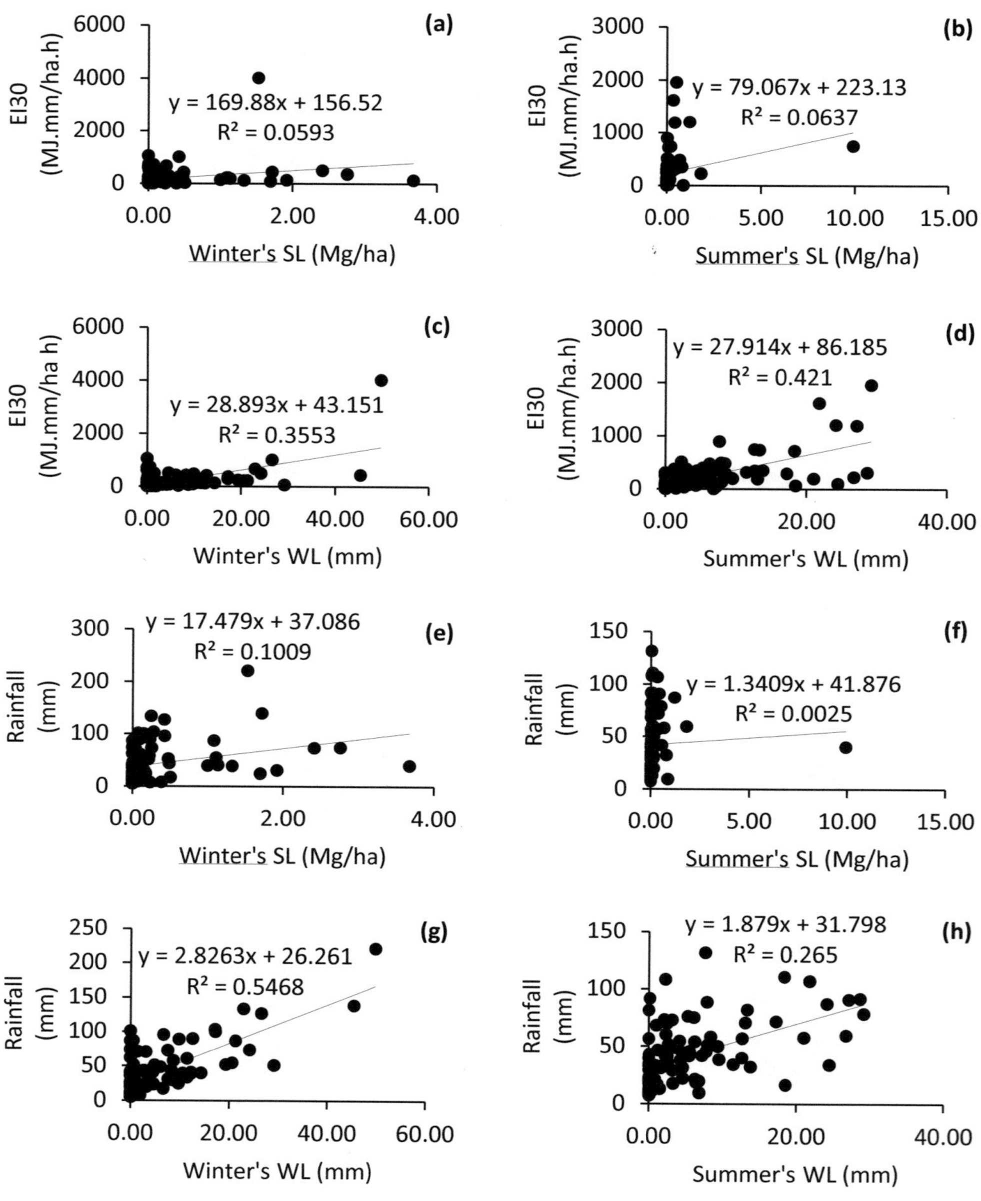

Figure 6: LCNT treatment correlations during winter and summer periods of 5 years of experimentation. **(a)** between EI_{30} and SL in winter; **(b)** between EI_{30} and SL in summer; **(c)** between EI_{30} and WL in winter; **(d)** between EI_{30} and WL in summer; **(e)** between rainfall and SL in winter; **(f)** between rainfall and SL in summer; **(g)** between rainfall and WL in winter; **(h)** between rainfall and WL in summer.

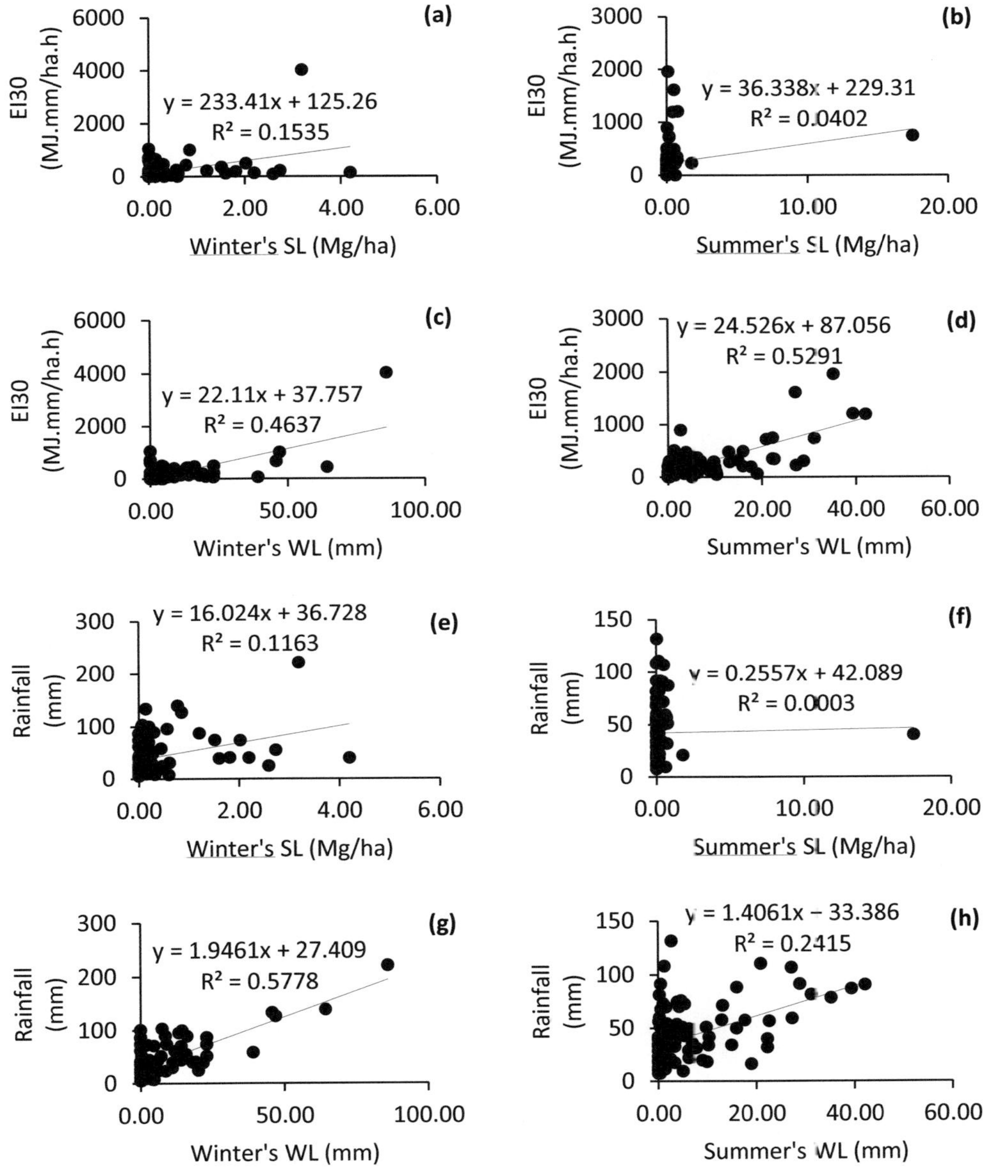

Figure 7: OSNT treatment correlations during winter and summer periods of 5 years of experimentation. **(a)** between EI_{30} and SL in winter; **(b)** between EI_{30} and SL in summer; **(c)** between EI_{30} and WL in winter; **(d)** between EI_{30} and WL in summer; **(e)** between rainfall and SL in winter; **(f)** between rainfall and SL in summer; **(g)** between rainfall and WL in winter; **(h)** between rainfall and WL in summer.

4 Discussion

As expected, total soil losses were higher in conventional tillage than in non-tillage management systems. Those results were also found in the studies of Stumpf [12] and Schmidt [14]. Conservation management systems, such as non-tillage, reduce water erosion losses. Coverage in direct contact with the soil surface dissipates the kinetic energy of rainfall near the soil surface and reduces runoff energy by increasing hydraulic surface roughness, increasing infiltration and decreasing runoff [18–20]. According to Dechen et al. [21], grasses are more efficient in forming and increasing aggregate stability than legume crops, providing greater soil resistance. In Table 1, for the non-tillage treatments, lupine and corn crop succession were more efficient in controlling soil losses due to erosion than oat and soybean crop succession. This is due to the high density of thin roots, which promote the approximation of particles by the pressure exerted as they grow as well as absorb large amounts of water from the soil [22]. The results of this research, which indicate that grasses are more efficient than legume crops in prevention from soil losses for non-tillage management systems, agree with the studies of Stumpf [12] and Dechen et al. [21].

When the total water losses in mm for the 5 years series are observed in Table 2, conventional management systems (CT) had more losses than non-conventional management systems (NT), following the same pattern as soil losses for crop types as well. However, when the correlations between water losses and EI_{30} index and rainfall amount for the total values are separated by seasons (winter and summer), the patterns between management systems types change (Figures 3 to 7). Water losses were better explained by EI_{30} index and rainfall magnitude in non-tillage systems than in conventional tillage systems. This is consistent with the studies of Bertol [9] and Lemos [10], which say that when the water dynamics is analyzed in these systems (NT), it is not possible to observe a standard behavior, in some cases, there is a greater loss of water in them. Deuschle et al. [11] state that the magnitude and intensity of the rainfall event will directly affect the soil and water losses in NT systems. These studies suggest that it is likely necessary to analyze rainfall events patterns that interfere more in the water dynamics that make NT systems more vulnerable to erosion and runoff losses. Wischmeier and Smith [17] state that EI_{30} values needs to be estimated over long time intervals (over 20 years) to include apparent cyclic rainfall patterns; so analyzing more years for this region (Santa Maria) would improve the possibility of solving doubts on that matter.

For the USA, Foster et al. [23] classified Factor K values as: *(a) low:* with values between 0.01 and 0.03 Mg.ha.h/MJ.mm.ha; *(b) medium:* with values between 0.03 and 0.06 Mg.ha.h/MJ.mm.ha, and *(c) high:* with values above 0.06 Mg.ha.h/MJ.mm.ha. In this study, the K Factor for an acrisol was estimated in a value of 0.0363 Mg.ha.h/MJ.mm.ha (Table 3), while in the other studies from the monitoring network a value of 0.0338 Mg.ha.h/MJ.mm.ha was estimated for an ultisol [12] and a value of 0.0091 Mg.ha.h/MJ.mm.ha for an oxisol [14]. Despite the oxisol's region has the lowest erodibility value, its EI_{30} index was 10,137.6 MJ.mm/ha.h, while in the ultisol's region the EI_{30} index was 5,607.3 MJ.mm/ha.h. In this study, the EI_{30} index was estimated in a value of 8,363.78 (Table 3). This information, derived from field-measured data, helps in decision-making and soil and water conservation planning measures that best suit for each location.

5 Conclusions

Soil losses were higher in conventional tillage than in non-tillage management systems for a 5 years series. Also, for the non-tillage treatments, lupine and corn crop succession were more efficient in controlling soil losses due to erosion than oat and soybean crop succession. Water losses were higher in conventional tillage than in non-tillage management systems for a five years series. Also, for the non-tillage treatments, lupine and corn crop succession were more efficient in controlling water losses due to runoff than oat and soybean crop succession. Water losses are better explained than soil losses when correlated with EI_{30} and rainfall during summer and winter periods. Information derived from field-measured data helps in decision-making and soil and water conservation planning measures that best suit for each location.

6 Acknowledgements

The author would like to thank EXCEED Swindon project and DAAD (German Academic Exchange Service) for supporting her participation at the International Expert Workshop on "Water on Agricultural Practices: Training the Trainers", in Rio de Janeiro, Brazil on September 15th to 21st of 2019.

7 References

[1]   R. Billia De Miranda, G. D'almeida Scarpinella, R. Siloto Da Silva, F.F. Mauad: Water Erosion in Brazil and in the World: A Brief Review. Mod. Environ. Sci. Eng., 2015, 1(1), 17–26.

[2]   V.G. Bahia: Fundamentos da erosão acelerada do solo (tipos, formas, mecanismos, fatores atuantes e controle). Inf. Agropecuário, 1992, 16(176), 16.

[3]   L.C. Hernani, P.L. de Freitas, F.F. Pruski, I.C. de Maria, C. Castro Filho, J.N. Landers: A erosão e o seu impacto,. In: Uso agrícola dos solos brasileiros, J.R.R. Manzatto, C.V. Freitas Junior, E. de Peres (Eds.), Rio de Janeiro: Embrapa Solos, 2002, 47–60.

[4]   G.H. Merten, J.P.G. Minella: The expansion of Brazilian agriculture: Soil erosion scenarios. Int. Soil Water Conserv. Res., 2013, 1(3), 37–48.

[5]   J.A.A. Anache, E.C. Wendland, P.T.S. Oliveira, D.C. Flanagan, M.A. Nearing: Runoff and soil erosion plot-scale studies under natural rainfall: A meta-analysis of the Brazilian experience. Catena, 2017, 152, 29–39.

[6]   IAPAR: Plantio direto no estado do Paraná. Londrina, 1981.

[7]   J. Schick: Fatores R e K da USLE e perdas de solo e água em sistemas de manejo sobre um Cambissolo Húmico em Lages, SC. Universidade do Estado de Santa Catarina, 2014.

[8]   A.L. Londero, J.P.G. Minella, D. Deuschle, F.J.A. Schneider, M. Boeni, G.H. Merten: Impact of broad-based terraces on water and sediment losses in no-till (paired zero-order) catchments in southern Brazil,. J. Soils Sediments, 2018, 18(3), 1159–1175.

[9]   I. Bertol: Conservação do solo no Brasil: Histórico, situação atual e o que esperar para o futuro. 2016.

[10] A. Minossi de Lemos: Matéria orgânica e perdas de solo, água e nutrientes por erosão em sistemas de preparo e de adubação orgânica e mineral em Argissolo Vermelho-Amarelo. Universidade Federal do Rio Grande do Sul, 2011.

[11] D. Deuschle, J.P.G. Minella, T. de A.N. Hörbe, A.L. Londero, F.J.A. Schneider: Erosion and hydrological response in no-tillage subjected to crop rotation intensification in southern Brazil. Geoderma, 2019, 340, 157–163.

[12] T.S. da Silva: Erodibilidade de um argissolo vermelho-amarelo e fator manejo e cobertura vegetal da equação universal de perdas de solo. 2016.

[13] E.A. Cassol, T.S. da Silva, F.L.F. Eltz, R. Levien: Soil erodibility under natural rainfall conditions as the K Factor of the Universal Soil Loss Equation and application of the nomograph for a subtropical Ultisol. Rev. Bras. Ciência do Solo, 2018, 1–20.

[14] M.R. Schmidt: Fatores erosividade das chuvas de Augusto Pestana (RS), cobertura e manejo do solo e erodibilidade de Latossolo Vermelho para uso na equação universal de perdas de solo. Universidade Federal do Rio Grande do Sul, 2017.

[15] J.A. Moreno: Clima do Rio Grande do Sul. Porto Alegre: Secretaria da Agricultura do Estado do Rio Grande do Sul, 1961.

[16] FAO: World reference base for soil resources: International soil classification system for naming soils and creating legends for soil maps, 3rd ed. Rome: FAO, 2014.

[17] W.H. Wischmeier, D.D. Smith: Predicting rainfall erosion losses: a guide to conservation planning. Washington, DC, 1978.

[18] N.P. Cogo: Effect of residue cover, tillage induced-roughness, and slope length on erosion and related parameters. Purdue University, 1981.

[19] P.R.C. Lopes, N.P. Cogo, R. Levien: Eficácia relativa de tipo e quantidade de resíduos culturais espalhados uniformemente sobre o solo na redução da erosão hídrica. Rev. Bras. Ciência do Solo, 1987, 11(1), 71–75.

[20] I. Bertol: Comprimento critico de declive para preparos conservacionistas de solo. Universidade Federal do Rio Grande do Sul, 1995.

[21] S.C.F. Dechen, F. Lombardi Neto, O.M. Castro: Gramíneas e leguminosas e seus restos culturais no controle da erosão em latossolo roxo. Rev. Bras. Ciência do Solo, 1981, 5, 133–137.

[22] J.C. Ramos, I. Bertol, F.T. Barbosa, J. Marioti, R.S. Wener: Influência das condições de superfície e do cultivo do solo na erosão hídrica em um Cambissolo Húmico. Rev. Bras. Ciência do Solo, 2014, 38(5), 1587–1600.

[23] G.R. Foster, D.K. Mccool, K.G. Renard, W.C. Moldenhauer: Conversion of the universal soil loss equation to SI metric units. J. Soil Water Conserv., 1981, 36, 355–359.

IDENTIFYING SURFACE RUNOFF USING NRCS-CN MODEL AT THE ARID REGION NILE DELTA - EGYPT

<u>M. Abu-hashim</u>[1], E. Mohamed[2], A. Belal[2]

[1]*Soil Science Department, Faculty of Agriculture, Zagazig University, 44511 Zagazig, Egypt;* <u>dr.mabuhashim@gmail.com</u>

[2]*National Authority for Remote Sensing and Space Sciences (NARSS), Cairo, Egypt*

Keywords: Soil Types, Surface Runoff, Curve Number, Hydrological Hazard, Forecasting

Abstract

Evaluation of surface runoff is an essential factor in the precision water and soil conservation management through their main extreme impacts on soil properties. Twenty eight soil profiles are prepared in Nile Delta, Egypt to cover different geomorphic units and hydrological soil groups in the study area. The Natural Resource Conservation Service - Curve Number model (NRCS-CN) is used to estimate the magnitude of runoff. Collected topographic data are used to explain the effects of slope variation on water retention and surface runoff. The results revealed that the highest values of surface runoff were distinguished close to the urban area and ranged between 40-50 mm. In urban areas, the surfaces are paved, and there is no infiltration of water. Consequently, the runoff water directly flow to the storm channels. Runoff values ranging between 30-40 mm occurred at north of the study area. The sloping surface and the nature of the clay soil contributed to generate more runoff than do the lowland areas. The study presented and tested the hydric runoff estimation based-model on the integration of hydric balance parameters. The GIS tools analyze and compose these parameters to perform an indirect method for the quantity of water that results from direct surface runoff flow. This method helps to gain a clear imaging of the surface runoff risks on the study area.

1 Introduction

Water requirements in arid and semi-arid countries are continuously increasing due to population increase and improving living level as well as due to the government policy to encourage industrialization [1]. Recently, water requirements have been increased in Egypt especially with increasing the population rate. With its loosening grip on the Nile River, water scarcity could endanger the country's stability and regional dominance [2]. Thus, precise water management practices consider one of the main issues and strategies to face the water scarcity and to convey the required water for the cultivated lands. Otherwise, soil runoff is considered another important parameter that should be considered in water management and land degradation [3-5]. Runoff occurs, when rainfall and/or the flooding irrigation reach the earth surface and the soil is saturated, then the excess water starts flowing on the surface. Thus, precise water management practices should be considered to find the most proficient technology to save water and to convey it to drought regions and to newly reclaimed regions. Furthermore, the flow rate is depending on various parameters: rainfall (intensity and distribution), soil characteristics (texture, infiltration rate, organic matter contents, porosity, soil aggregation), vegetation cover and the topographic parameters including slope and aspect [6, 7]. Soil water retention potential (S) and values of curve number NRSC-CN model consider the main parameters that affect identifying and computing the surface runoff [3, 8-10]. Density of vegetation covers effect the magnitude of surface runoff, where removing vegetation cover leads to an increasing tendency in soil surface runoff, thereby increase soil water erosion [4, 8, 11]. Also, land-use changes, urbanization and forested crops, an important factor in the runoff process, affect the soil water holding capacity [3-5]. Integration GIS techniques with remote sensing are used as effective tools to resolve and to monitor the potential flood water based on spatial hydrological modelling [4]. Though data of remote sensing were established to initiate the curve number by correlating hydrologic soil groups (HSG) and land use/ land cover (LULC) with data derived from the NRCS tables. GIS technique was used as an efficient tool for preparation and input date required by SCS-CN model for precision water management to understand the reactions of arid ecosystems and to support estimation of the returning water benefits [5, 12, 13]. Establishment of the hydrological soil groups (HSGs) was based on several parameters: soil texture, infiltration and potential water retention in each pedologic entity. The different land-use types had consequences in the surface hydric approach: water quantity, surface runoff, interception of the precipitated, infiltration, and the pedospheric cover. Although the CN model was mainly prepared for identifying storm runoff in small watersheds and agricultural basins, it was well modified for different land-uses such as urbanized watersheds.

The objective of this research is to integrate soil type data, remote sensing data, and hydrological based-model to estimate the direct earth surface runoff in the middle of the Nile Delta as well as to simulate the changes of surface runoff with changes in rainfall amount as a result of climate change.

2 Material and Methods

2.1 Study area

The study area lies between (30°45$^\backslash$ N, 30°55$^\backslash$ E), in middle of the Nile Delta at the catchment area of Tanta that aligned by Damietta and Rosetta branches (Figure 1). The study area is characterized by Mediterranean climatic conditions, where the rainfall is seasonal and intermittent, and its falls in the winter. However, rainfall recorded in November 2016 was about 58 mm according to the Climatic data of Tanta Meteorological Station.

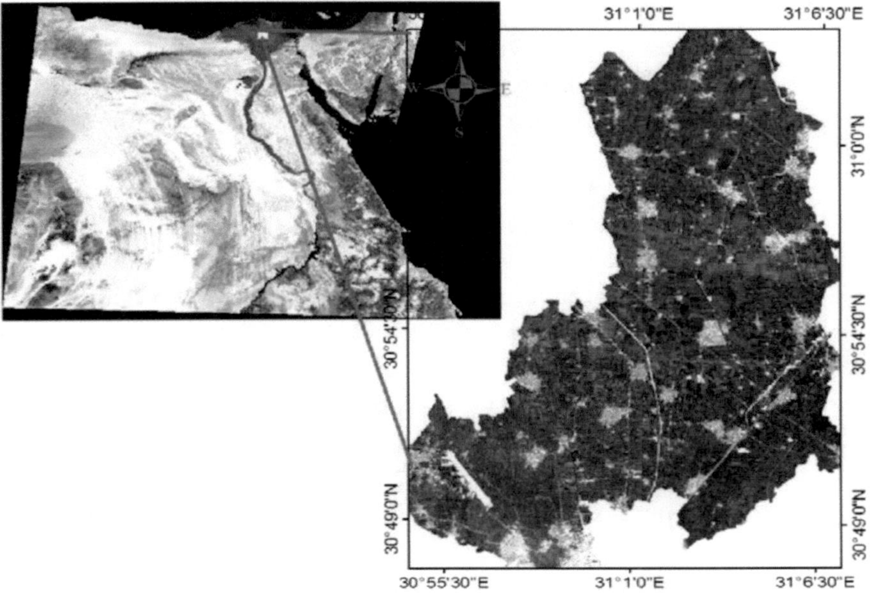

Figure 1: Location of the study area at Tanta catchment area in the middle of the Nile Delta

Mean and maximum temperatures were 18.9 and 24.9 °C, respectively. The main landscape of the investigated area is flood plain [5, 14]. Landsat 8 with spatial resolution 30 m was acquired in November 2016. Topographic maps and Digital Elevation Model (DEM) with 30 × 30 m resolution was derived using Shuttle Radar Topography Mission and elevation points were recorded during the field survey by GPS.

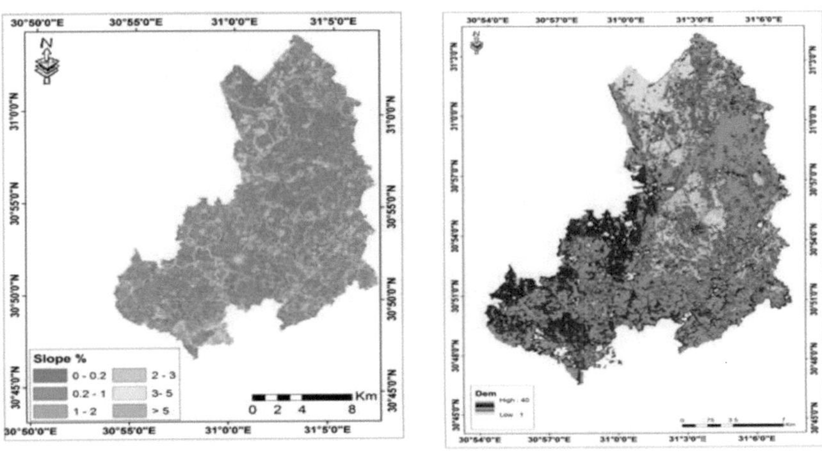

Figure 2: Digital elevation model and the slope distribution at the catchment area

2.2 Land use

Land use was determined using Support Vector Machine (SVM) as described by [5, 14] for the Nile Delta. The SVM technique was applied in the image acquired in November 2015 to determine the land use / land cover changes of the study area, as the SVM classifier provided an accurate method for discriminate LULC.

2.23Watershed delineation and soil sampling

GIS was used for determining watershed area and network of the streams in order to identify the sub-catchments that contribute to a single stream based on DEM considering the site morphological properties. Soil sample data cover geomorphological units of the study area (Figure 1). Suitable sampling sites were identified by computing hydrologic soil groups (HSG), soil type, and land-use for the region and combining them to initiate hydrologic response units (HRU$_s$) using GIS. Soil physical analyses were conducted according to [15]. The dominant soil types are loamy and clay loam soil. Soil field capacity varied from 22.8 to 41.8%, while dry bulk density ranged between 1.41 and 1.55 g/cm^3.

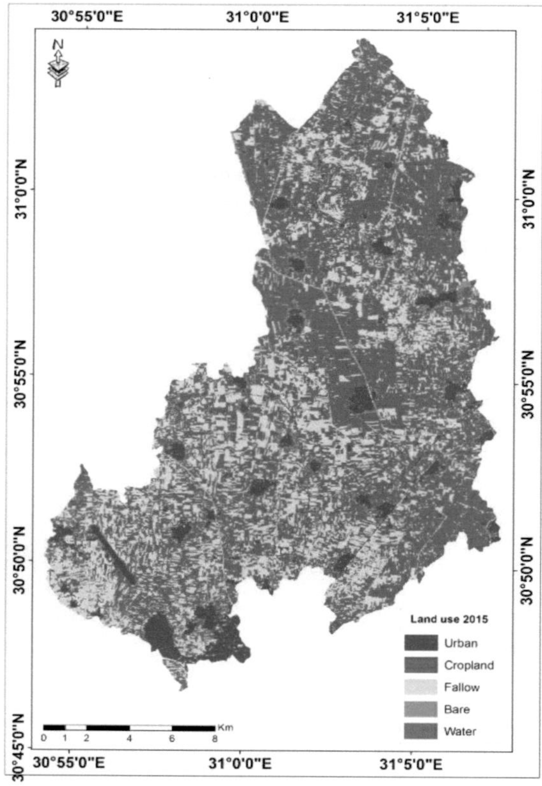

Figure 3: Characterization of land-use types at the catchment area

2.4 Surface runoff computation

SCS-CN model [4] was used to calculate runoff, where it depends on universal water balance, as follows:

$$P = Q + E + \Delta S \tag{1}$$

where are P: precipitation, Q: runoff, E: evapotranspiration, ΔS: storage term. Natural resource conservation service modified the water balance to the following equation:

$$Q = P\,(F/S) \tag{2}$$

Where are F: actual loss, S: potential loss. Evaporation of universal water balance equation and storage term has been included into the relation of actual (F) and water potential loss (S). By substituting F (actual loss):

$$F = P - Q \tag{3}$$

runoff was formulated as:

$$Q = P^2 / (P+S) \tag{4}$$

Equation 4 explains the runoff as a function of precipitation and water potential loss (S) that is identified as retention potential or soil water holding capacity. Since runoff is produced, if there is rainfall, the term initial abstraction (I_a) has been introduced [4], which is subtracted from the total rainfall to retrieve effective precipitation:

$$P_e = P - I_a \tag{5}$$

Where are P_e: effective precipitation, I_a: initial abstraction. I_a is all losses before runoff begins that includes water retained in surface depressions, water abstracted by vegetation, evaporation, and infiltration. NRCS-CN approach [4] is expressed as:

$$Q = (P - 0.2S)^2 / (P + 0.8S) \text{ for } P > Ia; \quad Q = 0 \quad \text{for } P \leq Ia \tag{6}$$

The relation of the potential water retention S is shown in the following equations:

$$S = (25400/CN) - 254 \tag{7}$$

Theoretically, CN ranging between 0 and 100, where $S = \infty$ and $S = 0$ respectively.

$$Q = \frac{(P-0.2S)^2}{(P+0.8S)} \qquad P \geq 0.2S, \geq 0.2S, \qquad Q = 0, \text{otherwise} \tag{8}$$

Where are Q: surface runoff (mm), P: precipitation (mm), and S: potential water retention (mm).

The computed monthly summarized direct runoff was an assumption that the precipitation is equal or higher than 0.2S [4]. Table 1 shows the curve number hand book values (TR55) that were published by [16, 17] and used in the context of this work. The CN was identified under the hydrologic soil groups (HSGs), and the land use and land management for each the soil types in the catchment area.

Table 1: Curve number values extracted from published tables listed in NRCS-CN, 1986 (TR-55) Handbook manual.

Land-use	Treatment	Hydrologic condition	Curve number for hydrologic Soil group			
			HSG A	HSG B	HSG C	HSG D
Bare soil	-	-	77	86	91	94
Fallow	Crop residue cover (CR)	Poor	76	85	90	93
		Good	74	83	88	90
Cropland	Straight row (SR)	Poor	72	81	88	91
		Good	67	78	85	89
Cropland	SR + CR	Poor	71	80	87	90
		Good	64	75	82	85
Cropland	Contoured (C)	Poor	70	79	84	88
		Good	65	75	82	86
Cropland	C + CR	Poor	69	78	83	87
		Good	64	74	81	85
Urban area	-	-	98	98	98	98
Water	-	-	100	100	100	100

Poor: Factors impair infiltration and tend to increase runoff; Good: Factors encourage average and better than average infiltration and tend to decrease runoff.

Figure 4 displays the flow chart of the method that was used to identify the direct surface runoff using the NRCS-CN model and the GIS technique.

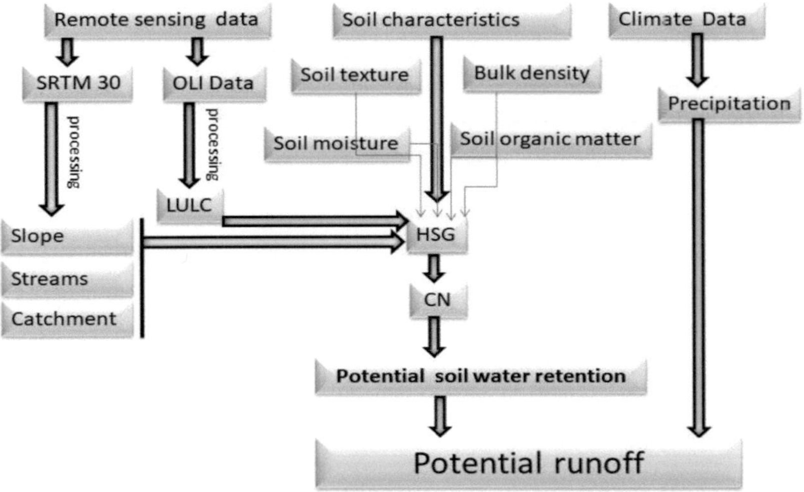

Figure 4: Flow chart of the method used to identify the direct surface runoff using the NRCS-CN model and the GIS technique

3 Results and Discussion

Several factors have been integrated using GIS techniques to model surface runoff in the study area, where land-use has distinct effects on infiltration rate and soils' water holding capacity [18]. Land use was determined using Support Vector Machine (SVM). Land use / land cover map showed that crop lands occupies an area about 59.3% of the total area, fallow land covers 27.2%, while urban areas occupy an area of 8% and bare soil 5.2%, as shown in Table 2 and Figure 3.

Table 2: Land-use distribution at the catchment area

Land-use	Area %
Cropland	59.3
Bare soil	5.2
Fallow	27.2
Urban area	8.0
Water	0.3

Soil types of the watershed varied between clay, clay loamy and loam soil. Saturated hydraulic conductivity varied from 3.9 to 49 cm/d. Soil field capacity ranged between 22.8 and 41.8%, while dry bulk density ranged between 1.41 and 1.55 g/cm^3. Soil organic matter varied from 1.2 to 1.6%. The area is described by flat to gently sloping, where the northern parts of the study area is attributed by flat ranging between 0- 0.2% except of some patches Figure 2, while the southern part is characterized by gradient slope varied from 0.2 to 1%.

NRCS-CN model values were calculated depending on the obtained data of land use / land cover, slope, texture, infiltration, and retention capacity of each pedologic unit (HRUs). Figure 5 shows the curve number map that represents the spatial distribution values of the study area. HSG was established for the basin based on soil parameters and land-use by using the text book of [19]. Thereby, spatial distributions (S) were mainly determined by the spatial heterogeneousness of LU/LC and soil types of the areas (Figure 6).

Direct surface runoff at the study area was calculated using the daily rainfall values. The mean of the rainfall amount was computed using the data of Tanta Meteorological Station. The obtained data shows a mean rainfall of 58 mm at the time of soil samples. This value considered as the highest rainfall amount comparing with the same time in the previous years, using climatic data of the meteorological station in the same region.

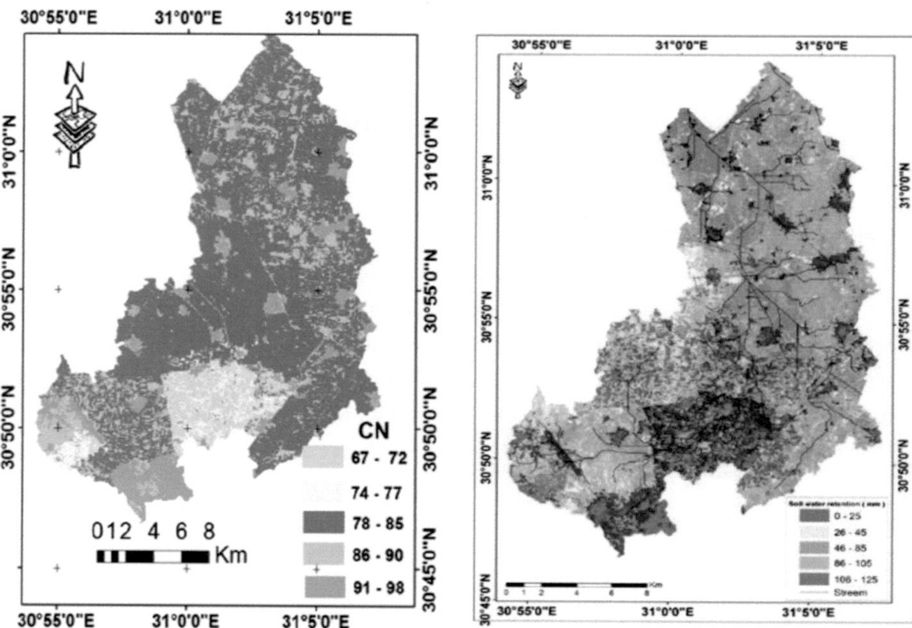

Figure 5: Spatial distribution of the curve number

Figure 6: Spatial distribution of potential water retention

The results illustrated that the highest values of runoff were distinguished around urban areas, where it ranged between 40-50 and 50-58 mm, respectively, and paved surfaces and rooftops did not allow water to penetrate into soil causing direct surface runoff. In addition, the runoff values that ranged between 30-40 mm in the cropland area located at the north of the study area, in which the slope tends to generate more runoff than the lowland areas do. These results were relevant with the findings of [4, 5]. That enrichment in urbanization activity results in decreasing the soil infiltration capacity, which leads to increase the land degradation. Furthermore, the main phenomena in the middle of the Nile Delta declared that urban area was sprawled by building industrial communities [20, 21]. Thereby, changes in LULC have distinct impact on direct surface runoff through their influences on the soil water holding capacity [18, 22]; urbanization, deforestation, and changes in land-use activities significantly affect the infiltration capacity and the distribution of surface flow [23]. Nevertheless, the most representative results were noticed in the north of the catchment area, in which low values of runoff was reported as 10-20 mm.

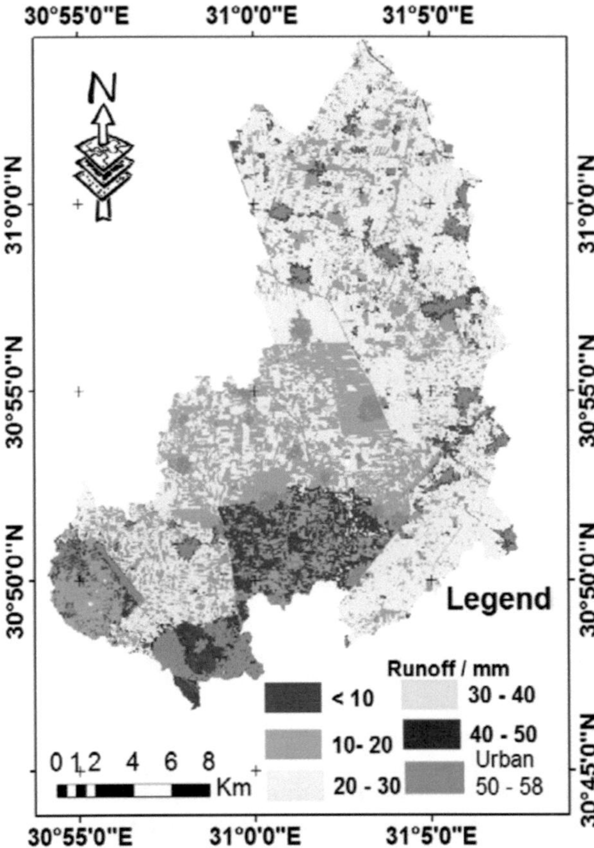

Figure 7: *Spatial distribution of runoff values for a rainfall of 58 mm/m²*

The obtained results confirm the fact that soil surface conditions in cropland through infiltration capacity have direct effect on the surface tendency of the region (Figure 2). These results correspond well with the findings of [1, 21-23], who stated that land-use controls soil infiltration, and the variation in infiltration capacity is produced due to the effect of the land-use systems on soil properties. Because of the reduced surface of the reception basin, the quantity of precipitation was considered to be uniform in the whole basin. The behavior of the basin was simulated in conditions of precipitation of 58 mm/m² as current status quo. By applying equation (8), two layers were obtained in the evaluation of the direct surface runoff water layer for each of the two rain categories (Figure 7). Expanding urban activities to cultivated soils results in an increased tendency of surface runoff with the possibility of climate change on local scale and increased precipitation.

4 Conclusions

Applying the NRCS-CN model combined with the GIS technique for analyzing the direct surface runoff at a catchment scale can be an efficient tool in the context of the demand of forecasting and projecting the hydric hazards. The strengths of these scenarios using this model are the precision, possibility and rapidity of obtaining results to simulate the hydric surface runoff either on a daily, monthly, seasonal base, or on annual scale. Increasing the urban activities in arable lands would result in increasing the tendency of these soils for surface runoff with the possibility of climate change. In addition, the results illustrated that the highest value of surface runoff were pronounced around the urban areas (50-58 mm) as a result of paved surfaces and rooftops that did not allow water to penetrate into the soil. Otherwise, the lower runoff values that ranged between 30-40 mm in the cropland area located at the north of study area. Thus, precise water management practices could be applied to find the most efficient technology to save water and to convey it to the drought regions. Nevertheless, the identified drawback in this model is ignoring certain parameters such as evapotranspiration. Remote sensing data provide essential information about land use/land cover and the topographic factors. Therefore, integration of these data with soil characteristics using GIS techniques can be used to model the surface runoff based on spatial hydrological modelling.

5 Acknowledgements

The authors would like to thank DAAD and Exceed SWINDON project for providing them financial support to participate at this workshop.

6 References

[1]   Elkhrachy, I.: Flash flood hazard mapping using satellite image and GIS tools: a case study of Najran city, Kingdom 306 of Saudi Arabia KSA. Egypt. J. Remote Sens. Space Sci., 2015, 18(2), 261–278

[2]   Ministry of Water Resources and Irrigation, Egypt: Water Scarcity in Egypt. Report 1-5, 2014, http://www.mfa.gov.eg/Site Collection Documents/Egypt %20 Water % 20 Resources %20 Paper_2014.pdf.

[3]   Mishra, S.K., Singh, V.P.: Another look at SCS-CN method, J. Hydrol. E.-ASCE, 1999, 4(3), 257–264.

[4]   Hawkins, R.H., Ward, T.J., Woodward, D.E., Van Mullem, J.A.: Curve Number Hydrology: State of the practice, Rev. Ed., U.S.D.A., Washington D.C., U.S.A., 2009

[5]   Abu-hashim, M., Mohamed, E.S., Belal A.A.: Identification of Potential Soil Water Retention using Hydric Numerical Model at Arid Regions by Land-use Changes. International Soil and Water Conservation Research, 2015, doi:10.1016/j.iswcr.2015.10.005.

[6]   Dunne, T., Leopold, L.B.: Water in Environmental Planning, W.H. Freeman Co., San Francisco, 1978, p. 818.

[7]   Morgan, R.P.C.: A simple approach to soil loss prediction: a revised Morgan–Morgan-Finney model. Catena, 2001, 44(4), 305–322

[8]   Michel, C., Andreassian, V., Perrin, C.: Soil Conservation Service Curve Number method: How to mend a wrong soil moisture accounting procedure, Water Resour. Res., 2005, 41, W02011.

[9]   Hawkins, R.H.: A symptotic determination of runoff curve numbers from data, J. Irrig. Drain., 1993, 15; E.-ASCE, 119(2), 334–345.

[10] Agnihotri, I., Punia M.P., Sharma J.R.: Estimation and Assessment of Spatial Variations in Water Availability in West Flowing River Basin of Kutch, Saurashtra and Marwar (WFR-KSM basin) using Geospatial Technology. Int. J. Adv. Remote Sens GIS, 2017, 6(1), 1994-2000

[11] Mantey, S., Tagoe, N.D.: Spatial modelling of soil conservation service curve number grid and potential maximum soil water retention to delineate flood prone areas: A case study. Res. J. Environ. Earth Sci., 2013, 5(8), 449–456.

[12] Soulis K.X., Valiantzas, J.D., Dercas, N., Londra, P.A.: Analysis of the runoff generation mechanism for the investigation of the SCS-CN method applicability to a partial area experimental watershed. Hydrol. Earth Syst. Sci. Discuss., 2009, 6, 373–400.

[13] Latha M., Rajendran, M., Murugappan, A.: Comparison of GIS based SCS-CN and Strange table Method of Rainfall-Runoff Models for Veeranam Tank, Tamil Nadu, India. Int. J. Sci. Eng. Res., 2012, 3(10), 1

[14] Mohamed, E.S., Belal, A., Shalaby, A.: Impacts of soil sealing on potential agriculture in Egypt using remote sensing and GIS techniques. Eurasian Soil Sci., 2015, 48(10), 1159-1169.

[15] Klute, A.: Water retention: Laboratory methods. In: Klute, A. (Ed). Methods of soil analysis. Part 1. 2nd edition. Agronomy Monograph 9, pp 635-662. ASA and SSSA, Madison, WI., Phys. Chem. Earth, 1986, 28, 1377-1387.

[16] Lyon, S.W., Walter, M.T., Gerard-Marchant, P., Steenhuis, T.S.: Using a topographic index to distribute variable source area runoff predicted with the SCS curve-number equation, Hydrol. Process., 2004, 18(15), 2757–2771.

[17] Zhan, X., Huang, M.: Arc CN-runoff: An Arc GIS tool for generating curve number and runoff maps. Environ. Model. Softw., 2004, 19(10), 875–879.

[18] Abu-hashim, M., Mohamed, E., Belal, A.E.: Land-use changes and site variables on the soil organic carbon pool: the potential application for the MENA region. Adv. Environ. Res., 2016, 65.

[19] Chow V.T., Maidment D.R., Mays, L.W.: Applied Hydrology. McGraw-Hill, New York, 1988.

[20] Mohamed, E.S., Saleh, A.M., Belal, A.A.: Sustainability indicators for agricultural land use based on GIS spatial modeling in North of Sinai-Egypt. Egypt. . Remote Sens. Space Sci., 2014, 17(1), 1-15.

[21] Abu-Hashim, M.S.D.: Impact of land-use and land management on water infiltration capacity on a catchment scale. PhD Thesis, Technical University Braunschweig, Germany, 2011.

[22] Mantey, S., Tagoe, N.D.: Spatial Modelling of Soil Conservation Service Curve Number Grid and Potential Maximum Soil Water Retention to Delineate Flood Prone Areas: A Case Study. Res. J. Environ. Earth Sci., 2013, 5(8), 449-456,

[23] Romero, P., Castro, G., Gomez, J.A., Fereres, E.: Curve number values for olive orchards under different soil moisture management. Soil Sci. Soc. Amer. J., 2007, 71(6), 1758-1769.

FISH FEED COMPOSITION AND ANIMAL MANURE APPLICATION AS FACTORS IN THE IMPAIRMENT OF AQUACULTURE POND WATER QUALITY - A PILOT STUDY

Abimbola Olumide Adekanmbi

Environmental Microbiology and Biotechnology Laboratory, Department of Microbiology, University of Ibadan, c/o El-Shaddai Baptist Church, P. O. Box 20000, U. I. Post Office, Ibadan, Oyo State, Ibadan, Nigeria; bimboleen@yahoo.com

Keywords: Aquaculture ponds, fish feed composition, animal manure application, physicochemical parameters, metal composition

Abstract

The role of aquaculture in food security, especially the provision of a cheap and affordable source of protein, cannot be completely overlooked. However, in an attempt to boost fish production and to enhance output, so many unwholesome practices such as the use of animal manure as fertilizers are being practiced. This exposes the environment especially water bodies to potential pollutants and pathogens upon discharge of this nutrient and bacterial laden water from fishing operations. This study aimed at determining the physicochemical and bacteriological qualities of water from selected aquaculture ponds in comparison with the water sources used for their operations. Water samples from selected aquaculture ponds were collected in pre-cleaned sample containers, labeled appropriately and transported from the site of collection to the microbiology laboratory on ice chest. Physicochemical and metal analyses of the samples were conducted using standard methods and atomic absorption spectrophotometry (AAS) respectively, while bacteriological analyses were carried out using the pour plate technique on MacConkey agar. The metal analyses showed that the concentrations of copper, zinc, lead, nickel and chromium were highest in the ponds with manure application and lowest in the reservoir ponds. Of the total coliforms obtained, 62.6% were isolated from the manure-applied ponds, while 28.3% and 9.1%, respectively, were isolated from the ponds without manure application and the reservoir ponds. The physicochemical parameters showed that all the parameters including pH, Biochemical Oxygen Demand (BOD), Chemical Oxygen Demand (COD), Total Dissolved Solids (TDS), Electrical Conductivity (EC), Salinity, Nitrate, Phosphate were higher in the ponds with animal manure application than in the ponds without manure application and the reservoir ponds. The only exception was the Hardness, which was highest the ponds without manure application (59.4 mg/L), with the ponds with manure application having 41.0 mg/L and the reservoir ponds with 56.4 mg/L. The study confirmed the negative impacts the application of animal manure in aquaculture ponds on the physicochemical and bacteriological qualities of water used in aquaculture ponds. Though further studies are going on in this area of research, there is a need for regulations regarding the use of animal wastes in production process in the interest of the environment and the final consumers. There is a need to put an effective treatment option in place for water generated from aquaculture to prevent the introduction of potentially pathogenic bacteria into receiving water bodies.

1 Introduction

Water is an indispensable resource for the continued existence of most life forms. However, growing population coupled with other factors such as leaching from septic tanks, improper disposal of wastes, improper management of wastes generated from animal husbandry and inadequate infrastructure have led to the impairment of the quality of most water sources especially of surface waters [1]. The use of animal manure in aquaculture coupled with the composition of fish feed has predisposed aquaculture ponds to a geometric rise in nutrient level, which could be a factor in the onset of eutrophication, thus impairing the physicochemical and microbiological characteristics of water from such ponds [2]. This is because fish feed contain a lot of proteinaceous materials, which could serve as nutrient for the proliferation of bacteria that coincidentally have been introduced into aquaculture ponds via the use of animal manure to promote the growth of certain phytoplankton and zooplankton [3]. The study aimed at determining the physicochemical and bacteriological qualities of water from selected aquaculture ponds in tandem with their water sources.

2 Materials and Methods

2.1 Description of study site

The study was carried in Ibadan, Southwest Nigeria. The aquaculture ponds sampled were located in an integrated farming system in Ibadan. Their main sources of water are boreholes and surface water. These water sources are stored in reservoir ponds prior to use in the ponds.

2.2 Sample collection

Water samples from the selected aquaculture ponds were collected in pre-cleaned sample containers, labeled appropriately and transported from the site of collection to the microbiology laboratory on ice chest. The water samples were analyzed within three hours after collection. The collection was done in three groups: (a) Ponds with manure application, (b) Ponds without manure application, and (c) Reservoir ponds.

2.3 Determination of physicochemical parameters and metal composition of water samples

Selected physicochemical parameters of the water samples, e.g., pH, Electrical Conductivity (EC), Total Solids (TS), Total Hardness, Total Suspended Solids (TSS), Total Dissolved Solids (TDS), anion concentrations, Chemical Oxygen Demand (COD), and Biochemical Oxygen Demand (BOD_5) were determined using standard methods [4], while the metal composition was determined using a flame atomic absorption spectrophotometry (FAAS) (UNICAM 929, London Atomic Absorption Spectrophotometer powered by SOLAAR software). Standards of the selected metals were prepared from 1000 mg/L stock solutions. Cathode lamps of each metal were used for the analysis of the respective mineral oils in the standards and metal concentrations in the sample filtrate [5].

2.5 Culture media used

MacConkey agar and Nutrient agar (Oxoid, UK) were used in the isolation of enteric bacteria (coliform) from the water samples obtained from the aquaculture ponds. MacConkey agar was used in the isolation of the enteric bacteria (coliforms), while Nutrient agar was used in the subculturing of the isolates obtained. The preparation of these media was done in accordance with the manufacturer's instruction, after which they were sterilized at 121 °C for 15 min inside the autoclave.

2.6 Isolation and characterization of coliforms from water samples
Isolation of coliforms from the water samples was carried out using the standard pour plate techniques on MacConkey agar [6]. The plates were incubated overnight at 35±2 °C, after which colonies with pink colour on the isolation medium were selected and sub-cultured to obtain pure cultures. The purified isolates were then stored on nutrient agar slants for further studies. Morphological, biochemical and sugar fermentation tests were used in confirming the identity of the isolates obtained [7].

3 Results and Discussion

3.1 Metal composition of the aquaculture ponds
In this study, the metal and physicochemical qualities of selected aquaculture ponds were determined to derive the effects of animal manure application on the qualities of the ponds under study. Table 1 is showing the mean metal concentration of the three groups of ponds. For all the metals, the concentration in the pond treated with animal manure is higher than the other two ponds. This trend might be attributed to the introduction of the waste, which might contain trace amounts of metals. The same trend was also supported by Nnaji et al. [8] in their study on the heavy metal risks on integrated chicken-fish farming. This is also evident in this study, where the ponds without manure application and the reservoir ponds had lower concentrations of metals. It should be noted, however, that nickel was below detection limits in the two ponds (< 0.005 mg/L).

Table 1: Mean metal concentrations (mg/L) of water obtained from the aquaculture ponds

Metal	Ponds with manure application	Ponds without manure application	Reservoir ponds
Copper (Cu)	0.04	0.01	0.01
Zinc (Zn)	2.55	0.28	0.13
Lead (Pb)	0.22	0.04	0.02
Nickel (Ni)	0.31	BDL	BDL
Chromium (Cr)	2.11	0.42	0.22

BDL: Below Detection Limit (0.005 mg/L)

3.2 Distribution of coliform bacteria in the aquaculture ponds
The use of organic fertilizers in the form of animal manure has been reported to result in a high level of potentially pathogenic microorganisms in the water generated from aquaculture operations [9]. In this study, a larger percentage of the coliforms obtained were from the manure application ponds, where 62.6% of coliforms were isolated. The other two categories of ponds had 28.3% and 9.1%, respectively, for the ponds without manure application and the reservoir ponds, respectively. The presence of coliforms and other bacteria could be linked to fecal origin, hence

the trend observed in Figure 1 [10]. El Saidy et al. [11] also reported a high microbial load for ponds treated with animal manure in comparism with the untreated ones.

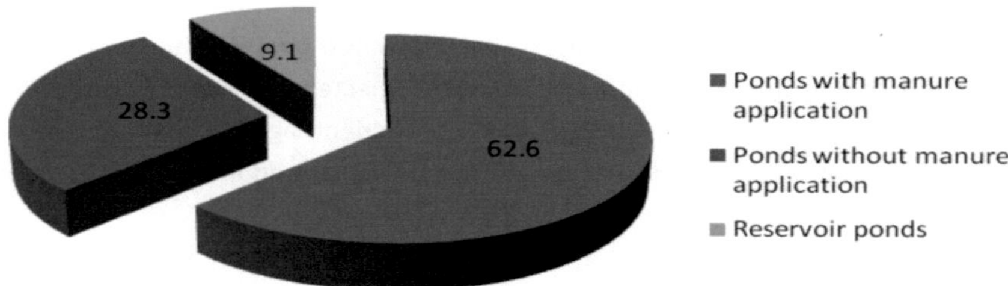

Figure 1: Distribution of the coliform bacteria from obtained from the aquaculture ponds (%)

In Table 2, the physicochemical parameters of the three groups of aquaculture ponds investigated in this study are shown. With the exception of total hardness, which is higher in the ponds without animal manure application (56.4 mg/L) and reservoir ponds (59.4 mg/L) in comparism with the ponds, where animal manure is applied (41.0 mg/L), all the other parameters were higher for the manure-applied aquaculture ponds. This could be due to the loading of the ponds with nutrients and other minerals especially from animal excreta, and this could be responsible for the higher values in comparism with the other two groups. The BOD_5 was very high in the manure-applied soil, suggesting a high level of contaminants or pollutants herein [3].

Table 2: Mean of selected physicochemical parameters of the aquaculture ponds

Parameters (mg/L)	Ponds with manure application	Ponds without manure application	Reservoir ponds
pH	7.7	7.2	7.4
BOD	323	120	2.7
COD	614	236	5.8
TDS	389	356	128
EC (µS/cm)	262	238	84.7
Salinity (g/L)	187	170	65
Hardness	41.0	56.4	59.4
TSS	257	4.93	2.8
Nitrate	20.7	1.00	0.40
Phosphate	2.90	0.70	0.10

Note: Values are a mean of five different aquaculture ponds;
BOD: Biochemical Oxygen Demand; COD: Chemical Oxygen Demand; TDS: Total Dissolved Solids; EC: Electrical Conductivity;
TSS: Total Suspended Solids.

4 Conclusions and Recommendation

There is a high occurrence of coliforms in the water samples studied, suggesting the presence of organic matter in the ponds. This is in addition to the physicochemical parameters and metal concentrations well beyond the acceptable limits in many of the ponds. There is, therefore, a need to put an effective treatment option in place for the treatment of water generated from aquaculture operations before eventual discharge into the environment. More research on this particular subject especially on the contribution of fish feed to aquaculture pond water is ongoing.

5 Acknowledgements

The author wishes to acknowledge Federal Ministry for Economic Cooperation and Development, Germany (BMZ), German Academic Exchange Service (DAAD), Technical University of Braunschweig (TUBS) and Exceed-Swindon project for sponsoring him to attend the International Expert Workshop on "Water on Agricultural practices: Training the trainers" in Brazil, which was organized by the Pontifical Catholic University of Rio de Janeiro, and Embrapa Soils, Rio de Janeiro (September 15-21, 2019), where this paper was presented. The efforts of my research collaborators and students are also acknowledged.

6 References

[1]   Huttly, S.R.A.: The impact of inadequate sanitary conditions on heath in developing countries. World health statistics quarterly. Rapport trimestriel de statistiques sanitaires mondiales 1990, 43(3), 118-26

[2]   Sloan, D.R., Kidder, G., Jacobs, R.D.: Poultry Manure as a Fertilizer. University of Florida, Gainesville, USA, 2003, 326 pp.

[3]   Espino, A.N.J., Barrog, A.J., Abucay, J.S., Cinense, M., Domingo C.J.: A GIS-aided study of integrated aquaculture and livestock/poultry production in the Philippines. Science City of Muños: Central Luzon State University, College of Fisheries and College of Veterinary Science and Medicine, 2008, pp135.

[4]   APHA/AWWA/WEF: Standard methods for the examination of water and wastewater. American Public Health Association, American Water Works Association, Water Environment Federation. (2012).

[5]   Hseu, Z.: Evaluating heavy metal contents in nine composts using four digestion methods. Bioresource Technology 2004, 95, 53-59.

[6]   Harrigan, W.F., McCance, M.E.: Laboratory methods in microbiology. Academic Press (1966)

[7]   Sneath, P.H.A.: Bergey's Manual of Determinative Bacteriology. William and Wilkins, Baltimore, 1996.

[8]   Nnaji, J.C., Uzairu, A., Gimba, C., Kagbu, J.A.: Heavy Metal Risks in Integrated Chicken-fish Farming. Journal of Applied Sciences, 2011, 11, 2092-2099.

[9]   Ansah, Y.B.: Characterization of pond effluents and biological and physicochemical assessment of receiving waters in Ghana. MSc thesis, Virginia Polytechnic Institute and State University, USA. 2010.

[10] Ampofo, J.A., Clerk, G.C.: Bacterial flora of fish feeds and organic fertilizers for fish culture ponds in Ghana. Aquaculture Research, 2003, 34(8), 677-680.

[11] Elsaidy, N., Abouelenien, F., Kirrella, G.A.K.: Impact of using raw or fermented manure as fish feed on microbial quality of water and fish. Egyptian Journal of Aquatic Research, 2015, 41, 93–100.

HYDROPONIC AS ONE OF WATER CONSERVATION TECHNIQUES
JORDAN CONTEXT

Ahmad Alulayyan

Jordan University of Science and Technology, Ali Kaied Hassan St. Building # 8, Shafa Badran, 11934, Amman, Jordan; alulayyan.ahmad@gmail.com

Keywords: Agriculture, Hydroponic, Water efficiency

Abstract

Jordan is a Mediterranean country that depends mostly on rain as its main water resource. Recent years have witnessed shortage in the rainfall in different parts of the country. As a result, numerous streams have dried out, underground water level has fallen to critical levels, and most water aquifers are experiencing high salinity as a result of over abstraction, which makes them unsuitable for domestic or irrigation uses. In addition, extreme weather conditions such as flash floods during winter and heat waves during summer are becoming more frequent in the region. These conditions are direct consequences of global climate change that has recently been affecting several locations around the world, which are dramatically impacting wide ranges of ecosystems. The climate change reflected on people's lives in Jordan, especially of people, who live in the rural areas that depend on rain-fed agriculture to make their livings. In addition, Jordan suffers from lack of rain fed cultivatable soil, where more than 80% of the country's land is desert (less than 200 mm rainfall) with a very harsh environment. The soil in these areas suffers from lack of plant nutrients, and cultivation in such environment requires a huge investment in fertilization and soil amending. Lack of sustainable water resources and deterioration of soil affect the agricultural production adversely. The quality of products is highly affected by water shortage, lack of nutrients and soil-borne diseases; consequently, the profitability of the agricultural sector has decreased significantly. Soilless culture and especially the hydroponic emerge as a technical solution to mitigate the consequences of water crisis and the unsuitability of soil for cultivation. Many projects are being implemented in Jordan to provide an evidence-based solution on the effectiveness of hydroponic as one of the adaptation technology to climate change impacts on water availability. The results of these projects are very promising in terms of water conservation, product quality and the environmental impacts and will be presented in this paper.

1 Introduction

Jordan is a food-deficit country and depends heavily on irrigated agriculture, for what it does produce. Agriculture comprises a relatively small share of gross domestic product (GDP). Jordan 2025 visions calls for increased share. Irrigated agriculture is considered as the biggest water consumer in Jordan, as it consumes more than 51% of the available water resources. Agriculture has serious socio-economic impacts and, at times, also a high political significance. It provides most of the agricultural production in the Kingdom and, therefore, contributes to the national food security and also offers high percentage of direct and indirect jobs. The production of food in semi-arid countries like Jordan is hardly possible without irrigation [1].

Severe water scarcity is responsible for the significant decline of water use in agriculture, while municipal use of water has increased steadily due to demographic and economic growth. Most irrigation occurs in two distinct areas; Jordan Valley and the highlands. About 60% of agriculture in Jordan depends on rainwater and 40% is irrigated agriculture in the highlands and the Jordan Valley. This 40% irrigated agriculture produces 90% of total agricultural products. Agriculture in the Jordan Valley produces 70% of total agricultural products, while consuming only 35% of the irrigation water. This demonstrates the greater productivity of lands cultivated with irrigation water and the significance of irrigation in the Jordan Valley. Table 1 shows the agriculture by type and value [2].

Table 1: Agriculture by type and value in Jordan

Type of Agriculture	Area		Value	
	Hectare	Percentage	JD* Million	Percentage
Rain fed	160,000	60	49	10
Irrigated	100,000	40	461	90
Total	260,000	100	510	100

JD = Jordan Dinar = 1.4 USD

The sever water scarcity and the increased demand on water for domestic uses as a result of the influx of refugees from the unstable neighbouring countries as well as other factor such as natural population increase, economic development and climate change led to a high pressure on the agricultural sector. The Jordanian Government represented by the Ministry of Water and Irrigation and the Ministry of Agriculture has taken many technical, institutional and economic measures for the management of demand on irrigation water and improved water use efficiency for supporting sustainability and stability of the agricultural sector.

2 Hydroponics

Hydroponics is a technique of growing plant in nutrient solutions with or without using inert medium such as gravel, vermiculite, rock wool, peat moss, saw dust, coir dust, coconut fibre, etc. for providing mechanical support. The term *hydroponics* was derived from the Greek words hydro' meaning water and ponos' labour; literally it means water work. The word *hydroponics* was coined by Professor William Gericke in the early 1930s; it describes the growing of plants with their roots suspended in water containing mineral nutrients. Researchers at Purdue University, USA developed the nutriculture system in 1940. During 1960s and 70s, commercial hydroponics farms were developed in Arizona, Abu Dhabi, Belgium, California, Denmark, German, Holland, Iran, Italy, Japan, Russian Federation and other countries. Most hydroponic systems operate automatically to control the amount of water, nutrients and photoperiod based on the requirements of different plants [3].

Due to rapid urbanization and industrialization not only the cultivable land is decreasing, but also conventional agricultural practices causing a wide range of negative impacts on the environment. To sustainably feed the world's growing population, methods for growing sufficient food have to evolve. Modification in growth medium is an alternative for sustainable production and to conserve fast depleting land and available water resources. In the present scenario, soilless cultivation might be commenced successfully and considered as an alternative option for growing healthy food plants, crops and vegetables [4]. Agriculture without soil includes hydro agriculture (Hydroponics), aqua agriculture (Aquaponics) and aerobic agriculture (Aeroponics) as well as substrate culture. Among them, hydroponics techniques are gaining popularity because of its efficient management of resources and food production. Various commercial and specialty crops can be grown using hydroponics including leafy vegetables, tomatoes, cucumbers, peppers, strawberries, etc. This paper covers different aspect of hydroponics, vegetables grown in hydroponics system and the global hydroponic market.

In hydroponic cultures, all essential plant nutrients are supplied via the nutrient solution, except carbon that taken up from air as CO_2. To prepare nutrient solutions containing all essential nutrients, inorganic fertilizers are used as nutrient sources, except for iron, which is added in chelated form to improve its availability for the plants. Most fertilizers used to prepare nutrient solutions in soilless culture are highly soluble inorganic salts, but some inorganic acids are also used.

2.1 Hydroponic types

There are two kinds of hydroponic systems; Circulating Methods (closed systems) and Non-Circulating method (open systems). In the *closed system*, drained nutrient solution is recovered, replenished and recycled (Figure 1). Compared with the open-loop system, it requires more precise and frequent control of the nutrient solution; technical know-how is needed, as it is more sensitive to operational mistakes. The returned nutrient solution has to be treated to restore its original nutrient element composition and to remove any foreign substances. Moreover, spreading of root-borne diseases may occur, thus sterilization of the solution must be provided to kill pathogens. In *open-systems*, water and nutrients are supplied as for a conventional on-soil crop and the drained nutrient solution is thrown out of the system (Figure 2).

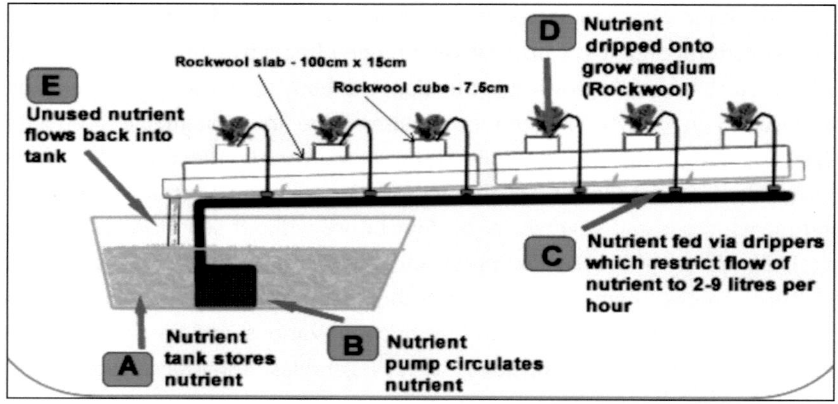

Figure 1: Circulating Hydroponic System

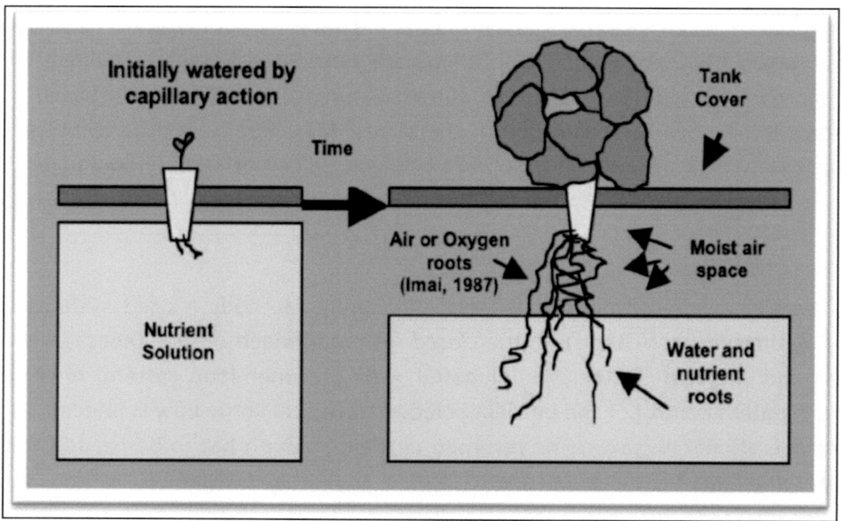

Figure 2: Non-Circulating Hydroponic System

There are a variety of techniques used in the closed systems these include:

- *Nutrient Film Technique (NFT)*

In this system, the plant roots are directly exposed to nutrient solution. The nutrient solution flows constantly, so no timer is required for the submersible pump. The nutrient solution is pumped into the growing tray (usually a tube) and flows over the roots of the plants, and then drains back into the reservoir. There is usually no growing medium used other than air, which saves the expense of replacing the growing medium after every crop. Normally, the plant is supported in a small plastic basket with the roots dangling into the nutrient solution. It is very crucial to monitor the pump performance, because it is very important to keep the nutrient solution flow constant.

- *Deep flow technique (DFT)/pipe system*

In this technique, 2-3 cm deep nutrient solution flows through 10 cm diameter PVC pipes, to which plastic net pots with plants are fitted. The plastic pots contain planting materials, and their bottoms touch the nutrient solution that flows in the pipes. Plants are established in plastic net pots and fixed to the holes made in the PVC pipes. Old coir dust, or carbonized rice husk, or mixtures of both may be used as planting material to fill the net pots. A small piece of net is placed as a lining in the net pots to prevent the planting material falling into the nutrient solution. When the recycled solution falls into the solution in the stock tank, the nutrient solution gets aerated. The PVC pipes have a slope of drop of 1 in 30-40 to facilitate the flow of nutrient solution.

The Non-Circulating system also has different techniques that include:

- *Root dipping technique*

In this technique, plants are grown in small pots filled with little growing medium. The pots are placed in such a way that the lower 2–3 cm of the pots is submerged in the nutrient solution. Some roots are dipped in the solution, while others hang in the air above the solution for nutrient and air absorption, respectively. This 'low tech'growing technique is inexpensive to construct and needs little maintenance. Importantly, this technique does not require expensive items such as electricity, water pump, channels, etc.

- *Floating Technique*

In this technique, shallow containers (10 cm deep) can be used. Plants established in small pots are fixed to a styrofoam sheet or any other light plate and allowed to float on the nutrient solution, filled in the container, and the solution is artificially aerated.

- *Capillary Action Technique*

Planting pots of different sizes and shapes with holes at the bottom are used. The pots are filled with an inert medium, and seedlings/seeds are planted in inert medium. These pots are placed in shallow containers filled with the nutrient solution. Nutrient solution reaches inert medium by capillary action. Aeration is very important in this technique. Therefore, old coir dust mixed with sand or gravel can be used. This technique is suitable for ornamental flowers and indoor plants.

2.2 Advantages and disadvantages

There are many advantages of growing plants hydroponically over soil-based culture [6]. Soilless culture offers opportunities to provide optimal conditions for plant growth and, therefore, higher yields can be obtained compared with open field agriculture; gardening is clean and extremely easy, requiring very little effort [7]. Soilless culture averages compared with ordinary soil yields are given in Table 2. Hydroponics offers a means of control over soil-borne diseases and pests, which is especially desirable in the tropics, where the life cycles of these organisms continues uninterrupted and so does the threat of infestation. It is also effective for the regions of the world having scarcity of arable or fertile land for agriculture [8]. It reduces the costs and time taken for various tasks, which are avoided in soilless culture of cultivation. It offers a clean working environment and thus hiring labour is easy.

Table 2: Soilless culture averages compared with ordinary soil yields

Name of crop	Hydroponic Equivalent (kg/ha	Agricultural Average (kg/ha)
Wheat	918	110
Oats	551	156
Rice	2,203	138-165
Maize	1,468	276
Soybean	275	110
Potato	28,000	3,000
Beet root	3,674	1,653
Cabbage	3,307	2,388
Peas	2,572	367
Tomato	73,000	2,000-4,000
Cauliflower	5,511	1,837-2,753
French bean	7,716 pods for eating	-
Lettuce	3,858	1,653
Cucumber	5,144	1,286

However, application of hydroponic techniques on commercial scale requires technical knowledge and higher initial capital expenditure. This will be further high if the soilless culture is combined with controlled environment agriculture (8). High degree of management skills is necessary for solution preparation, maintenance of pH and EC, nutrient deficiency judgment and correction, ensuring aeration; maintenance of favourable condition inside protected structures, etc. Great care is required with respect to plant health control. Finally energy inputs are necessary to run the system (9). Considering the significantly high cost, the soil-less culture is limited to high value crops of cultivated area.

4 Quality of hydroponically grown vegetables

Literature provides contrasting evidences. Nevertheless, hydroponics generally enables the production of vegetables of quality comparable with those grown in soil [6]. Some species, such as tomato, cucumber, pepper, strawberry and almost all leafy plants (lettuce, rocket salad, basil, and celery) seem to respond better to hydroponic culture than other crops, such as eggplant and muskmelon. For example, reduced sweetness and firmness of hydroponically grown muskmelons is often observed [7]. Moreover, leafy and fruit vegetables produced hydroponically do not contain residues of chemicals used for soil disinfection or other soil pollutants, and usually, they are very clean, a crucial factor for strawberry and ready-to-eat leaf and shoot vegetables (microgreens, baby leaves), the consumption of which is rapidly increasing in many countries. Cultivation of strawberries in suspended bag culture reduces labour costs for planting and harvesting, and significantly reduces susceptibility to grey mould (*Botrytis cinerea*). Proper manipulation of the culture solution can substantially improve nutritional or sanitary characteristics, for example, to reduce the level of nitrates.

5 Hydroponics and water use efficiency

Barbosa *et al.* stated that hydroponic lettuce yielded 41 ± 6.1 kg/m^2.yr per greenhouse unit (815 m^2) and had water demands of 20 ± 3.8 L/kg.yr. In comparison, conventional lettuce production yields 3.9 ± 0.21 kg/m^2.yr with water demands of 250 ± 25 L/kg.yr [11].

Production of tomatoes in a drip-irrigated field is in the order of 3-8 kg/m^2.yr, while in a hydroponic greenhouse it is 50-80 kg/m^2.yr. Thus, the production rate is 10 to 20 times higher in a hydroponic greenhouse than in the field. Water use in the field varies a lot too; it can be, for instance, 300 L/m^2 over the growing season (part of the year), while water use in hydroponic greenhouses is in the order of 1,000 L/m^2 over the growing season (a whole year). This means that Produce Water Use in the field is 300 L/m^2 divided by 5 kg/m^2 = 300/5 = 60 L/kg. In a greenhouse, Produce Water Use varies from 1,000/50 = 20 L/kg to 1000/80 =12.5 L/kg. This is on average a Produce Water Use of about 15 L/kg in a greenhouse versus 60 L/kg in the field, a four-fold improvement in Produce Water Use by moving from drip-irrigated field production to hydroponic greenhouse production [12].

5 Conclusions

Reducing the irrigation water consumption is crucial to maintain the agricultural sector functional in Jordan because of the increasing demand on the limited water resources. Hydroponics is seen as a very promising opportunity for semi-arid and arid regions like Jordan to overcome the chronic water crisis, to promote the quality of agricultural products and to support the sustainability of the agricultural sector, which is one of the national security pillars. In near future the hydroponic industry is expected to grow exponentially. The Jordan Government and other stakeholders should create the enabling environment to encourage this industry to grow commercially. It is important to develop low-cost hydroponic technologies that reduce dependence on human labour and lower overall startup and operational costs.

6 Acknowledgements

The Author would like to thank DAAD and Exceed Swindon project for providing this opportunity to present this work in the International Expert Workshop on Water on Agricultural Practices in Rio de Janeiro, Brazil, 15-21 September 2019, and for the associated financial support they provided. The Author would also like to thank Dr. Nabeel Bani Hani and Mr. Faisa Zyadat for their contributions and feedback.

7 References

[1]   Ministry of Water and Irrigation, National Water Strategy (2016-2025), http://www.mwi.gov.jo/sites/enus/Hot%20Issues/Strategic%20Documents%20of%20%20The%20Water%20Sector/National%20Water%20Strategy(%202016-2025)-25.2.2016.pdf.

[2]   Ministry of Agriculture, Annual Report (2012), http://moa.gov.jo/Portals/0/annual%20reports/%D8%A7%D9%84%D9%83%D8%AA%D8%A7%D8%A8%20%D8%A7%D9%84%D8%B3%D9%86%D9%88%D9%8A%202012%20(Repaired).pdf.

[3]   Resh, H.M.: Hydroponic Food Production: a Definitive Guidebook for the Advanced Home Gardener and the Commercial Hydroponic Grower. CRC Press, Boca Raton, FL. (2013)

[4]   Butler, J.D., Oebker, N.F.: Hydroponics as hobby growing plants without soil. Circular 844, Information Office, College of Agriculture, University of Illinois, Urbana, IL. 2006, 6180p.

[5]   Gruda, N.: Do soilless culture systems have an influence on product quality of vegetables? J. Applied Botany & Food Quality, 2009, 82, 141-147.

[6]   Savvas, D.: Nutrient solution recycling in hydroponics. In: Hydroponic Production of Vegetables and Ornamentals (Savvas, D., Passam, H.C., eds), Embryo Publications, Athens, Greece 2002, pp 299–343.

[7]   Silberbush, M., Ben-Asher, J.: Simulation study of nutrient uptake by plants from soilless cultures as affected by salinity buildup and transpiration. Plant and Soil, 2001, 233, 59–69

[8]   Sonneveld, C.: Effects of salinity on substrate grown vegetables and ornamentals in greenhouse horticulture. PhD Thesis, University of Wageningen, The Netherlands (2000)

[9]   Van Os, E.A., Gieling, Th.H., Ruijs, M.N.A.: Equipment for hydroponic installations. In: Hydroponic Production of Vegetables and Ornamentals (Savvas, D., Passam, H.C., eds), Embryo Publications, Athens, Greece, 2002, pp 103 141.

[10]   Pardossi, A., Malorgio, F., Incrocci, L., Tognoni, F.: Hydroponic technologies for greenhouse crops. In D. Ramdane, ed. Crops: Quality, growth and biotechnology, WFL Publisher, Helsinki, Finland. 2006, 360-378.

[11]   Barbosa, G. Gadelha, F. Kublik, N. Proctor, A. Reichelm, L. Weissinger, E. Wohlleb, G., Halden, R .: Comparison of Land, Water, and Energy Requirements of Lettuce Grown Using Hydroponic vs. Conventional Agricultural Methods. Environmental research and public health, 2015, Vol. 12(6), pp 6879–6891

[12]   Anonymous: Practical Hydroponics & Greenhouses UR Greenhouse Horticulture, Wageningen, 2010, p. 52 - 59. ISSN 1321-8727

REMOVAL STUDY OF A CONTAMINANT OF EMERGING CONCERN, HYDROQUINONE FROM WASTEWATER BY ADSORPTION PROCESS ON ACTIVATED CARBON

Dègninou Houndedjihou[1], Ephraim Vunain[2], Tomkouani Kodom[1]

[1]*Laboratoire d'Hydrologie Appliquée et Environnement, Faculté Des Sciences, Universités de Lomé, BP : 1515 Lomé, Togo ;* *christophehound@yahoo.fr*

[2]*Research Laboratory, Chemistry Department, Science Faculty, Chancellor College, University of Malawi P.O Box 280, Zomba, Malawi;*

Keywords: Activated carbon, Hydroquinone, Sunflower seed hull, Synthetic wastewater

Abstract

The removal of a contaminant of emerging concern (CEC), hydroquinone in aqueous solution has been studied using activated carbon prepared from agricultural waste. Activated carbon from sunflower seed hull has been used in this study. This activated carbon has a high volume of wide micropores and presence of small mesopores, high carbon yield, high surface area, and a large pore volume. Physicochemical activation at two ratios (1:1.5 and 1:2) has been performed using zinc chloride ($ZnCl_2$) as activating agent. FTIR, SEM-EDS, XRD and BET analyses have been performed for the characterization of the synthetized adsorbent. Batch experiments were performed to study the removal of target pollutant by adsorption process. Hydroquinone concentrations in synthetic wastewater were monitored by a High-Performance Liquid Chromatography (HPLC) system equipped with a phenyl Xbridge column and Photodiode Array (PDA) detector. The results show a maximum adsorption rate of 250 mg/g hydroquinone on activated charcoal. The equilibrium has been reached at 120 min. Langmuir isotherm and second order kinetic fit well the removal of hydroquinone by adsorption on this adsorbent, while the thermodynamic calculation show that the adsorption is endothermic. Some parameters such as initial hydroquinone concentration, contact time, solution temperature and adsorbent amount that influence the adsorption of hydroquinone in aqueous media have been studied in order to find the best conditions for an optimal removal.

1 Introduction

Hydroquinone (1,4-dihydroxybenzene, HQ) is one of phenolic compounds widely used in the industry [1]. It is mainly used in the production of dyes, photo stabilizers, plasticizers, cosmetics, pesticides and some pharmaceuticals [2]. So, phenol and its derivatives (catechol, hydroquinone, resorcinol, etc.) widely used in different syntheses processes are classified among contaminants of emerging concern (CECs) [3, 4]. Thus, hydroquinone is considered to be one of the commonly encountered organic pollutants in industrial effluents discharged into the environment causing severe environmental problems [5] and polluting a large part of water resources. These phenolic compounds, being non-biodegradable, can be found in the aquatic environment at very weak concentration, but they are hazardous and toxic to humans and the ecosystem [6, 7]. Some tests of subcutaneous injection of hydroquinone in rats revealed fertility reduction in male and estrus cycle disturbance with prolongation in female rats [8]. Dark urine, vomiting, abdominal pain, tachycardia, tremors, convulsions, coma and deaths have been reported after ingestion of photographic developing agents containing hydroquinone as effects on humans [8]. Thus, kinds of phenolic compounds are classified as priority contaminants by the European Union (EU). The US EPA regulations propose a maximum concentration of 1 ppm for all phenolic compounds in wastewater [1]. In drinking water, the 80/778/EC directive states a maximum concentration of 0.5 µg/L for total phenols, while individual concentration should be under 0.1 µg/L [9].

The removal of phenolic compounds from water has become a global environmental concern. Several technologies or processes including advanced oxidation processes AOPs and adsorption on activated carbon (AC) are developed for their elimination from wastewater. Heterogeneous photocatalytic [7, 13, 14], electrochemical [2, 12], heterogeneous catalytic ozonation [10], and adsorption processes [1, 11, 15-18] were used. So, many technologies have been developed for this purpose, and amongst all, the adsorption process by activated carbon is one of the most efficient techniques [19]. The use of advanced oxidation processes AOPs (heterogeneous photocatalysis, ozonation, H_2O_2/UV, Fenton, photo-Fenton, etc.) have been reported to be efficient on such organic chemicals. But the main problem of AOPs is the generation of by-products that are sometimes more dangerous. Adsorption on activated carbon has the advantage of not generating by-products. A number of researches have been devoted to the enhancement of adsorption process using different adsorbents [14]. Activated carbon (AC) is recognized to be the most common and efficient adsorbent material for the removal of low molecular weight organic compounds like hydroquinone, catechol and resorcinol from water because of its high adsorption efficiency [1]. This efficiency of AC has gained increased interest, and different biomasses from agroforestry waste are used as precursors, e.g., palm kernel and coconut shell [20, 18]; rattan sawdust [15]; sunflowers seeds shell [21]; sunflower seed hull [1]; pine cone [22]; coffee residues [23]; sunflower seed husk [19]; *Canarium schweinfurthii* nutshell [24].

Sunflower (*Helianthus annuus*) seeds are among the world's most important oil seeds with an estimated global production in million metric tons [1]. In Malawi, a large quantity of sunflower seed is produced annually with a large amount of sunflower seed hull (SSH) generated as wastes or (by-product) from the production of sunflower seed oil. In our previous work in Malawi, the SSHAC has been found efficient in the removal of two (catechol and resorcinol) of the three dihydroxybenzene isomers with a maximum capacity that can be taken up per mass of adsorbent different [1]. The third isomer hydroquinone has not been studied in our previous work because of the lack of this compound in the lab during our stay.

The aim in this work was to study the efficiency and the conditions of optimal removal of a contaminant of emerging concern, hydroquinone from water on activated carbon adsorption process.

2 Materials and Methods

The sunflower seed hull activated carbon (SSHAC) used in this work was obtained from Vunain et al. [1]. Different kinds of SSHAC used in this work are shown in Table 1.

Table 1: Ash content, moisture content, volatile matter and carbon content of SSHAC used [1]

Samples and impregnation ratio	Ash content (%,)	Moisture content (%,)	Volatile matter (%,)	Carbon content (%,)
Inactivated biomass	10.00	14.98	50.12	24.90
AC-500; 1:2	7.77	14.10	51.37	32.76
AC-500; 1:1.5	7.76	14.01	44.42	33.81
AC-600; 1:2	5.84	10.80	30.88	52.48
AC-600; 1:1.5	5.66	10.82	31.02	52.50
AC-700; 1:2	3.97	8.54	28.61	58.88
AC-700; 1:1.5	3.19	8.52	28.07	60.22
AC-800; 1:2	0.89	7.13	24.08	67.90
AC-800; 1:1.5	0.32	4.30	21.78	73.60

Elemental analyses of SSHAC (AC-800; 1:1.5) sample and precursor material (SSH) gave their compositions (Table 2). Fourier transform infrared (FTIR) spectroscopic analysis using a Perkin-Elmer Frontier FTIR-spectrometer, the surface morphologies examined using a JEOL JSM-IT300 SEM instrument coupled with EDX, Nitrogen adsorption/desorption isotherms of the carbon samples measured at 77 K using a Micromeritics ASAP 2020 surface area and porosity analyzer, thermal behavior measured with a Discovery TGA 550 TA instrument were performed on SSHAC and SSH [1].

Table 2: Elemental analysis of SSHAC (AC-800; 1:1.5) sample and precursor material (SSH) [1]

Sample	(C) (%, wt/wt)	(O) (%, wt/wt)	(H) (%, wt/wt)	(N) (%, wt/wt)	(S) (%, wt/wt)	(Others) (%, wt/wt)
SSH	21.80	44.20	4.32	0.89	0.63	28.16
(AC-800; 1:1.5)	72.20	22.70	4.17	0.31	0.27	0.35

Pore structure parameters of raw SSH biomass and SSHAC used in this work are illustrated in Table 3. All chemicals and reagents were used without purification; hydroquinone was obtained from Sigma Aldrich. Batch experiments were performed at room temperature accorcing to the method of Vunain et al. [1] to study adsorption equilibrium and different parameters affecting the removal of HQ fixing the pH at 10.

pH 10 has been chosen according to the work of Suresh et al. [25] and Vunain et al. [1] on phenolic compounds; when pH ≥ pKa, the adsorption process is favorable. The initial pH of hydroquinone solution was adjusted using 0.1 mol/L NaOH. The influence of initial concentration of pollutant (50; 100; 150 and 200 mg/L), adsorbent dosage (0.1–1 g/L), contact time (10–420 min) and temperature (25, 35 and 50 °C), were investigated. HPLC system (Alliance Waters 2690, Waters) coupled to a diode array detector (Waters 2996, PDA) with an Xbrigde column of phenyl type (150 mm x 4.6 mm, 5 mm) was used for the hydroquinone concentration monitoring. Kinetic and isotherm experiments were carried out by agitating 200 ml of synthetic wastewater's concentration 100 mg/L with different constant adsorbent amount (25; 50; 100; 150 and 200 mg) at a constant agitation speed of 200 rpm, 25 °C and pH = 10. HANNA pH-meter with a combined pH electrode has been used for measuring. Three different models: pseudo first-order, pseudo second-order, intra-particle diffusion and four isotherm models: Langmuir, Freundlich, Dubinin–Radushkevich (DR), Temkin isotherms have been studied in this work.

Table 3: Pore structure parameters of raw SSH biomass and SSHA [1]

Sample	Specific surface area (m^2/g)	Pore diameter (nm)	Pore volume (cm^3/g)	
			Micropore volume (cm^3/g)	Total pore volume (cm^3/g)
Raw SSH biomass	206	1.65	0.7780	0.9884
SSHAC	1374	1.92	1.9780	2.0463

The amount of adsorbate adsorbed at equilibrium, Qe (mg/g), was calculated from the following formula:

$$Q_e = \frac{C_0 - C_e}{m} \times V \quad (1)$$

where, Co and Ce (mg/L) are the initial and equilibrium concentration of hydroquinone, m is the mass of adsorbent (SSHAC) (g) used and V is the volume of the solution(L).

The percentage removal (%) of hydroquinone at each time was calculated as follows:

$$\%\textbf{Removal} = \frac{C_0 - C_t}{C_0} \times 100 \quad (2)$$

C_t is the concentration at time t, determined using the calibration curve plot based on the pic area (Figure 1).

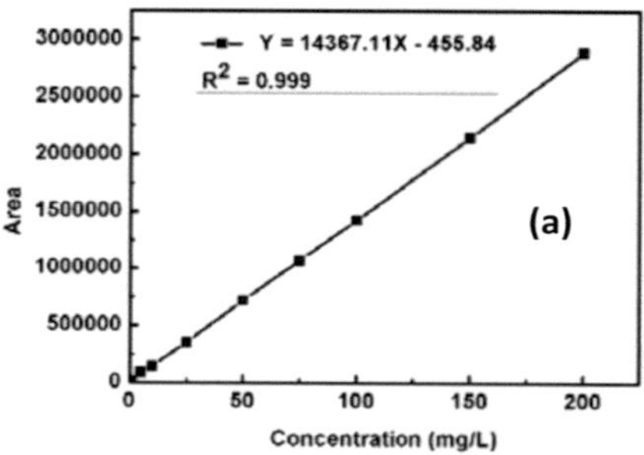

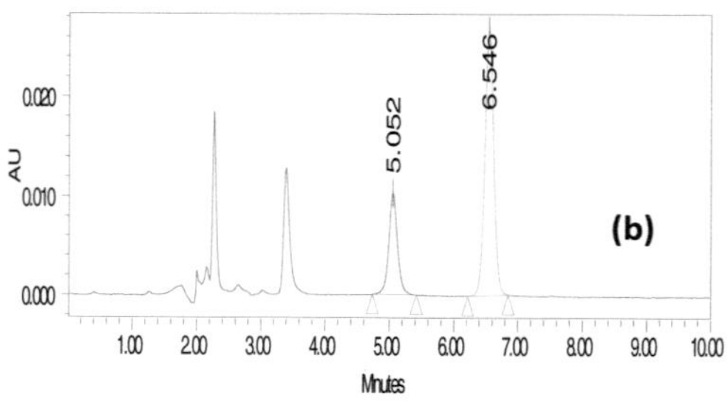

Figure 1: (a) Hydroquinone calibration curve; (b) hydroquinone chromatogram

3 Results and Discussion

3.1 Effect of hydroquinone initial concentration and contact time

Figure 2 shows the quantity of adsorbate removed per mass unit as function of initial concentration of pollutant and contact time. The adsorbed amount per mass unit by the adsorbent increased when the contact time increased and also when the initial concentration of hydroquinone increased. The adsorption capacity at equilibrium increased with increasing of the initial hydroquinone concentration from 100 to 250 mg/g, but the equilibrium time was insignificantly affected by the initial hydroquinone concentration.

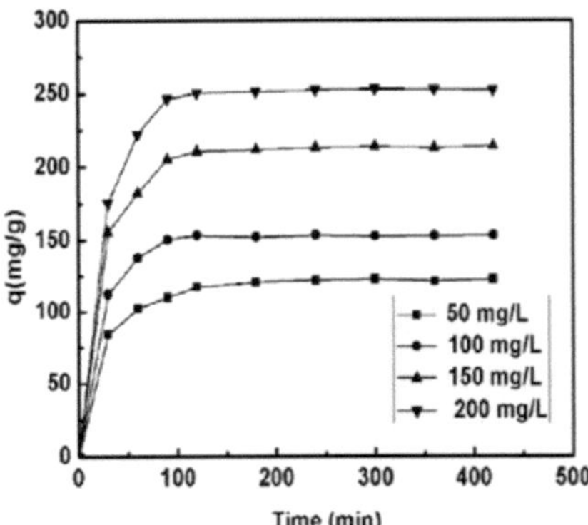

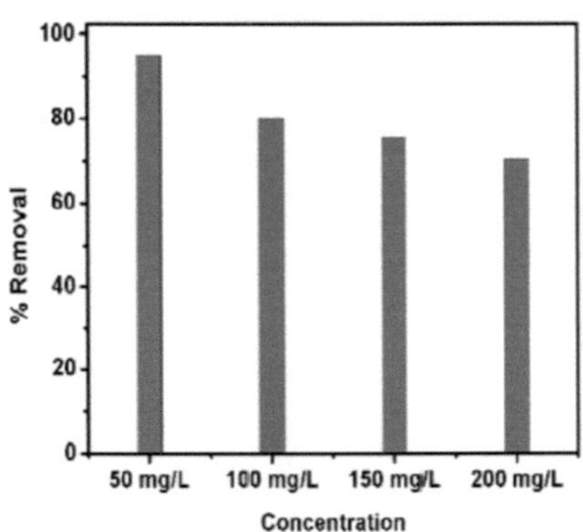

Figure 2: Effect of initial concentration and contact time
(adsorbent 0.2 g; pH = 10, temperature: 25 °C and time 7h)

The adsorption process is rapid the first thirty minutes for all initial concentrations. The adsorption rate decreases after this period and reaches equilibrium at about 120 min. This can be explained by the availability of the surface sites. In fact, the susceptible sites of the AC surface to adsorb the pollutant molecules are more available at the first stage of the adsorption, and after this period these sites are occupied by hydroquinone molecules [1]. The amount of phenolic compound removed and the capacity (q) increased with the increase of initial concentration using activated carbon produced, but the per cent removal of the phenolic compound decreased with increasing the initial concentration. The same trend was reported by Jain et al. [26] on chromium (VI) removal from aqueous systems and also by Zoweram et al. [27] on the removal of the dye Acid Red 18.

3.2 Effect of adsorbent dose

The adsorption of hydroquinone on SSHAC is a function of the adsorbent dose. Figure 3 shows that the increasing of the adsorbent dose increases the amount of adsorbed pollutant, and the percentage removal (adsorption efficiency) increases, but on the other hand, the amount adsorbed per unit mass decreases. Similar trend has been reported by Aljeboree et al. [18] on the adsorption of the dyes GRL and DY 12.

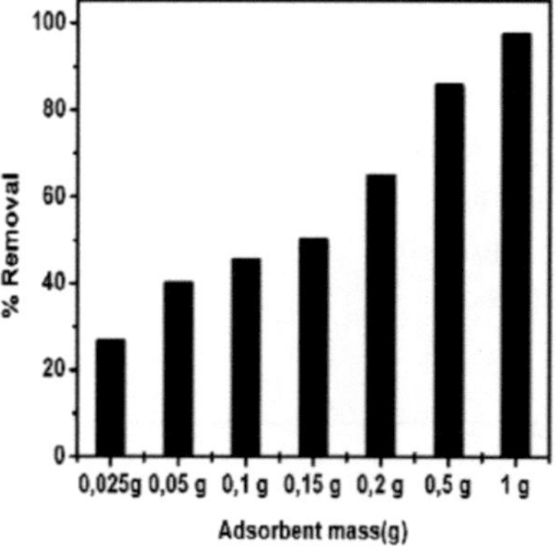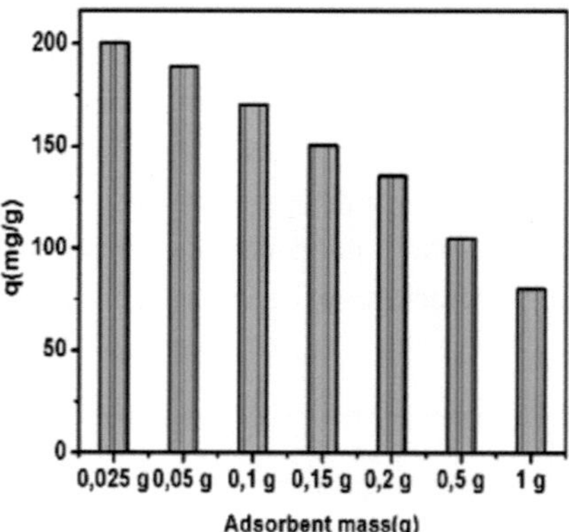

Figure 3: Effect of adsorbent dose on (a) Adsorption efficiency (% removal), (b) Adsorbate adsorbed per unit mass (g) (C = 100 mg/L, pH = 10, temperature: 25 °C and time 150 min)

The increase of percentage removal can be explained by the fact that by increasing the adsorbent dose, the number of available adsorption sites and surface area increase, therefore, more hydroquinone molecules are adsorbed [28]. The decrease of the adsorbed amount per unit mass, when the adsorbent dose increases, can be mainly attributed to unsaturation of the adsorption sites [29].

3.3 Effect of temperature

Hydroquinone concentration of 100 mg/L had been in contact with 0.2 g of SSHAC during 150 min at three different temperatures (25; 35 and 50 °C). The result shows that the adsorption of hydroquinone on SSHAC is affected by the solution temperature (Figure 4). The adsorbate amount removed increases when the temperature increases during the contact time. One can conclude that the reaction of adsorption process of hydroquinone on SSHAC is endothermic. Ong et al. [19] reported in their work concerning the removal of basic dye by sunflower seed husk a similar trend; however, they noticed a slight effect of temperature on the basic dye from 30 to 70 °C. The increase the of the diffusion rate of the adsorbate molecules across the external boundary layer and in the internal pores of the adsorbent particle could explain the increase of the adsorbate amount removed when the temperature increases [21].

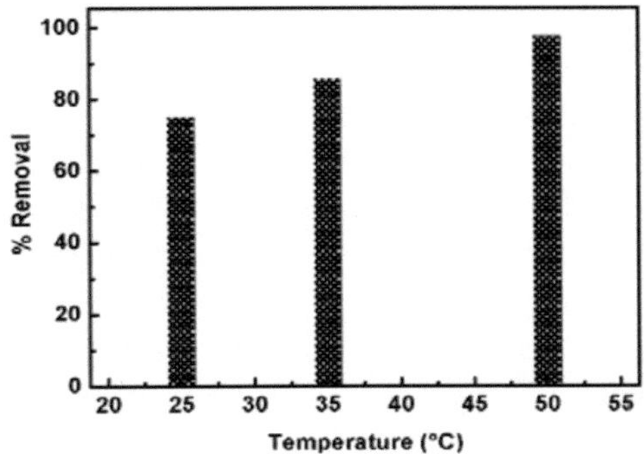

Figure 4: Effect of temperature on the removal percentage of hydroquinone
(C = 100 mg/L, AC: 0.2 g, pH10)

3.4 Hydroquinone adsorption isotherm

Four different isotherms have been studied. The adsorption data has been analyzed using the linear forms of the Langmuir, Freundlich, Temkin and Dubinin-Radushkevich isotherms. These linear equations are:

Langmuir Isotherm: $\dfrac{C_e}{q_e} = \dfrac{1}{q_0 b} + \dfrac{C_e}{q_0}$ $\qquad$ (3)

Freundlich Isotherm: $\log(q_e) = \log K_F + \dfrac{1}{n}\log(c_e)$ $\qquad$ (4)

Dubinin-Radushkevich Isotherm: $\ln(q_e) = \ln(q_0) - B\mu^2$ $\qquad$ (5)

Temkin Isotherm: $q_e = \beta \ln K_T + \beta \ln(c_e)$ $\qquad$ (6)

where are Ce the concentration of adsorbate solution at equilibrium (mg/L), q_e the amount of adsorbate adsorbed per mass of adsorbent (mg/ g), b the equilibrium constant related to the sorption energy between the adsorbate and adsorbent (L/mg), and q_0 limiting amount of adsorbate, the monolayer or maximum capacity that can be taken up per mass of adsorbent; K_F and n are the Freundlich constants, which represent adsorption capacity and degree of non-linearity between solution concentration and adsorption [30]; B is the D-R isotherm constant related to mean free sorption energy E according to relationship $B = \dfrac{1}{2E^2}$ (7) $(mol^2.kJ^{-2})$; μ is the Polany potential defined by $\mu = RT \ln\left(1 + \dfrac{1}{C_e}\right)$ (8) [31] (Suteu et al., 2012), R is the universal gas constant (J/mol.K), and T is the absolute temperature (K); $\beta = \dfrac{RT}{b}$ (9) with b Temkin constant related to the heat of adsorption (J/mol) and K_T empirical Temkin constant related to the equilibrium binding constant related to the maximum binding energy (L/mg) [18].

The different constants associated with each isotherm model were evaluated from the intercepts and slopes of the corresponding linear plots of each model (Figure 5). These different parameters are shown in Table 5. It can be seen that Langmuir isotherm has the highest R^2 (0.997) and the lowest root mean square error RMSE (0.005), indicating that the Langmuir isotherm gives the best fit for the hydroquinone adsorption with a high maximum adsorption of 250 mg/g. The separation factor (R_L), which is an important characteristic of Langmuir isotherm, can be defined by **$R_L = 1/(1+bC_0)$** (10), where b is the Langmuir constant and C_0 is the highest phenol concentration (mg/L) [19]. This value of R_L indicates the type of the isotherm to be either unfavorable ($R_L > 1$), linear ($R_L = 1$), favorable ($0 < R_L < 1$) or irreversible ($R_L = 0$). The dimensionless separation factor, R_L was calculated from the Langmuir constant b of every initial concentration (Table 4). The separation factor values, which are decreased when the initial concentration of the pollutant increased, were comprised between 0 and 1 indicating favorable on the activated carbon (Table 4 and Figure 6). Similar observations have been reported of the adsorption of phenol on hazelnut bagasse [28].

The Freundlich constant 1/n obtained from the slope of the plot of log (q_e) vs log (c_e) (Figure 5) is smaller than 1, which can be attributed to the molecular interaction between adsorbent and adsorbate, and this indicates a favorable process [32]. The value of n is above unity; the adsorption of hydroquinone on SSHAC is a favorable physical process [30] and is also valid for many organic contaminants such as hydroquinone [33].

The constant B in Dubinin-Radushkevich isotherm model gives an idea on free energy E (KJ/mol). It can be seen in Table 5 that this free energy gives information about the sorption mechanism and calculated from equation (10) is less than 8KJ/mol, indicating that the hydroquinone adsorption on SSHAC is a physical process [31]. The maximum removal capacity based on Dubinin-Radushkevich isotherm model is higher than that valuated in the Langmuir isotherm model. Similar trend has been reported by Jain et al. [26], and this can be related to the fact that Dubinin-Radushkevich isotherm follows a pore filling mechanism [31].

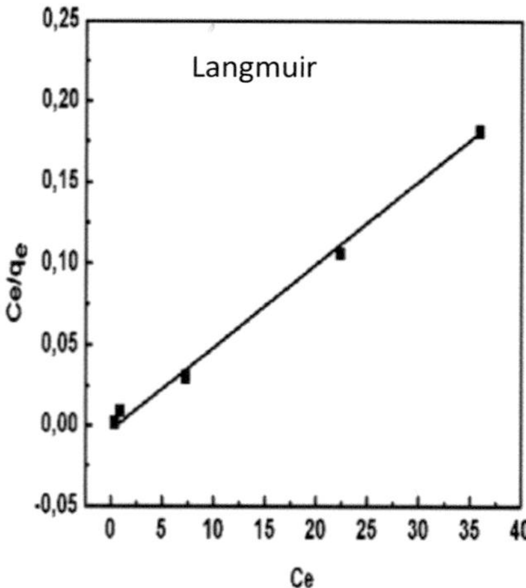

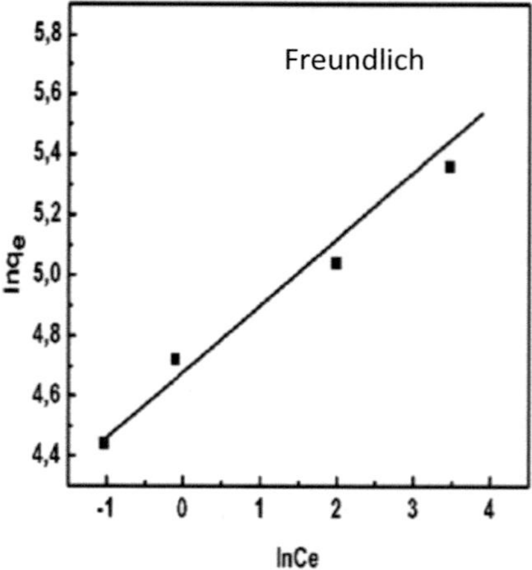

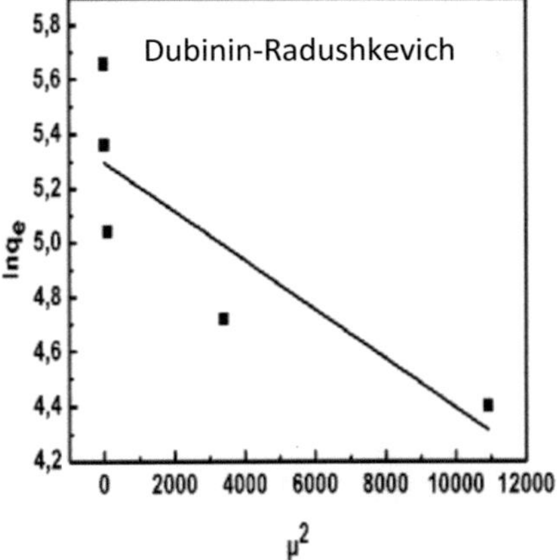

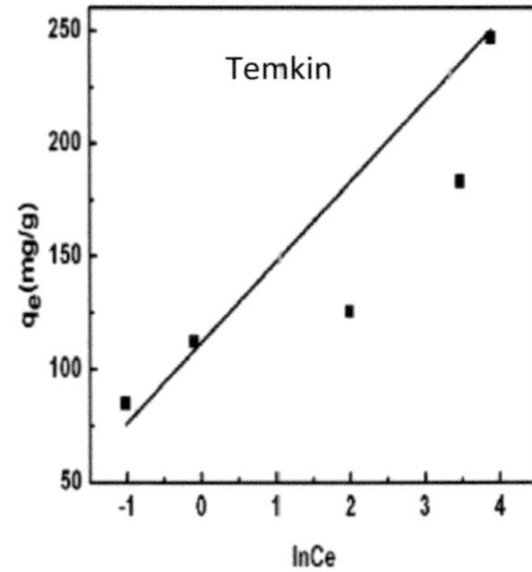

Figure 5: Different isotherms models for the hydroquinone adsorption study at 25 °C (t: 150 min)

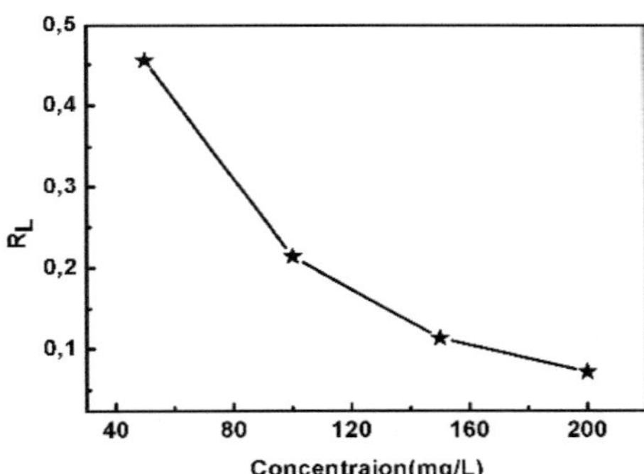

Figure 6: Evolution of R_L versus hydroquinone corcentration

Table 4: Value of Langmuir separation factor (RL) for different concentration

Initial Concentration	R_L values
50 mg/L	0.455
100 mg/L	0.214
150 mg/l	0.113
200 mg/L	0.072

Based on the R-squared (R^2) value, the classification of the models according to the simulation of the adsorption isotherms is: Langmuir >Freundlich > Temkin > Dubinin-Radushkevich.

Table 5: Parameters for different isotherm models for the hydroquinone adsorption study at 25 °C

Isotherms	Parameters	Values
Langmuir	b (L /mg)	0.72
	q_o (mg/g)	249.78
	R^2	0.997
	RMSE	0.005
Freundlich	K_F (mg/g) $(L/mg)^{1/n}$	10.776
	n	1.46
	R^2	0.961
	RMSE	0.11236
Dubinin - Radushkevich	B (mol^2/kJ^2)	9.65×10^{-3}
	E (KJ/mol)	7.2
	q_o (mg/g)	259.4
	R^2	0.721
	RMSE	0.30459
Temkin	K_T (L/g)	23.04
	β (J/mol)	35.63428
	R^2	0.898
	RMSE	29.868

3.5 Adsorption kinetic

Three of the most widely used kinetic models pseudo first-order, pseudo-second-order and intra-particle diffusion equation were used to study the adsorption kinetic behavior of hydroquinone onto adsorbent. So the Lagergreen model for the pseudo first-order with the linear equation $\ln(q_e - q) = \ln q_e - K_1 t$ (11), where k_1 is the rate constant of pseudo-first order sorption model, min^{-1}; Ho model for the pseudo second–order with the equation: $\dfrac{t}{q} = \dfrac{1}{K_2 q_e^2} + \dfrac{t}{q_e}$ (12), where K_2 is the rate constant of pseudo-second order sorption model, g/mg min and qe2k2 is initial sorption rate, mg/g min; and finally Webber and Morris model for intra-particle diffusion with the equation $q = K_p \cdot t^{\frac{1}{2}} + C$ (13), where Kp is the rate constant for intra-particle diffusion (mg g^{-1}min$^{-0.5}$) and c is the intercept [31] have been used.

The kinetic data of the sorption of phenolic compounds on the AC from sunflower seed hull were summarized in Table 6. They were calculated from the intercepts and slopes of the corresponding linear plots (Figure 7). The analysis of the different parameters in the Table 6 shows that the pseudo–first kinetic has a low RMSE, but the correlation coefficients are low. Furthermore, the q_e calculated from this model is largely different form the experiment value (q_{exp} = 250 mg/g). This kinetic model is not suitable to describe the adsorption kinetic of hydroquinone. The different parameters of the pseudo-second order in the same Table 6 show good correlation coefficients, very low RMSE values, and the limiting amount of adsorbate calculated agrees well with the experimental values, confirming that the rate of the sorption process is well described by a pseudo second order equation. Similar observation has been reported by Osma et al. [34] on the adsorption of Reactive Black 5 on sunflower seed shell activated carbon, and also by L. Li et al. [11] on the adsorption of hydroquinone on graphene oxide.

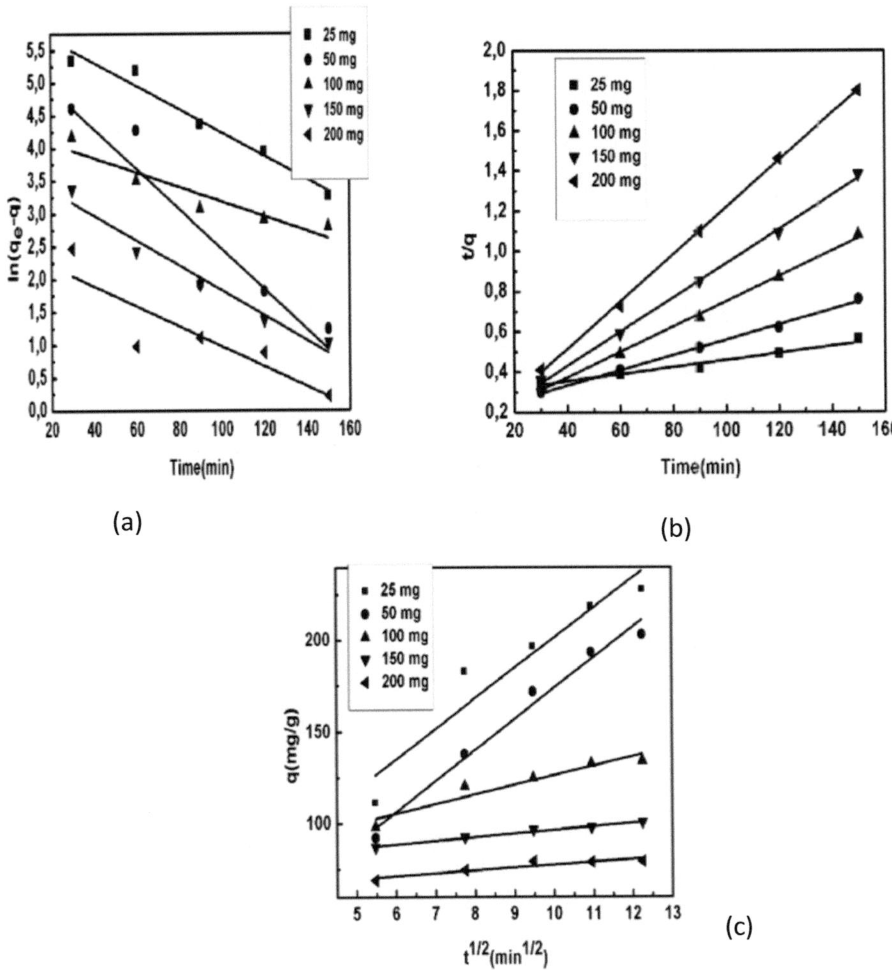

Figure 7: (a) Pseudo first-order, (b) pseudo-second-order, and (c) intra-particle diffusion kinetic of hydroquinone adsorption (pH = 10; time: 150 min)

The intra-particle diffusion model is used to describe the mechanism and the rate controlling steps affecting the adsorption kinetics [30]. The linear plots of q versus t$^{1/2}$ are used to determine K_p and C as slope and intercept. These linear plots do not pass through the origin, and all constants C are different from zero. One can assume that the rate limiting process is not only controlled by intraparticle diffusion [35, 36]; some other mechanisms alongside with intraparticle diffusion might be involved as well [37].

Table 6: Kinetic parameters for hydroquinone adsorption on SSHAC (pH 10, C = 100 mg/L)

Parameter	25 mg	50 mg	100 mg	150 mg	200 mg
Pseudo First-order kinetic					
K_1 (min^{-1})	0.0179	0.0307	0.01117	0.01913	0.01533
q_e (mg/g)	319.05	253.4	73.92	42.35	12.32
R^2	0.969	0.879	0.889	0.968	0.781
RMSE	0.17386	0.62404	0.21528	0.19005	0.44463
Pseudo Second-order kinetic					
K_2 (g mg^{-1}min^{-1})	$1.01{*}10^{-5}$	$7.77{*}10^{-5}$	$3.17{*}10^{-4}$	$7.8{*}10^{-5}$	$2.89{*}10^{-3}$
q_e (mg/g)	388.23	305.25	249.48	118.06	85.47
R^2	0.969	0.996	0.997	0.998	0.999
RMSE	0.01673	0.0114	0.01862	0.01789	0.01402
Intra-particle diffusion					
Kp (mg g^{-1}min-$^{0.5}$)	20.219	12.775	9.567	6.723	3.364
C	74.9	50.7	42.3	39.5	27.8
R^2	0.986	0.952	0.858	0.839	0.708
RMSE	9.91012	10.2162	8.8549	5.00905	2.6924

3.6 Adsorption thermodynamic

Thermodynamic parameters have been studied to evaluate the energetic aspects, the thermodynamic feasibility and the spontaneous nature of hydroquinone adsorption process on SSHAC. The spontaneity and energetic aspects of the sorption are given, respectively, by Gibbs free energy and enthalpy. The classical van't Hoff equation (14) was used to evaluate the different thermodynamic parameters: $\ln K_D = \frac{-\Delta H^o}{RT} + \frac{\Delta S^o}{R}$ **(14)**, where are K_D = q_e/C_e, T the absolute temperature (K), ΔH^o the enthalpy (KJ/mol), ΔS^o the entropy (J/mol), and R the universal gas constant (8.314 x 10^{-3} KJ/mol.K) [25]. Gibbs free energy at different temperature is determined using the following equation: $\Delta G^o = \Delta H^o - T\Delta S^o$ (15). By plotting the classical van't Hoff equation ($\ln K_D$ versus $\frac{1}{T}$), enthalpy and entropy are determined as slope and intercept.

Table 7: Thermodynamic parameters of hydroquinone adsorption on SSHAC

Temperature (K)	$\Delta G°$ (KJ/mol)	$\Delta H°$ (KJ/mol)	$\Delta S°$ (J/mol)	R^2
298	-4.6	1.47	20.62	0.998
308	-4.9			
323	-5.2			

The different thermodynamic parameters calculated are shown in Table 7. It can be seen that the enthalpy is positive and Gibbs free energy is negative for the three temperatures showing that the hydroquinone adsorption is endothermic and spontaneous on SSHAC surface. Similar trend has been reported by other authors [31, 19, 1 ,38]. The negative values of ΔG_0 at all temperatures suggest the spontaneous nature of the adsorption. The decrease (more negative) in the value of ΔG_0 with increase in temperature suggests that higher temperatures would make the adsorption of hydroquinone onto SSHAC easier than lower temperatures (adsorption more favorable at higher temperatures) [1]. The positive value of ΔH_0 confirmed the endothermic nature of adsorption of hydroquinone over the SSHAC. The positive value of ΔS_0 suggested that there was increased randomness between the adsorbent/adsorbate interfaces during the adsorption of hydroquinone onto SSHAC. This also supported the affinity of sunflower seed hull derived AC for hydroquinone molecules.

4 Conclusions

Adsorption process of removal of contaminants of emerging concern, hydroquinone in wastewater has been investigated using AC prepared from agroforestry waste, sunflower seed hull. The effect of some parameters such as contact time, initial pollutant concentration, adsorbent dose and solution temperature have been investigated to find the best condition of optimal removal of hydroquinone by adsorption on SSHAC. The amount adsorbed per unit mass increases when pollutant concentration increases and decreases when adsorbent dose increases, while the removal percentage decreases for increasing pollutant concentration and increases for an increase of adsorbent dose. Isotherm and kinetic models studied showed that Langmuir isotherm and pseudo-second kinetic fit well to describe the hydroquinone adsorption, while the thermodynamic parameters studied in the temperature range of 25 to 50 °C showed that the Gibbs free energy ΔG^O and enthalpy ΔH^O are negative and positive, respectively. The negativity of ΔG^O and the positivity of ΔH^O confirmed the spontaneous nature and feasibility of the adsorption process, and proved the hydroquinone adsorption as endothermic. The percentage of removal increases with the temperature.

5 Acknowledgements

Many thanks to the EXCEED Swindon program, the German Academic Exchange Service (DAAD) and the German Federal Ministry for Economic Cooperation and Development (BMZ) for an award of a three months research grant, which allowed us to synthetize the adsorbent used in this study. The corresponding authors would also like to express his gratitude to the EXCEED Swindon project

Water Perspectives in Emerging Countries
Water in Agricultural Practices: Training the Trainers
September 15-21, 2019 - Rio de Janeiro, Brazil
Jose Araruna Jr., Aluisio Granato, Julia Stuchi, Rodolfo Silva (Eds.)

and DAAD for making possible his participation at the International Expert Workshop on *"Water on Agricultural Practices: Training the Trainers"* in Rio de Janeiro, Brazil from 15 – 21 September 2019.

The authors deliver their gratitude also to the « Laboratoire d'Hydrologie Appliquée et Environnement », Faculté Des Sciences, Universités de Lomé.

6 References

[1] E. Vunain, D. Houndedjihou, M. Monjerezi, A.A. Muleja, B. Kodom: Adsorption, Kinetics and Equilibrium Studies on Removal of Catechol and Resorcinol from Aqueous Solution Using Low-Cost Activated Carbon Prepared from Sunflower (*Helianthus annuus*) Seed Hull Residues. Water, Air, Soil Pollution, 2018, 229(11), 46

[2] Y. Zhang, S. Xiao, J. Xie, Z. Yang, P. Pang, Y. Gao: Simultaneous electrochemical determination of catechol and hydroquinone based on graphene–TiO2 nanocomposite modified glassy carbon electrode. Sensors and Actuators B: Chemical, 2014, 204, 102–108

[3] S. Sauvé, M. Desrosiers: A review of what is an emerging contaminant. Chemistry Central Journal, 2014, 8(1), 15

[4] D.E. Vidal-Dorsch, S.M. Bay, K. Maruya, S.A. Snyder, R.A. Trenholm, B.J. Vanderford: Contaminants of emerging concern in municipal wastewater effluents and marine receiving water. Environmental Toxicology and Chemistry, 2012, 31(12), 2674–2682

[5] C.-H. Chiou, R.-S. Juang: Photocatalytic degradation of phenol in aqueous solutions by Pr-doped TiO_2 nanoparticles. Journal of Hazardous Materials, 2007, 149(1), 1–7

[6] Y.A. Shaban, M.A. El Sayed, A.A. El Maradny, R.Kh. Al Farawati, M.I. Al Zobidi: Photocatalytic degradation of phenol in natural seawater using visible light active carbon modified (CM)-n-TiO_2 nanoparticles under UV light and natural sunlight illuminations. Chemosphere, 2013, 91(3), 307–313

[7] S. Ahmed, M.G. Rasul, R. Brown, M.A. Hashib: Influence of parameters on the heterogeneous photocatalytic degradation of pesticides and phenolic contaminants in wastewater: A short review. Journal of Environmental Management, 2011, 92(3), 311–330

[8] International Programme on Chemical Safety, Ed.: Hydroquinone: health and safety guide. World Health Organization, Geneva, (1996)

[9] I. Rodríguez, M.P. Llompart, R. Cela: Solid-phase extraction of phenols. Journal of Chromatography A, 2000, 885(1-2), 291–304

[10] J. Xu, Y. Yu, K. Ding, Z. Liu, L. Wang, Y. Xu: Heterogeneous catalytic ozonation of hydroquinone using sewage sludge-derived carbonaceous catalysts. Water Science and Technology, 2018, 77(5), 1410–1417

[11] L. Li, L. Fan, M. Sun, H. Qui, X. Li, H. Duan, C. Luo: Adsorbent for hydroquinone removal based on graphene oxide functionalized with magnetic cyclodextrin–chitosan. International Journal of Biological Macromolecules, 2013, 58, 169–175

[12] J.S. Hu, J.L. Dong, Y. Wang, L. Guan, Y.Y. Duan: Hydroquinone Wastewater Treatment by Means of Electrochemical Oxidation in Three-Dimensional Bipolar Cell. Advanced Materials Research, 2012, 518–523, 2539–2542

[13] D. Houndedjihou, T. Kodom, A.A. Dougna, I. Tchakala, G. Djaneye-Boundjou, L.M. Bawa: Oxidation of Catechol and Hydroquinone in Aqueous Media by Heterogeneous Photocatalysis Using Thin Layer of TiO_2 P25', Chemical Science International Journal, 2018, 24(2), 1–10

[14] W.T. Tsai, M.K. Lee, T.Y. Su, Y.M. Chang: Photodegradation of bisphenol-A in a batch TiO_2 suspension reactor. Journal of Hazardous Materials, 2009, 168(1), 269–275,

[15] B.H. Hameed, A.A. Rahman: Removal of phenol from aqueous solutions by adsorption onto activated carbon prepared from biomass material. Journal of Hazardous Materials, 2008, 160(2-3), 576–581

[16] K.N. Gupta, N.J. Rao, G.K. Agarwal: Adsorption of Toluene on Granular Activated Carbon. International Journal of Chemical Engineering and Applications, 2011, 2(5), 310–313

[17] M.C. Gérard, J.P. Barthélemy: An assessment methodology for determining pesticides adsorption on granulated activated carbon. Biotechnology, Agronomy, Society and Environment, 2003, 7(2), 79–85

[18] A.M. Aljeboree, A.N. Alshirifi, A.F. Alkaim: Kinetics and equilibrium study for the adsorption of textile dyes on coconut shell activated carbon. Arabian Journal of Chemistry, 2017, S3381–S3393

[19] S.T. Ong, P.S. Keng, S.L. Lee, M.H. Leong, Y.T. Hung: Equilibrium Studies for the Removal of Basic Dye by Sunflower Seed Husk (*Helianthus annuus*). International Journal of Physical Sciences, 2010, 5(8), 1270–1276

[20] A.R. Hidayu, N. Muda: Preparation and Characterization of Impregnated Activated Carbon from Palm Kernel Shell and Coconut Shell for CO_2 Capture. Procedia Engineering, 2016, 148, 106–113

[21] M.M. el-Halwany: Kinetics and Thermodynamics of Activated Sunflowers Seeds Shell Carbon (SSSC) as Sorbent Material. Journal of Chromatography & Separation Techniques, 2013, 4(5), 183

[22] Ü. Geçgel, H. Kolancılar: 'Adsorption of Remazol Brilliant Blue R on activated carbon prepared from a pine cone. Natural Product Research, 2012, 26(7), 659–664

[23] J.S. Aznar: Characterization of activated carbon produced from coffee residues by chemical and physical activation. Activated Carbon, 2011, p. 66.

[24] S.O. Adegboyega, A.A. Olusegun, S.O. Michael, T.I. Mku, S.A. Sam: Preparation of phosphoric acid activated carbons from *Canarium schweinfurthii* Nutshell and its role in methylene blue adsorption. Journal of Chemical Engineering and Materials Science, 2015, 6(2), 9-14

[25] S. Suresh, V.C. Srivastava, I.M. Mishra: Adsorption of catechol, resorcinol, hydroquinone, and their derivatives: a review. International Journal of Energy and Environmental Engineering 2012, 3(32), 19

[26] M. Jain, V.K. Garg, K. Kadirvelu: Chromium (VI) removal from aqueous system using *Helianthus annuus* (sunflower) stem waste. Journal of Hazardous Materials, 2009, 162(1), 365–372

[27] A.N. Zoweram, M. Majlesi, M.R. Masoudinejad: Remove food dye (Acid Red 18) by using activated carbon of sunflower stalk modified with Iron nanoparticles Fe_3O_4 from aqueous solutions. International Journal of Advanced Biotechnology and Research, 2016, 7(3), 289-297

[28] B. Karabacako: Liquid Phase Adsorption of Phenol by Activated Carbon Derived From Hazelnut Bagasse. Journal of International Environmental Application & Science, 2008, 3(5), 373-380

[29] R. Malik, D.S. Ramteke, S.R. Wate: Adsorption of malachite green on groundnut shell waste based powdered activated carbon. Waste Management, 2007, 27(9), 1129–1138

[30] P. Senthil Kumar, S. Ramalingam, C. Senthamarai, M. Niranjanaa, P. Vijayalakshmi, S. Sivanesan: Adsorption of dye from aqueous solution by cashew nut shell: Studies on equilibrium isotherm, kinetics and thermodynamics of interactions. Desalination, 2010, 261(1-2), 52–60

[31] D. Suteu, C. Zaharia, T. Malutan: Equilibrium, kinetic, and thermodynamic studies of Basic Blue 9 dye sorption on agro-industrial lignocellulosic materials. Open Chemistry, 2012, 10(6), 1913-1926

[32] C. Ding, Y. Sun, Y. Wang, Y. Lin, W. Sun, C. Luo: Adsorbent for resorcinol removal based on cellulose functionalized with magnetic poly(dopamine). International Journal of Biological Macromolecules, 2017, 99, 578–585

[33] A. Mohammad-Khah, R. Ansari: Activated Charcoal: Preparation, characterization and Applications: A review article. International Journal of ChemTech Research, 2009, 1(4), 859-864

[34] J.F. Osma, V. Saravia, J.L. Toca-Herrera, S.R. Couto: Sunflower seed shells: A novel and effective low-cost adsorbent for the removal of the diazo dye Reactive Black 5 from aqueous solutions. Journal of Hazardous Materials, 2007, 147(3), 900–905

[35] A.O. Ifelebuegu: Removal of Steriod Hormones by Activated Carbon Adsorption - Kinetic and Thermodynamic Studies. Journal of Environmental Protection, 2012, 3(6), 469–475

[36] M.I. Konggidinata, B. Chao, Q. Lian, R. Subramaniam, M. Zappi, D.D. Gang: Equilibrium, kinetic and thermodynamic studies for adsorption of BTEX onto Ordered Mesoporcus Carbon (OMC). Journal of Hazardous Materials, 2017, 336, 249–259

[37] E. Vunain, T. Biswick: Adsorptive removal of methylene blue from aqueous solution on activated carbon prepared from Malawian baobab fruit shell wastes: Equilibrium, kinetics and thermodynamic studies. Separation Science and Technology, 2019, 54(1), 27–41

[38] D. Sun, Z. Zhang, M. Wang, Y. Wu: Adsorption of Reactive Dyes on Activated Carbon Developed from *Enteromorpha prolifera*. American Journal of Analytical Chemistry, 2013, 4(7), 17–26

A STUDY ON FARMERS' PERCEPTION OF IMPACTS OF FARM WASTES ON SOIL AND WATER QUALITY: A CASE STUDY OF OYO STATE, NIGERIA

I.O. Kunlere[1,2,3], P.A. Ogar[2,4], A.S. Kunlere[3,5], A.A. Mumuni[6]

[1]*Department of Microbiology, University of Ibadan, Ibadan, Nigeria;*

[2]*Environmental Quality Control Unit, National Environmental Standards and Regulations Enforcement Agency (NESREA), South-West Zonal Headquarters, Eleyele, Ibadan, Nigeria;*
idowu2nice@yahoo.com

[3]*Wastesmart Green-Project Foundation, Ibadan, Nigeria;*

[4]*Institute of Ecology and Environmental Studies, Obafemi Awolowo University, Ife-Ife, Nigeria;*

[5]*Agricultural Science Department, Federal Government College (under the Federal Ministry of Education), Ikirun, Nigeria;*

[6]*Ecotoxicology and Conservation Unit, Department of Zoology, University of Lagos, Lagos, Nigeria;*

Keywords: Farm wastes, environmental impacts, perception, soil quality, water quality

Abstract

For many years before the discovery of oil in commercial quantity in Nigeria, agriculture was the mainstay of Nigeria's economy. Although agriculture was displaced as Nigeria's major income earner following years of concentration on the oil industry, it continues to provide food, jobs and livelihoods for millions of Nigerians. However, every year, farms in Nigeria produce millions of tons of wastes. Every point along the food supplies chain (pre-planting, planting, harvesting, and post-harvest operations, and acquisition by the final consumer) produces one form of waste or the other. Meanwhile, sustainable waste management remains a challenge in many developing countries, including Nigeria. In Nigeria, over 50% of wastes end up indiscriminately in the environment, resulting in water, air and soil pollution. As Nigeria continues to experience increase in population growth and agricultural activities, the accompanying increase in volume of farm wastes, which are large caches of untapped biomass resources, has been linked to incessant cases of pollution that threat food security, environmental sustainability and national development. Despite the threats that farm wastes pose, little is known about how aware of these critical issues, Nigerian farmers whose farms produce these enormous wastes, are. A qualitative approach was employed for the study, which saw fifteen farmers engaged in a semi-structured interview format. Data obtained from the interviews were interpreted using thematic analysis. The study showed low knowledge and poor perception of respondents on farm wastes and their impacts. One of the three farms in the study exclusively either burned its wastes or allowed them decay over time on the field and has had no contacts with municipal waste management authorities since its inception. The study concludes by recommending intense and sustained public awareness, targeted capacity building training on sustainable waste management and environmental reporting for farm owners, employers, and residents of host communities.

1 Introduction

Annually, farming activities generate millions of tons of farm wastes globally [1]. Meanwhile, recent population figures show the world has continued to witness significant increase in population growth [2]. By 2050, 68% of the world's population is expected to reside in urban areas; of this increase, 35% would live in India, China and Nigeria alone, while a whopping 90% of that increase would occur in only two continents, Asia and Africa [3]. Meanwhile, Africa is the world's fastest urbanizing region [4]. Amongst other things, these figures mean that there would be steady, higher demands for food to meet the needs of the teaming populace. Thus, as Nigeria continues to experience rapid population growth and urbanization, agricultural (farming) activities have also continued to increase to up food production [5]. However, these increases also mean an increase in the volume of generation of farm wastes.

Agriculture and Nigeria's economy
Agriculture was the pillar of the Nigerian economy for many decades before the advent of oil exploration. Although the intervening years after Nigeria exported its first oil shipment following the advent of commercial exploration of petroleum in Nigeria in the 1950s saw petroleum quickly displaced agriculture as Nigeria's major foreign exchange earner [6], agriculture has remained a significant pillar of the Nigerian economy, contributing 19.3% and 17.4% of Nigeria's nominal GDP in the second quarter of 2017 and first quarter of 2018, respectively [7-9]. Agriculture is also the single biggest employer of labour, particularly in rural areas in Nigeria. With Nigeria's population rapidly growing, agricultural activities have also been on the increase [10]. However and inadvertently so, as agricultural activities increase, farm wastes (wastes produced by farms) have also been increasing; this increase is projected to continue as population and economic growth continue [2].

Agriculture and farm wastes
Although agriculture is a strategic sector in Nigeria's economy, the enormous volumes of wastes that farming activities continue to generate is a big concern. In some places, large volumes of edible or processed food continue to be lost annually as food wastes. Like in many developing countries, waste management remains a big challenge in Nigeria [1, 11, 12]. Waste management has both global dimensions and implications, and the harmful environmental and public health consequences of indiscriminate waste dumping necessitate that more attention is paid to it. In the past few years in Nigeria, the growing volumes of farm wastes make the situation even worse. While unplanned and unregulated increases in the volume of farm wastes may adversely impact the profitability of farms, indiscriminate disposal and poor management of farm wastes are inimical, both to the environment and the public as well as to food security, environmental sustainability and national development [10]. In recent times, awareness of environmental sustainability issues has been on the rise in Nigeria, however, more attention has been inadvertently paid to management of municipal waste than to management of farm waste; thus, many farms do not properly manage the wastes they produce. Although how much farmers, farm managers and farm workers know about farm wastes, how they perceive farm wastes, its adverse

impacts, methods of sustainable management of farm wastes etc. are key determinants of whether farms sustainably manage their farm wastes or not, little is known about these critical points. This study, therefore, investigates the general perception that farmers have about the impact of farm wastes from their wastes on soil and water quality.

What are farm wastes?

Agricultural waste is waste produced as a result of various agricultural operations. It includes manure and other wastes from farms, poultry houses and slaughterhouses; harvest waste; fertilizer run-off from fields; pesticides that enter into water, air or soils; and salt and silt drained from fields [13]. According to [14], farm wastes are non-product outputs of production and processing activities of agricultural products, which may be beneficial, but whose economic value may be less than the actual cost of collection, transportation, and processing into useful materials or products. The definition of [14], which introduces the angle of economics (economic value) to the definition of farm waste, is an important one, because at the end of the day, economics determines what is regarded as waste or not, and how much value people are willing place on or exchange for it. Indeed, what constitutes waste to one may not be waste to another. Economics also determines, if the inherent value in a "farm waste" is high enough or makes economic sense for that value to be exploited through reuse, recovery or recycling.

Farm wastes can also be defined as by-products or residual materials, which directly or indirectly result from the cultivation and/or processing of raw agricultural products on the farm. Oftentimes, farm waste, agricultural wastes, and food waste are used interchangeably, although depending on use or circumstance, they may not exactly mean the same thing. For example, while farm waste may mean all types of residual materials (wastes) produced on the farm (before, during or after cultivation), during transportation, processing or before it gets to the final consumer, food waste specifically refers to edible organic products, which are originally cultivated or produced for human consumption but do not get consumed due to various reasons, including excess production, underuse, spoilage or sub-quality. On the other hand, when broadly used, agricultural wastes (or agro-waste) may refer to both farm and food wastes. Farm wastes are generated during pre-cultivation, cultivation, pre-harvesting, harvesting and processing phases on the farm. In essence, farm wastes tend to be produced throughout the lifecycle of the farming season or cultivation process. In addition, for farm wastes, emphasis is placed on waste generated before the agricultural product gets to the final consumer. Agricultural wastes that are generated after the agricultural product gets to the final consumer is more generally referred to as food waste.

According to [14], farm wastes may include animal waste (such as manure, animal carcasses), food processing waste (for example, while only 20% of maize is canned, 80% of maize, including its cob goes to waste), crop waste (corn stalks, sugarcane bagasse, drops and culls from fruits, vegetables, and pruning), and hazardous and toxic agricultural waste (such as pesticides and herbicides), etc. Examples of farm wastes include wastes from fruits, vegetables, meat, poultry, dairy products, and crops. Farm wastes are made up of poultry wastes, dairy wastes, green wastes, crop wastes, etc.

They include various categories of wastes, produced on the farm or in activities related to livestock, animal husbandry or crop cultivation. Farm wastes include animal feces, urine, litter, beddings, animal carcass, dairy parlor washings, waste milk, wasted feed, run-off from feed lots and holding area, paunch waste, abattoir waste water, animal viscera, horns, feather, bones blood, fur, placenta, birth tissue, fetal membranes, aborted fetus, groundnut shells, cereal husks, palm kernel shells, sugar cane bagasse, waste wood, spoilt packaging products like boxes, bags and plastics, worn-out machinery and equipment components such as tires and batteries [15]

Types of agricultural wastes

Farm wastes may be classified on the basis of their physical state, source, biodegradability, recyclability, toxicity or infectiousness (ability to cause or serve as conduit/dispersal for pathogens and diseases). On the basis of physical source, farm wastes may be in the form of solid, liquid (or slurries) and gas, while on the basis of source, farm wastes may also classified as organic or inorganic. They may also be classified based on their toxicity or infectiousness as toxic or non-toxic wastes, infectious or non-infectious wastes. Table 1 summarizes the classification of farm wastes.

Table 1: Classification of farm wastes

Organic waste	Inorganic waste	
Examples: Feces, tree stumps, weeds, harvest remains, spoilt fruits, vegetables and other edibles, carcasses, hair, bones, hyde and skin, urine, blood, methane gas, manure, slurry, straw, shells, sugarcane trash/bagasse, molasses, cotton sticks, rice bran, wood remains, etc.	Agrochemicals (phytosanitary, pesticide and veterinary products), plastic and metallic packages of chemicals, used tires, used lubricant oils, asbestos, tiles, bricks, concrete, etc.	
Infectious waste	**Non-infectious waste**	
Contaminated bodily fluids, carcasses of animals that died from diseases, contaminated or spoilt fruits, vegetables, plant remains, residues of pesticides and other agrochemicals, etc.	Leaf droppings, animal droppings from healthy animals, papers, plastics etc.	
Solid (or slurry) waste	**Liquid waste**	**Gas waste**
Feces, tree stumps, weeds, harvest remains, spoilt fruits, vegetables and other edibles, carcasses, hair, bones, hyde and skin.	Urine, blood.	Methane gas

Volume of annual farm wastes generation

Annually, billions of tons of agricultural, industrial and municipal wastes are produced globally [2]. While some reports say about 1 billion tons of agricultural waste is generated annually [14], close to 5 million tons of agricultural waste was produced between 2012 and 2017 in the Emirate of Abu Dhabi, United Arab Emirates, UAE [16]. About 1.7% (or approximately 98 million tons) of the 2.6 billion tons of waste produced across the EU in 2008 were agricultural wastes [17]. Although data on exact volume of farm waste generated in Nigeria are hard to come by as record keeping by farms is not strictly adhered to and supervision by regulatory authorities remains largely weak, more than 60% of municipal solid waste generated in Nigeria is organic, mostly from food waste. As the number of farms in Nigeria and their production capacities continue to increase in response to the growing population, it is estimated that the volume of waste they produce will be the rise. In fact, from public records, as population grows in areas once sparsely-populated and only occupied by farms, one of the leading cases of public complaints in such areas is environmental pollution resulting from the wastes and activities of such farms. It has been noted that a single hog produces three times the amount of waste as a human produces, while a dairy cow produces 20 times the amount of waste of a human [18].

Impact of farm wastes on soil and water quality
There have been growing concerns about the risks that poor management of farm wastes in Nigeria poses to the environment and public health, amongst other things. This section discusses some of the direct and indirect environmental impacts of farm wastes, particularly from the perspective of water and soil quality. The huge amounts of pathogens, drug-resistant microorganisms and nutrients that farm wastes (for example, animal carcasses, effluents, pesticides, etc.) often contain means that when farm wastes are indiscriminately left on farmlands, they could contaminate underground water or nearby surface water systems, which if used without proper treatment, could lead to serious disease outbreaks.

Figure 1 shows some of the major means through which farm wastes contaminate natural water resources. In Nigeria, clean potable water from municipal facilities is hardly available, particularly in rural areas, where majority of the population are farmers and where rural dwellers mostly directly depend on local streams and other water bodies for water, hence, environmental pollution from farm wastes poses serious risks [19-21]. Disease-causing organisms could be easily spread through poor handling of farm wastes. Although personal protection equipment is recommended to protect farm workers and to reduce the risk of accidents or spread of pathogens, they are hardly used by farmers [22]. In cases of epidemics or where farm crops or animals are destroyed by contagious diseases, poor management of the resulting farm waste of animal carcasses or plant remains could cause easy spread of the disease to other animals or plants on the same or other plants or pollute surrounding water bodies [21].

According to [23], agriculture is one of the biggest users of fresh water resources. In fact, on a global scale, agriculture averagely uses about 70% of all water supplies. Unsurprisingly, agriculture, particularly through farm wastes, has also been implicated in the pollution of groundwater and surface water [24]. Indiscriminate or overuse of manure, fertilizers and agrochemicals such as pesticides or release of untreated farm waste directly into the environment could result in chemicals being washed into nearby streams, rivers, etc., causing a sudden surge in the concentration of nutrients, which then results in rapid and unrestricted growth of microalgae or cyanobacteria, a phenomenon that is called eutrophication [25].

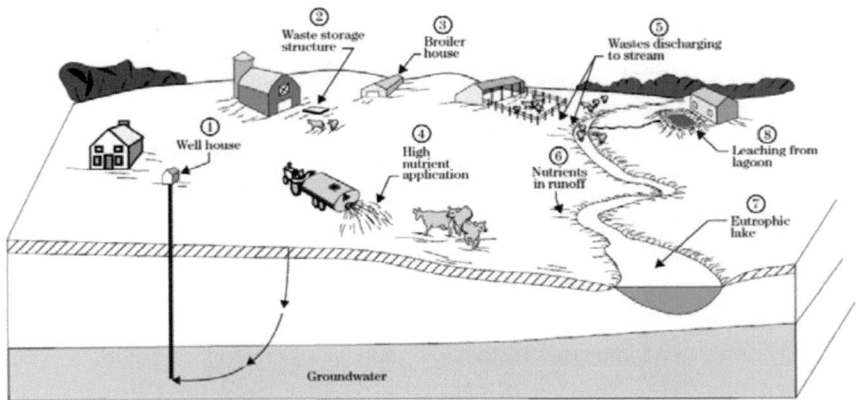

Figure 1: Possible danger points in the environment from uncontrolled animal waste [26]

According to [27], eutrophication is a process, whereby a water body acquires excessive amounts of nutrients, particularly phosphates and nitrates, which then results in overgrowth of algae, depletion of oxygen in the water body and subsequent death of aquatic animals. Eutrophication has harmful effects on the environment, because as algae species rapidly and excessively grow in a water body (algal boom) due to excess nutrients, they use up more dissolved oxygen for their metabolism, leading to discoloration of the water. This also leads to a gradual and drastic reduction in the net concentration of dissolved oxygen available in the affected water body for other aquatic life to use. If this continues unchecked, the other aquatic lives are unable to withstand extreme low concentrations of dissolved oxygen and will begin to die off. As dead aquatic life in the water body decays, their decaying remains further contaminate the water body; thus, the water body starts to stink due to malodours, rendering the water unsafe both for man and animal use. The offensive, putrefying odour from decaying dead aquatic plant and animal species also makes life unbearable for the residents of the community.

Globally, a disproportionate number of people, particularly in rural areas, who are affected by the impacts of indiscriminate disposal of farm wastes, are women and children. In Nigeria for example, more people live in rural areas than in urban areas. Majority of residents in rural areas engage in one form of agriculture or the other, however, a significant number of farmers are women. This is also true for the USA, where various surveys reported more women working in farms than men and that the number of women working on farms across the USA continues to grow [28, 29]. These findings thus mean that people in rural areas as well as women and children are particularly at more risk to unsafe practices such as indiscriminate management of farm wastes, which could result in the pollution of soil and water resource, and epidemics. A research by [30] found out that incidents of indiscriminate disposal of farm wastes by big farms occurred more in minority and low income communities with vulnerable women and children, who have no means of fighting back to protect their homes and ancestral lands. That situation was described as "environmental racism", where the rich suppress the poor in poor, dehumanizing environmental conditions [30].

Indiscriminate burning of farm wastes is also a common practice. However, indiscriminate burning releases dioxins and furans into the environment, which are persistent pollutants and suspected carcinogens. Figure 2 shows air pollution from burning of wastes at a dumpsite, a common feature in many farms. Sadly, dioxins produced in one area can be transported via air and water to thousands of miles away, where they can cause new pollutions. Prolonged exposures to dioxins cause a variety of health implications. For example, in women of reproductive age, dioxins could lead to neurological damages to the fetus during pregnancy, abnormal menstruation, disruptions in hormone production, chronic acnes, cysts and cancers [31].

Because Nigeria lacks a comprehensive national policy on waste management, there is no specific national strategy, which encompasses sustainable management of all or common types of farm wastes. However, there are a number of piecemeal national regulations that may cover some areas such sanitation, waste control, management of animal carcasses, etc. However, enforcement of the existing piecemeal regulations has not been effective. Also significantly, in terms of supervisory roles of farms by environmental regulators, in the past few years, the focus of regulators has been more on big farms, almost overlooking small and medium ones, whose impacts could be just as significant. Farms also register and maintain administrative offices at locations different from where their farms are located at. So, while regulators may visit such farms' administrative offices which are neat and well organized, they are often not able to visit or effectively inspect, monitor or regulate the main farms, where the real issue of poor management of farm wastes occurs.

Figure 2: Indiscriminate burning of farm wastes is common in Nigeria. (Google Image)

Qualitative research and perception of farmers of impact of farm wastes
While the number of farms and volume of farm wastes in Nigeria have been increasing, few studies have focused on the perception of Nigerian farmers on farm wastes and their impacts on environmental mediums such as water and soil. Yet, farmers' perception of such critical issues influences their awareness of the impacts such issues have, and determine the approaches the

farmers adopt in managing them. On the other hand, right perception of these issues can inspire sustainable management and produce key benefits both to the farm and the larger society. In developing effective strategies to curb the rising pollution problem from farm wastes, it is, therefore, imperative to investigate how farmers perceive farm wastes and their impacts, and how they currently manage their farm wastes. Qualitative research methods are an important research tool as they give more elaborate insights than conventional quantitative research tools [32]. Examples of qualitative research tools are interviews and focus groups. Amongst other advantages, qualitative tools give insights into social issues in their true natural states (rather than in experimental states) and prioritize the perspectives of the respondents [33]. Qualitative research tools may also give deeper clues to areas that should be further investigated using quantitative research tools [34, 35].

Since farmers are major players, who determine what ends up as farm wastes or not, it is important to understand their perception of the subject matter and the tools they currently employ in minimizing the volume of the farm waste they generate ensuring sustainable management. Various qualitative researches had been carried out into farm and food wastes [32, 36-38]. However, majority of them employed closed-ended questions. By nature, however, closed-ended questions may be suggestive, pre-empt the interviewees and also limit the kind of responses they give [32]. This approach limits getting the true perspectives of the interviewees on the subject matter in "their own words". Furthermore, qualitative researches, which employed open-ended questions in investigating perception of farmers on farm waste, are rare.

In his study, [39] reported that some farmers reflect zero environmental awareness on important issues. If such farmers are unaware of the impacts of poor management of farm wastes, they may inevitably unsustainably manage their farm wastes (Figure 1). This, in turn, would adversely impact the environment. Perception of these issues, therefore, is key to the management methods that farmers adopt. Farmers' perception is influenced by various factors, including level of formal/informal education, experience/exposure, location of their farms, age, access to information, technical know-how, etc. These factors may act singly or in combination [39-41]. However, there is currently little information on the perception of farmers in Oyo State on farm wastes and their impacts on the environment, hence, this study.

This study has two specific objectives: (i) To determine the perception (level of awareness) of farmers and workers at selected farms in Oyo State on the dangers that unsustainable management of farm wastes poses to soil and water environments, and (ii) To ascertain current practices adopted by the farmers in the management of the farm wastes they generate.

Objectives
This study aims to investigate farmer's knowledge and perception of impact of farm wastes, types of waste each farm produced, available waste management methods and how farmers currently dispose of their wastes.

2 Materials and Methods

Study Site

Three farms were selected for the study, one from each of the 3 senatorial districts in Oyo state (Table 2).

Table 2: Details of Study Sites

	Products	Location of farm / LGA	Senatorial District	State
Farm A	Mango, cashew, livestock, fish pond	Ogbomoso	Oyo North	Oyo
Farm B	Livestock, oil palm, yam, cassava, maize	Lagelu	Oyo Central	Oyo
Farm C	Poultry, fish pond, cassava, maize	Ibarapa East	Oyo South	Oyo

Figure 3 shows the map of Oyo State. Oyo State is located in the South-West region of Nigeria. Oyo State is divided into 3 Senatorial Districts, 33 Local Government areas and 29 Local Council Development Areas. It has about 6,000,000 residents, many of whom live in sub-urban/rural areas and are subsistent farmers; many of these places lack access to clean, portable water. Oyo State occupies a total landmass of 28,454 km^2 and is bounded by the Republic of Benin and Ogun to the West, and by Osun, Kwara, and Ogun States to the East, North, and South, respectively. Its topography is characterised by gentle rolling low land in the south, with rivers that flow north-south. Its climate consists of dry and wet seasons, while its vegetation is predominantly rain forest in the south and guinea savannah in the north. Some of its most cultivated crops cultivated are maize, yam, cassava, plantain, cocoa tree, palm tree and cashew [42].

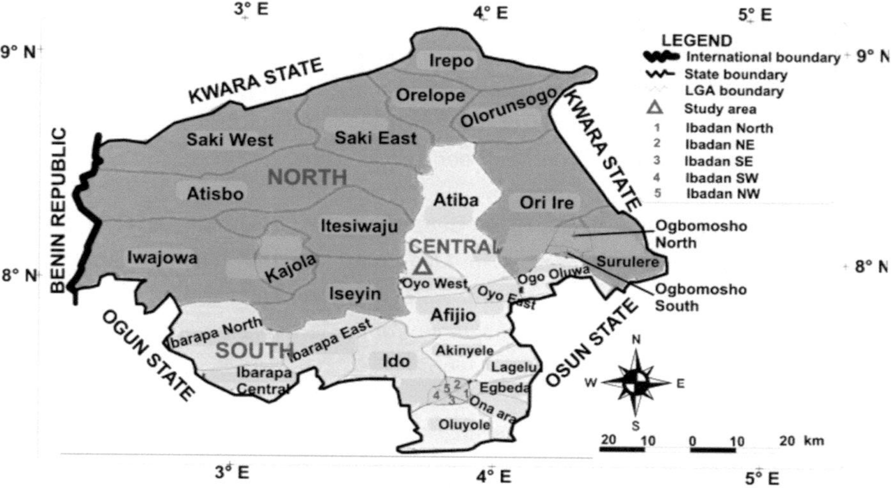

Figure 3: Map of study areas in Oyo State, Nigeria [45]

Methods

This study employs a qualitative approach to investigate farmers' perception of farm waste and their impacts. Semi-structured interviews were conducted with fifteen farmers. In total, each interview lasted between five to twenty minutes and was conducted in English, Yoruba and Hausa languages as well as Pidgin English. While key notes were taken during the interviews, in addition, with the permission of the interviewees, the interviews were recorded on tapes, which were later listened to multiple times and transcribed. During the interview, the interviewees were allowed to react to the meanings that interviewers read into their responses and were allowed to reject, confirm, or modify such meanings [43]. Attention was also paid to non-verbal communication of the interviewees such as body language for cues. At least, one interviewee from each of the three farms was in managerial or operations supervisory role, a position, which suggests that the interviewee is more conversant with the policies of the farm. Before the interview, a general observation of the farms was made and key details about the farms and the respondents were collected [44]. The observations made then formed the basis of open-ended questions asked during the main interviews like: Types of wastes generated on the farm; storage facilities for farm produce; waste management infrastructure available on the farm, workers' awareness of them, and frequency of use; awareness of impacts of unsustainable waste management practices; how often workers on the farm fall sick; number of complaints lodged by residents in the host community on pollution emissions from the farm. The questions data obtained from the interviews were interpreted using thematic analysis.

3 Results and Discussion

Table 3 shows the profile of the fifteen respondents interviewed for this study. An adapted method of the thematic analysis method of [46 and 47] was used to analyze qualitative data for this research. To identify the major themes in the transcribed interviews, thematic analysis was used. According to [47], a theme is an important idea in the interviewee's response, which bears close relationship to one or more of the research questions and which is conveyed in a repeated pattern. Through active listening and reading, repeated patterns were watched out for; these were then represented using simple codes. The codes were then arranged into themes based on relatedness. Based on various similar studies by other researchers, there were themes identified as a priori, while the rest were identified during the listening and reading process [48]. As the process continued, the themes in the farmers' responses were fully coded and exhausted. Figure 3 shows the themes and sub-themes identified in the responses of the farmers.

The study showed low knowledge and poor perception of respondents on farm wastes and their impacts. This is directly linked to lack of awareness and environmental education. None of the farms practised sorting of waste at source or before disposal, as all wastes on the farms were comingled. None of the farms had conducted an environmental awareness program, health talk or waste management training program on safe handling and sustainable management of farm wastes in the past one year before the study.

Table 3: Profile of respondents

	Farm A	Farm B	Farm C
Number of Interviewees	5	5	5
Type of employment of interviewees	Casual Staff: 4 Full Staff: 1	Casual Staff: 3 Full Staff: 2	Casual Staff: 4 Full Staff: 1
Gender of interviewee	Female: 4 Male: 1	Female: 3 Male: 2	Female: 1 Male: 4
Workforce of interviewees' farm	20	55	10
Age range	20-35	25-50	14-50
Marital status	Single: 0 Married: 5	Single: 0 Married: 5	Single: 2 Married: 3
Household size	>2 member-family: 1 >3 member-family: 4	>3 member-family: 5	> 3 members--family: 3 1 member--family: 2
Region of origin of Interviewees	South-West:1 South-East: 0 South-South: 0 North-West: 2 North-East: 0 North-Central: 2	South-West:1 South-East: 0 South-South: 0 North-West: 1 North-East: 1 North-Central: 2	South-West: 0 South-East: 0 South-South: 0 North-West: 1 North-East: 0 North-Central: 4
Level of Education	Primary: 4 Tertiary: 1	No education: 3 Secondary: 1 Tertiary: 1	Primary: 4 Tertiary: 1
Years of experience in farming	> 5 years: 3 >10years:1 >20 years: 1	> 5 years: 3 >10years:1 >20 years: 1	> 5 years: 0 >10years:4 >20 years: 1
Farm size	8 ha	20 ha	12 ha
Number of hospitalisation in the past one year	3	1	1
Number of deaths in the past one year	1	0	0
Quantity of wastes	Unknown (No records kept)	Unknown (No records kept)	Unknown (No records kept)

Themes
Sub-themes

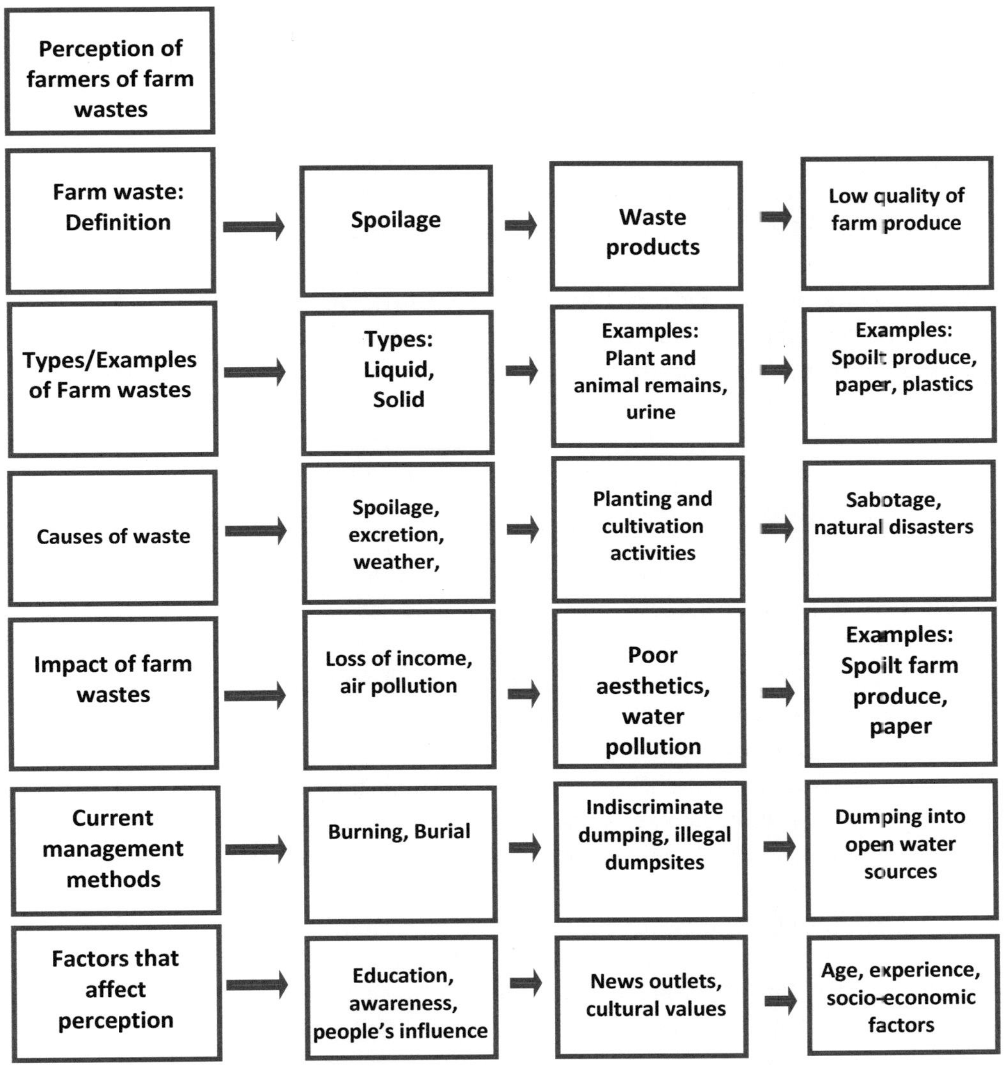

Figure 3: Coding framework for thematic analysis, including themes and sub-themes

One of the three farms in the study (Farm B) exclusively either burned its wastes or allowed them decay over time on the field and has had no contacts with municipal waste management authorities since its inception five years ago. None of the three farms is registered with any waste collector accredited by the Oyo State Government in the past one year. This observation with the three farms could be linked to its location in a remote and sparsely populated area. But interestingly, there are hundreds of other small and medium-sized farms (5 to 15 ha) like them that are currently off the radar of regulatory agencies, but whose activities and wastes may be polluting the environment, posing serious risks to the environment and to public health.

One farm manager (Farm A) explained that while unpacking animal dungs during harmattan the past three years ago, three workers suffered respiratory episodes that might have been asthmatic episodes. However, one of the affected workers eventually died a couple of days after. But at the time of the death, the workers believed it was a "spiritual attack". The surviving affected workers could not be interviewed for this study, as they had left the employ of the farm. As a result of the incidents, the Farm A stopped the practice unpacking animal dungs and thus, wastes were less emphasized and the farm's wastes have grown into heaps. Each of Farms B and C also reported that they had one case of hospitalization of farm workers in the past one year. None of the workers on the three farms in the study indicated that they had used personal protection equipment (PPE) during their farming operations in the past one year, and none was seen using any PPE during the period of the interviews.

Two of the farms (Farms A and B) export some of their products abroad. Respondents from both explained that more attention is placed on the quality of products meant for export. And those products not meeting the standards for exports are usually rechanneled into the local market, where they are easily sold off rather than being further processed or disposed of. This practice can also be linked to weak oversights in local markets. None of the farms has a waste management policy. None of the respondents could mention any waste management related-policy that the farms adhere to. As during this study in October 2018, despite the unsustainable farm waste management practices at all three farms sampled, the respondents acknowledged that there have been no public complaints of environmental pollution by residents in the host community against any of the three farms to regulatory authorities. This may be linked to the fact that all three farms are located in sparsely populated rural area and outskirts of major towns.

One of the three farms (Farm A) released its untreated effluent into the neighbouring water channels. Although in the case of Farm C, there is a stabilisation pond, which it also directly uses as a fish pond, no laboratory testing is done to ensure pollutants are within statutory limits before used water is released into the environment. All three farms, though large by characterisation, have no environmental permits on waste and toxic substances. Over 75% of the farmers interviewed revealed that they believe improper wastes may affect their health or those of neighbouring residents, but did not how to go about it. Over 50% of the workers interviewed at the three farms said, they think that there could actually be links between the poor management of farm wastes on their farm and the frequency of medical emergencies that had been experienced on the farm in the past two years. All three farms were indicated, they would be interested in trainings on how to make compost from their organic wastes and other sustainable methods of managing their farm wastes. The study showed that perception of respondents on farm wastes and their impacts was low. Although the farms surveyed had each, in the last 5 years, experienced at least 2 cases of environmental pollution linked to poor waste management, waste management was not a priority for the farms. One of the farms exclusively either burned its wastes or allowed them decay over time on the field. Majority of the farms' employees interviewed were unaware of the 3-Rs of waste management and lacked sustainable waste management skills but expressed enthusiasm to learn.

From this study, some of the problems identified for poor management of farm wastes include: poor awareness of impacts of unsustainable waste management practices; poor storage facilities for farm products; poor transport and social amenities and poor awareness of sustainable waste management practices. The study concludes by recommending intense and sustained public awareness, targeted capacity building training on sustainable waste management and environmental reporting for farm owners, employers, and residents of host communities.

4 Recommendations and Conclusion

Countries seeking to be more resource-efficient must improve their waste management system, of which management of farm waste is a critical segment. Finding economically viable ways of resource recovery from waste is key to ensuring that farms effectively cut down on waste generation and carbon emissions. In cutting down on carbon emissions and waste generation rather than focusing on a single method, attention should be on harnessing a combination of methods. Also, emphasis should be put on the waste hierarchy with priorities in waste prevention, reuse, recovery and recycling ahead of disposal of wastes such as in landfill [49]. The study concludes by recommending intense and sustained public awareness, targeted capacity building training on sustainable waste management and environmental reporting for farm owners, employers, and residents of host communities. Some of the solution to improper farm waste management include: better and more sustainable farming methods, more intense environmental education, particularly at the grassroots, adoption of recycling skills, upgrade of municipal facilities, sustained public sensitisation, more effective regulation of local market for low-quality farm products, and strengthening of environmental regulators to more effectively carry out their supervisory functions.

5 Acknowledgements

The authors acknowledge Wastesmart Green-Project Foundation, Nigeria for providing funding for this research project, and NESREA, for its support. The financial grant from DAAD and the Exceed Swindon Project for participation at this Expert Workshop is also deeply appreciated.

6 References

[1]   Gontard, N., Sonesson, U., Birkved, M., Majone, M., David, B., Celli, A., Angellier-Coussy, H., Jang, G., Verniquet, A., Jan, B., Schaer, B., Batista, A.P., Sebok, A.: A research challenge vision regarding management of agricultural waste in a circular bio-based economy, Critical Reviews in Environmental Science and Technology, 2018, 48,6, 614-654.

[2]   EPA: Municipal Solid Waste Generation, Recycling, and Disposal in the United States: Facts and Figures for 2010. United States Environmental Protection Agency Press, 2011, 12pp.

[3]   UNECA (United Nations Economic Commission for Africa): 2018 Africa Sustainable Development Report: Towards a transformed and resilient continent. (2018). https://www.uneca.org/publications/2018-africa-sustainable-development-report.

[4]   UNDESA (United Nations Department of Economic and Social Affairs): Urban safety and good governance. (2018). https://www.un.org/development/desa/en/news/population/2018-revision-of-world-urbanization-prospects.html

[5]   Vanguard: In Nigeria's steady progress in agriculture, potential history in the making. (2017). https://www.vanguardngr.com/2017/03/nigerias-steady-progress-agriculture-potential-history-making/

[6]   Okoruwa, V.O., Ogundare, G.O., Yusuf, S.A.: Determinants of traditional agricultural exports in Nigeria: an application of co integration and error correction model. Quarterly Journal of International Agriculture, 2003, 43(4).

[7]   Bola, O.: Nigeria Agricultural Sector. Oxford University Press. (2007).

[8]   Ufiobor, K.A.: Nigeria Agriculture and Sustainability: Problems and Solutions. Bachelor's Thesis, Sustainable Coastal Management, University of Applied Sciences, Novia. (2017).

[9]   NBS, National Bureau of Statistics: Nigerian Gross Domestic Product Report (Q2 2018) (Report Date: August 2018).

[10] Ugwuoke, C.U., Monwuba, N., Onu, F.M., Shimave, A.G., Okonkwo, E.N., Opurum, C.C.: Impact of Agricultural Waste on Sustainable Environment and Health of Rural Women. Civil and Environmental Research, 2018, 10(9), ISSN 2224-5790 (Paper) ISSN 2225-0514 (Online),

[11] Marshall, R.E., Farahbakhsh, K.: Systems approaches to integrated solid waste management in developing countries. Waste Management, 2013, 33, 988-1003.

[12] Kunlere, I.O., Fagade, O.E., Nwadike, B.I.: Biodegradation of low density polyethylene (LDPE) by certain indigenous bacteria and fungi. International Journal of Environmental Studies, 2019, 76(3), 428-440.

[13] GES, Glossary of Environment Statistics: Studies in Methods, Series F, No. 67, United Nations, New York, 1997.

[14] Obi, F.O., Ugwuishiwu, B.O., Nwakaire, J.N.: Agricultural Waste Concept, Generation, Utilization and Management. Nigerian Journal of Technology (NIJOTECH), 2016, 35(4), 957–964.

[15] SEPA, Scottish Environmental Protection Agency: A guide to agricultural waste (2005). www.sepa.org.uk.

[16] Waste Statistics: Waste generation in the Emirate of Abu Dhabi. Statistics Centre, Abu Dhabi Emirate. 2017, pp. 1-13.

[17] Eurostat: Generation and treatment of waste in Europe 2008, Steady reduction in waste going to landfills. Eurostat Press, 2011, 8 pp.

[18] Robin, M.: How factory farm lagoons and spray fields threaten environmental and public health. (2001). nrdc.org/water pollution/cesspools.pdf

[19] Davies, R.H.: A two year study of Salmonella typhimurium DT104 infection and contamination on cattle farms. Cattle Practice, 1997, 5, 189 – 194.

[20] Fischer, T.K., Steinsland, H., Mølbak, K., Ca, R., Gentsch, J.R., Valentiner-branth, P., Aaby, P., Sommerfelt, H.: Genotype profiles of Rotavirus strains from children in a suburban community in Guinea-Bissau, Western Africa. Journal of Clinical Microbiolology, 2000, 38, 264–267.

[21] Cameron, R., Mohammed, H., Ward, D.: Area-wide integration (AWI) and ts implications for animal and human health. Discussion Keynote Paper, Session 3 of Electronic Conference on Area Wide Integration of Specialized Crop and Livestock Production, 2000, 5 pp. http://www.virtualcentre.org/en/ele/awi_2000/3session/3paper.htm http://www.virtualcentre.org/en/ele/awi_2000/3session/3plenary.txt.

[22] Mackenzie, W.R., Hoxie, N.J., Proctor, M.E., Grad us, M.S., Blair, K.A., Peterson, D.E., Kazmierzak, J.J., Addis, D.G., Fox, K.R., Rose, J.B., Davies, J.P.: A massive outbreak in Milwaukee of Cryptosporidium infection transmitted through the public water supply. New England Journal of Medicine, 1994, 331, 161-167.

[23] USEPA: National Pesticide Survey: Update and Summary of Phase II results, Office of Water and Office of Pesticides and Toxic Substances, United State Environmental Protection Agency Report EPA570/9-91-021, Washington D.C. (1993).

[24] FAO: An overview of Pollution of water by Agriculture, J. A. Sagarday. In: Prevention of Water Pollution by Agriculture and Related Activities. Proceeding of the FAO Expert Consultation, Santiago, Chile 20 – 23 Oct., 1992. Water Report 1, FAO, Rome, 1993, pp 19-26.

[25] IAEA/FAO: International Atomic Energy Agency/ Food and Agriculture Organisation Guidelines for Sustainable Manure Management in Asian Livestock Production Systems. A publication prepared under the framework of the RCA project on Integrated Approach for

Improving Livestock Production Using Indigenous Resources and Conserving the Environment. (2008).

[26] USDA, United States Department of Agriculture, Natural Resources Conservation Service: Chapter 3, Agricultural Wastes and Water, Air, and Animal Resources In: Part 651 Agricultural Waste Management Field Handbook. (2012).

[27] Lawrence, I., Jackson, B.: Environmental Health: Toxic substances. (1998) www.toxics.usgs.gov/definitions/eutrophication.html

[28] National Agricultural Workers Survey (NAWS): United States Department of Labour Employment and Training Administration. (1995).

[29] USDA (United States Department of Agriculture): Characteristics of women farm operators and their farm. (2013). www.ers.usva.gov/media/1093194/eib/pdf

[30] Devries, M.: Speciesism. (2012). www.factoryfarmsdrones.com.

[31] CEC, Commission for Environment Co-operation: Burning agricultural waste: A source of dioxins. (2014). www.cec.org.

[32] Graham-Rowe, E., Jessop, D., Sparks, P.: Identifying motivations and barriers to minimising household food waste. Resources Conservation and Recycling, 2014, 84, 15-23, https://doi.org/10.1016/j.resconrec.2013.12.005.

[33] Pope, C., Mays, N.: Qualitative Research: Reaching the parts other methods cannot reach: an introduction to qualitative methods in health and health services research. BMJ (British Medical Journal), 1995, 311(6996), 42-45.

[34] Bryman, A.: Integrating quantitative and qualitative research: how is it done? Qualitative Research, 2006, 6(1), 97-113.

[35] Newenhouse, S., Schmit, J.: Qualitative methods add value to waste characterization studies. Waste Management and Research, 2000, 18(2), pp.105-114,

[36] Bonadonna, A., Matozzo, A., Giachino, C., Peira, G.: Farmer behavior and perception regarding food waste and unsold food. British Food Journal, 2019, 121(1), 89-103.

[37] Ofei, K., Holst, M., Rasmussen, H., Mikkelsen, B.: How practice contributes to trolley food waste. A qualitative study among staff involved in serving meals to hospital patients. Appetite, 2014, 83, 49-56.

[38] Hoek, A., Pearson, D., James, S., Lawrence, M., Friel, S.: Shrinking the food-print: A qualitative study into consumer perceptions, experiences and attitudes towards healthy and environmentally friendly food behaviours. Appetite, 2017, 108, 117-131.

[39] Rahman, S.: Environmental impacts of modern agricultural technology diffusion in Bangladesh: An analysis of farmers' perceptions and their determinants. Journal of Environmental Management, 2003, 68(2), 183-91.

[40] Adesina, A.A., Baidu-Forson, J.: Farmers' perception and adoption of new agricultural technology: evidence from analysis in Burkina Faso and Guinea, West Africa. Agricultural Economics, 1995, 13, 1 – 9.

[41] Negatu, W., Parikh, A.: The impact of perception and other factors on the adoption of agricultural technology in the Moret and Jiru Woreda (district) of Ethiopia. Agricultural Economics, 1999, 21, 205 – 216.

[42] OYSG (Oyo State Government): About Oyo State. (2017). https://oyostate.gov.ng/about-oyo-state/

[43] Travers, M.: Qualitative Research through case studies. SAGE Publications, London. (2001).

[44] Cohen, D., Crabtree, B.: Qualitative Research Guidelines Project. Robert Wood Johnson Foundation. (2006). Available at: http://www.qualres.org/HomeSemi-3629.html.

[45] Adagunodo, T.A., Sunmonu, L.A., Oladejo, O.P., Hammed, O.S., Oyeyemi, K.D., Kayode, O.T.: Site characterization of Ayetoro Housing Scheme, Oyo, Nigeria. *IOP Conf. Ser.:* Earth Environmental Sciences, (2018). 173 012031.

[46] Alhojailan, M.I.: Thematic Analysis: A Critical Review of its Process and Evaluation. West East Journal of Social Sciences, 2012, 1(1), 39-47.

[47] Braun, V., and Clarke, V.: Using thematic analysis in psychology. Qualitative Research in Psycholology, 2006, 3(2), pp. 77-101.

[48] Kalcic, M., Prokopy, L., Frankenberger, J., Chaubey, I.: An In-depth Examination of Farmers' Perceptions of Targeting Conservation Practices. Environmental Management, 2014, 54(4).

[49] EEA: Managing municipal solid waste — a review of achievements in 32 European countries. European Environment Agency Press, 2013, 40 pp.

EFFECTS OF PARACETAMOL ON SEED GERMINATION AND ROOT ELONGATION OF *LYCOPERSICON ESCULENTUM* AND *LACTUCA SATIVA*

Ferreira, J.W.F.[1], **Nascimento, J.C.M.**[1], **Costa, J.L.**[1], **Napoleão, D.C.**[2], **Barros, K.K.**[1]

[1]*Federal University of Pernambuco, Department of Environmental and Civil Engineering, 55014-900, Av. Campina Grande, s/n - Km 59 - Nova Caruaru, Caruaru - PE, Brazil;*
johnny.wander@yahoo.com.br

[2]*Federal University of Pernambuco, Department of Chemical Engineering, Recife, Brazil;*

Keywords: Paracetamol, phytotoxicity, hormesis, seed germination

Abstract

Many Pharmaceutically Active Compounds (PhACs) enter agroecosystems during reuse of treated wastewater, biosolids and manure, presenting a potential impact on plant development. Paracetamol (PCM), one of the most frequently used pharmaceuticals was tested to explore its role on initial growing effects in crop plants. This study aimed to test effects upon germination and development of two crop species, namely tomato (*Lycopersicon esculentum*) and lettuce (*Lactuca sativa*) after exposure to different PCM concentrations (1, 10, 100, 1000 mg/L). Various growth parameters such as germination percentage, root and shoot length, germination rate index (GRI), seedling vigor index (SVI) and Phytotoxicity Index (PI) were evaluated. The results showed that no significant effects were found on germination frequency. However, significant response for shoot and root elongation was found. It was observed that 1 mg/L PCM induced stimulus on tomato growth, and the same response was verified for lettuce, when its concentration was 100 mg/L, whereas doses above stimulating values caused phytotoxic effects. Remarking hormetic effect for both crops was found. It was concluded that the different concentrations of PCM showed different responses in the species, and PCM can be classified as shoot and root elongation stressor at high concentrations (ranging from 100 to 1000 mg/L), but at concentrations that are environmentally realistic (up to 1 mg/L) it can induce stimulation or non-significant response to tomato (*L. esculentum*) and lettuce (*L. sativa*).

1 Introduction

Over the past few decades, the pharmaceutical industry has grown remarkably, developing and providing a variety of medicines that are used for human and veterinary therapeutic purposes. Pharmaceutically Active Compounds (PhACs) may be released into the environment through different sources such as improper disposal or via wastewater [1, 2]. Once consumed, it is estimated that about 50% of drugs are excreted with feces and urine in an unchanged or slightly modified way [3, 4]. Besides, if these organic contaminants eventually reach the wastewater treatment plants (WWTPs), low efficiencies are pointed out in removing these compounds [5]. Thus, PhACs are present in treated wastewater, biosolids or manure [6 - 8], which are used and studied as sources of water and nutrients in agriculture. In addition, PhACs can be transported by runoff during rainfall, reaching surface waters or infiltrate and reach groundwater [9, 10]. Due to the continuous entry into the environment, PhACs are considered contaminants of emerging concern, classified as pseudo persistent, as they are ubiquitous in several environmental matrices, whose researches on their effects and fate on agricultural ecosystems are still scarce [1].

Paracetamol (PCM; N-(4-hydroxyphenyl) acetamide) is a non-steroidal, anti-inflammatory (NSAID) chemical of intensive use, and it is the most detected PhAC in water bodies and WWTPs [11]. PCM has analgesic and antipyretic properties, and in most countries, it may be purchased freely without a prescription. Currently, PCM is one of the most widely used painkillers in the world. However, it is not an absolutely safe medicine; it can cause liver necrosis, nephrotoxicity, extrahepatic lesions and even death in humans and animals when overdosed [12, 4]. Doses lower than 0.5 mM PCM can be metabolized in conjunction with cytochrome P450, which introduces N-acetyl-p-benzoquinone imine (NAPQI) in the body system [13], which is a chemical compound capable of participating in free radical reactions [14]. Consequently, PCM may lead to a number of unfavorable consequences on target (e.g., humans and fishes) and non-target (e.g., plants and micro fauna) organisms.

PCM may potentially affect all development stages of crop plants, from germination to reproduction [15]. As example, An et al. [4] observed inhibition of root elongation of wheat seeds (*Triticum aestivum L.*), when PCM concentration reached 668.8 mg/L, followed by chronic growth inhibition also at lower concentrations (1.4–22.4 mg/L), alterations of photosynthetic pigments and soluble protein fractions as well as damages of the antioxidant defense system. PCM has been classified as harmful to (*Copepod Tisbe battagliai*) [16] and has shown toxicity to a variety of aquatic species such as unicellular algae (*Pseudokirchneriella subcapitata*), cyanobacteria (*Cylindrospermopsis raciborskii*), mycrophytes (*Lemna minor*) and crustaceans (*Daphnia magna*) [17, 18].

Critical stages that defines all development and even the crop yield is mainly determined by the early stages of ontogenesis (seed germination, root system formation and growth of young seedlings). It can be assumed that in this relatively short period, when seeds by mobilizing energy reserves (e.g., polysaccharides including starch and proteins storage forms) convert heterotrophic nutrition into autotrophic [19], young seedlings still do not have sufficient detoxification capacity or active defense mechanisms [4]. Thus, by absorbing water from the substrate, they also absorb contaminants that are solubilized in the water, which can cause toxicity, and consequently, physiological disturbances to seeds and seedlings, reducing the processes of germination and crop vegetative development, for example [15].

Generally, germination of different species is inhibited only by treatments with high PhAC concentrations (several hundreds of mg/L) [4]. In contrast, primary root elongation may be inhibited at concentrations at the order of tens of mg/L, especially when NSAIDs act in synergy [8]. Other researches also suggest the effects of medicines on biochemical and physiological processes, both qualitative and quantitative, which are reflected in biomass production. By reporting that active pharmaceutical ingredients induce hormesis in plants that is a capacity of low doses increasing plant health, while high doses may suppress plant vigor [20]. Such effects may include changes in enzymatic or non-enzymatic antioxidant defense, respiration, nutrient absorption or in the activity of plant growth regulators [15]. According to Calabrese [21], hormesis occurs in many organisms and with numerous chemical agents. This type of over compensatory response is represented by an adaptive reaction that involves additional plant activity (e.g., increased resource allocation) to deal with a stress situation. For example, Pan & Chu [22] found that the use of antibiotics Tetracycline and Sulfamethazine at 0.01 mg/L caused positive hormesis effects in the plant aerial parts and root elongation on the germination of four cultivated species: lettuce (*Lactuca sativa*), carrot (*Daucus carota*), cucumber (*Cucumis sativus*) and tomato (*Lycopersicon esculentum*).

The aims of this study were: (1) to evaluate the relative sensitivity of PCM at different concentrations in germination and shoot and root elongation during the initial growth of tomato (*Lycopersicon esculentum)* and lettuce (*Lactuca sativa)*; (2) to determine, at which concentrations PCM may be an indicator of stress; and (3) to check the response to the possibility of stimuli and rapid growth of individuals by hormesis in the presence of significant PCM dosages.

2 Materials and Methods

2.1 Materials

Paracetamol (PCM) (98-101% purity) was purchased from Sigma-Aldrich, USA; CAS No. 103-90-2, and used without further purification. The chemical and physical properties of PCM are presented in Table 1. The variety of seeds tested were obtained by Isla Sementes Ltd., in Porto Alegre, Brazil.

Table 1:   Selected chemical and physical properties of paracetamol

Chemical Formula	$CH_3CONHC_6H_4OH$
Structure	
Molecular weight (g/mol)	151.163
Water solubility (g/L)	12.7^a
log K_{ow}	0.49^b

[a] Barros, 2014. [23]; [b] ILO, 2008. [24]

2.2 Experimental design

The experiments were conducted within a randomized completely block design (RCBD) with three replications. Ten seeds of each selected crop were placed on filter papers (42 to 44 gsm) and placed in glass petri dishes with dimensions (10 cm × 2 cm), which were moistened with a volume of 5 mL (added using a Transferpette S brand, 10 mL micropipette); the aqueous solution of PCM used were prepared using distilled water at five concentrations: 0, 1, 10, 100 and 1000 mg/L. A stock solution was prepared (1000 mg/L) and diluted to set the other concentrations (100, 10, 1 mg/L) and 0 mg/L as control. After moistened, glass petri dishes were covered by the lid, wrapped by plastic wrap and kept in the culturing box (shaker incubator/ INNOVA brand) and incubated in the dark at 25 ± 2 °C.

The interspecies sensitivity issue was investigated by determining threshold responses for two plant species from different families: tomato (*Lycopersicon esculentum*) and lettuce (*Lactuca sativa*). Between them, *L. esculentum* has been shown to be particularly able to take-up contaminants, besides being one of the main wastewater irrigated crop worldwide, where its parameters are mostly studied for human and ecosystems safety and productivity in agroecosystems [25]. *L. sativa* is considered a "standard" species due to its moderate sensitivity and elevated frequency of use in phytotoxicity tests [26].

Counting and measurement campaigns were conducted daily, and the germination experiment was considered finished after 7 days, when was checked the first true leaves. Seeds were only considered germinated, when radicle reached to 2 mm of length.

2.3 End-point measurements
Germination percentage (G%) is an estimate of seeds population viability, which was calculated by the rate between the number of germinated seeds and the number of seeds sown in the substrate [27] according to Equation (1):

$$(1) \quad G \% = \frac{S_G}{S_T} \times 100$$

where S_G is the number of seeds germinated at the end of experiment, and S_T is the number of seeds sown.

Germination rate index (GRI) was based on Wang [28] (Equation 2):

$$(2) \quad GRI = \frac{G1}{T1} + \frac{G2}{T2} + \cdots + \frac{Gi}{Ti}$$

where GRI is the germination speed index (germinated seeds per day). G1 to Gi is the number of germinated seedlings occurring each day, and T1 to Ti is the period of the experiment (days).

The seedling vigor index (SVI) was calculated according to the following equation, where the higher the SVI the more vigorous is the growth (Equation 3):

$$(3) \quad SVI = SL \times G\%$$

where *SL* is seedling length in cm (root length + shoot length).

Traditional phytotoxicity bioassays rely on measuring germination and root elongation [27]. Phytotoxicity index (PI) is estimated based on root elongation according to the Equation (4):

$$(4) \quad PI = 1 - \frac{RLT}{RLC}$$

where RLT is the root length in the treated seeds and RLC the root length in the control treatment. The values of the PI range between (-1) and (1), in which a higher value means a negative (i.e., toxic) effect and a lower value a positive (i.e., stimulatory) effect.

2.4 Statistical analysis

For statistical evaluation, the software STATISTICA (StatSoft, Inc.®) was used. The results were presented as mean ± SD (standard deviation). The significance of the differences was evaluated by the one-way analysis of variance after preceding verification of data normality and homogeneity of data variance (ANOVA). Statistical significance was accepted, when the probability of the result assuming the null hypothesis (p) was less than 0.05. Tukey's test was used to identify statistically significant differences between treatments.

3 Results and Discussion

3.1 Characteristics of control and PCM solutions

The major physicochemical characteristics of the control and prepared solution of PCM prepared from stock solution are summarized in Table 2.

Table 2: Physicochemical properties of the prepared solution for experiments

PCM Concentration	TKN[a] (mg/L)	pH	EC[b] (µS/cm)	salinity (mg/L)	NO_3^- (mg/L)	PO_4^{3-} (mg/L)	SO_4^{2-} (mg/L)	NH_3-N (mg/L)
0 mg/L	0.00	8.50	2.00	0.00	0.00	0.01	3.41	0.00
1 mg/L	0.11	7.90	2.00	0.00	0.90	0.01	3.42	0.00
10 mg/L	0.17	7.89	2.50	0.00	0.90	0.01	3.37	0.00
100 mg/L	1.08	7.47	3.40	0.00	1.30	0.01	3.36	0.00
1000 mg/L	8.61	6.69	11.40	0.00	1.90	0.01	3.38	0.00

[a]Total Kjeldahl Nitrogen; [b]Electrical Conductivity

Presence of nitrogen in PCM solutions were not found in the ammonia (NH_3-N) form, thus excluding the possibility of toxic action by this compound. Moreover, comparing the results of TKN and nitrate (NO_3^-), it was observed that the greatest presence of nitrogen in the dilutions is in organic form, which could be occurred from PCM metabolites. The observed phosphate (PO_4^{3-}) and sulfate (SO_4^{2-}) concentrations indicate that these compounds are probably derived from the distilled water and not from PCM composition. The values found for EC and salinity indicate the low presence of dissolved ions. Regarding the pH, it was observed that there was a decrease over the concentration range, and despite pH influencing the germinative enzymatic activity, the values resulted from different PCM do not influence negatively seeds germination, according to Corsato et al. [29].

3.2 Effects of PCM on the germination

The statistical analysis showed that there was no significant difference in the germination frequency (G%) of tested seeds exposed to experimental concentrations of PCM. This is consistent with most of the studies on the topic, which demonstrated weak effects of NSAIDs on germination of seeds from several crops, including wheat, onions, lettuce, pea and tomato [4, 15], Despite the low effect of PCM on G%, it is important to mention that seed coats play an important role on embryo protection, as reported by Araujo & Monteiro [30]. PCM could be adsorbed by seed coats, which would not affect the growth of first roots [4]. However, it was found by statistical analysis that the concentration of PCM had significant ($P<0.05$) effects on shoot and root elongation (Figure 1).

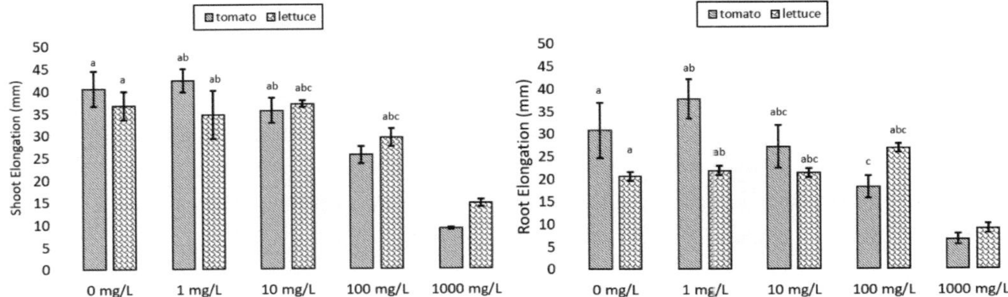

Figure 1: Bar plots of shoot and root elongation of tomato (*L. esculentum*) and lettuce (*L. sativa*) exposed to 0, 1, 10, 100, 1000 mg/L PCM concentrations. Different letters, if present, indicate significant differences between the doses of each treatment.

Few phytotoxicity studies have reported the potential effects of PhACs on crop plants [31], but within the few studies available, root length and root elongation have been used commonly as end-point measurements. Root cultures (as opposed to whole plants) have been used for detection of rapid phytotoxic responses after pharmaceutical exposure.

Results shown at Figure 1 after the 7-day exposure indicate that 1.0 mg/L PCM induced a better development in shoot and root elongation in tomato (105 ± 12%; 127 ± 33%, respectively), for lettuce roots increased (106 ± 17%) comparing to control, but these results were not statistically significant. On the contrary, the 10, 100, 1000 mg/L concentrations decreased significantly the shoot and root elongation gradually. It was verified that the concentration of 1000 mg/L were the most detrimental one to shoot and root growth and resulted in a rate decrease of 77± 3%; 78 ± 2% for tomato seeds, and 60 ± 3; 56 ± 4% for lettuce seeds, respectively. It was also indicated that the toxic effect of PCM on tomato and lettuce roots was higher than that on shoots (Figure 2).

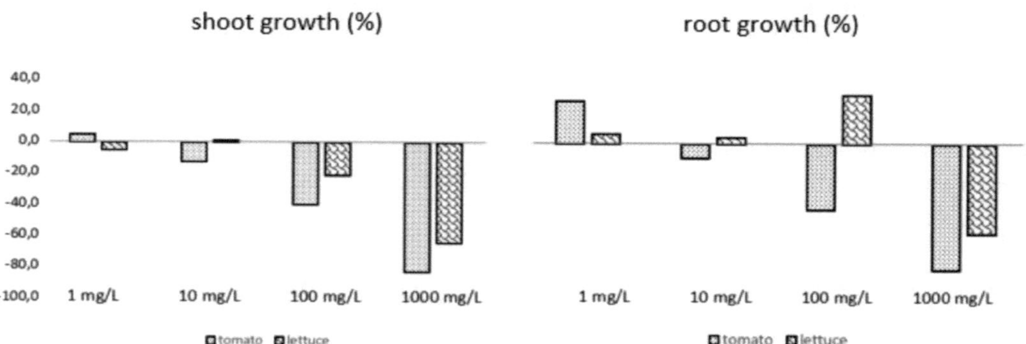

Figure 2: 7-day development data for tomato (*L. esculentum*) and lettuce (*L. sativa*), with the rate of shoot and root elongation percentages. The line represents the treatment control response. A positive (enhancement) effect of the treatment is shown by bars above the line, while negative effects (inhibition) are displayed by bars below the line.

For tomato and lettuce, there was not a significant increase in shoot growth (Figure 2). However, an increase of root growth at the end of the experiment was observed; greater root stimulation for tomato was found to occur at 1 mg/L PCM, while the other concentrations led to a decrease. For lettuce, there was no statistically significance at concentrations 1 and 10 mg/L PCM, but at 100 mg/L, a significant increase of 30 ± 8% was noted. According [32], the effects of PhACs on *L. sativa* are most beneficial on the root elongation than to shoot. In addition, Rede et al. [1] observed a similar response for *L. sativa*, where roots became thinner and longer compared with the control. Furthermore, *L. sativa* demonstrated to be particularly resistant to high dosages of PCM. Pino et al. [8] reported only inhibition of root elongation of *L. sativa*, when PCM concentrations reached up to 2.82 mg/L. Therefore, it could indicate a relatively low acute toxicity of PCM to *L. sativa*.

Regarding tomato root growth, the results showed a significant difference from those concluded by Zezulka et al. [15], who obtained an increase response at concentrations of 5 to 10 mg/L PCM by approx. 40%.

Results of the germination rate index (GRI) are depicted in Figure 3. This parameter can be also used as an indicator for phytotoxicity, were the higher values represent a more rapid rate of germination and show a better performance of germination [28].

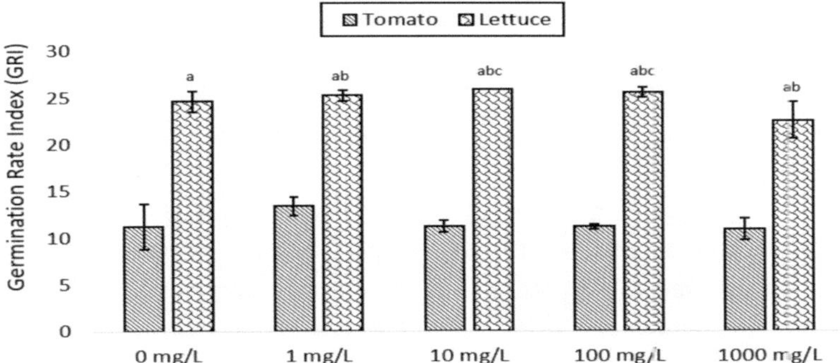

Figure 3: Bar plots of Germination Rare Index (GRI) elongation of tomato (*L. esculentum*) and lettuce (*L. sativa*) in relation to the 0, 1, 10, 100, 1000 mg/L PCM concentrations. Different letters indicate significant differences between the doses of each treatment

However, no statistical significance was found in tomato GRI. For lettuce GRI, treatments showed similar response with significant difference among treatments; but comparing with the control, they are non-significant among themselves. Which mean, PCM did not step in the daily germination speed.

The lowest Seedling Vigor Index (SVI) value was observed for the highest PCM concentration treatments (Figure 4). SVI was almost the same for lettuce until it reached 100 mg/L, after this concentration it became relatively low. For tomato, one can see this decreasing after 1 mg/L. This variation would be attributed to an adaptative reaction that involves additional plant activity, however, there is a limit that these seedlings can deal with, and when it is exceeded, it could be understood that seedlings are being poisoned.

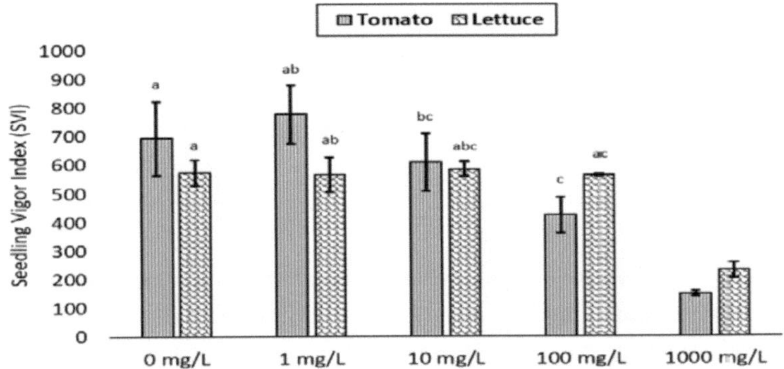

Figure 4: Bar plots of Seedling Vigor Index (GRI) where the higher the SVI the more vigorous is the growth. Tomato (*L. esculentum*) and lettuce (*L. sativa*) were tested in relation to the 0, 1, 10, 100, 1000 mg/L PCM concentrations. Different letters indicate significant differences between the doses of each treatment.

Results of the phytotoxicity index (PI) are depicted in Figure 5. Results of the study showed that the most efficient and stimulatory concentration of PCM for tomato was 1 mg/L, and for lettuce the highest efficiency occurred, when concentration was 100 mg/L. Afterwards, treatment enhanced phytotoxicity. Among the PCM concentrations studied, 1000 mg/L resulted in the highest (most toxic) PI, (0.78) for tomato and (0.56) for lettuce. Lettuce showed to be more resistant to PCM concentrations than tomato. These results are also in agreement with those reported for other plant species [4, 13, 22], suggesting that plants may be susceptible to phytotoxicity, when exposed to Paracetamol at high levels (ranging 100 to 1000 mg/L), which could mean that environmentally realistic PCM doses would not cause damage to plant development in the ontogenesis phase.

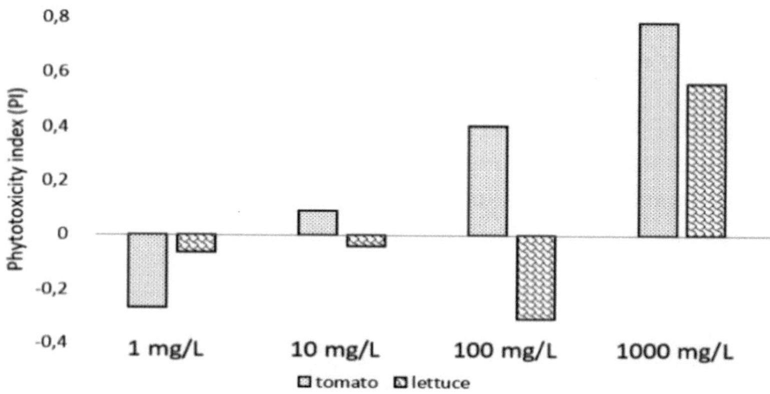

Figure 5: Phytotoxicity index (PI) for tomato (*L. esculentum*) and lettuce (*L. sativa*). The horizontal line represents the treatment control response. A positive value means a negative effect (i.e., toxicity) and a negative value means a positive effect (i.e., stimulatory).

Similar experimental phenomenon and results on other pollutants or plants had been reported by other researchers [20, 21, 25]. An et al. [4] described this singularity as a probably stress response of the plant, which can result in abnormal hormone secretion or induction of several stress proteins.

4 Conclusions

In the present study, it is concluded that different concentrations of paracetamol showed varied responses in a concentration dependent manner. And although not all tested PCM concentrations are environmentally realistic, the essay has a meaningful theoretical significance that can help to understand the toxic action and extent of PCM on the germination of plants. Out of the concentrations tested, 1.0 mg/L PCM increased growth biomarkers for tomato (*L. esculentum*) and 100 mg/L had an adverse effect on growth and development of lettuce (*L. sativa*). No rapid growth on germination was found, and higher PCM concentrations tested have shown to bring stress on shoot and root elongation.

5 Acknowledgments

The authors would like to thank EXCEED Swindon project and DAAD (German Academic Exchange Service) for support to participate at the *International Expert Workshop on Water on Agricultural Practices* held in Rio de Janeiro, Brazil on 15-21 September, 2019.

6 References

[1] Rede, D., Santos, L.H., Ramos, S., Oliva-Teles, F., Antão, C., Sousa, S.R., Delerue-Matos, C.: Individual and mixture toxicity evaluation of three pharmaceuticals to the germination and growth of *Lactuca sativa* seeds. Science of The Total Environment 2019, 673, 102-109.

[2] Singh, V., Pandey, B., Suthar, S.: Phytotoxicity of amoxicillin to the duckweed *Spirodela polyrhiza*: growth, oxidative stress, biochemical traits and antibiotic degradation. Chemosphere 2018, 201, 492–502.

[3] Heberer, T.: Occurrence, fate, and removal of pharmaceutical residues in the aquatic environment: a review of recent research data, Toxicological Letters 2002, 131, 5–17.

[4] An, J., Zhou, Q., Sun, F., Zhang, L.: Ecotoxicological effects of paracetamol on seed germination and seedling development of wheat (*Triticum aestivum* L.). Journal of Hazardous Materials 2009, 169(1-3), 751-757.

[5] Jelic, A., Gros, M., Ginebreda, A., Cespedes-Sánchez, R., Ventura, F., Petrovic, M., Barcelo, D.: Occurrence, partition and removal of pharmaceuticals in sewage water and sludge during wastewater treatment. Water Research 2011, 45(3), 1165-1176.

[6] Zorita, S., Martensson, L., Mathiasson, L.: Occurrence and removal of pharmaceuticals in a municipal sewage treatment system in the south of Sweden. Science of The Total Environment 2009, 407, 2760–2770.

[7] Martin, J., Camacho-Munoz, D., Santos, J.L., Aparicio, I., Alonso, E.: Occurrence of pharmaceutical compounds in wastewater and sludge from wastewater treatment plants: removal and ecotoxicological impact of wastewater discharges and sludge disposal. Journal of Hazardous Materials 2012, 23, 40–47

[8] Pino, M.R., Muñiz, S., Val, J., Navarro, E.: Phytotoxicity of 15 common pharmaceuticals on the germination of *Lactuca sativa* and photosynthesis of *Chlamydomonas reinhardtii*. Environmental Science and Pollution Research 2016, 23(22), 22530-22541.

[9] Hurtado, C., Parastar, H., Matamoros, V., Piña, B., Tauler, R., Bayona, J.M.: Linking the morphological and metabolomic response of *Lactuca sativa* L. exposed to emerging contaminants using GC×GC-MS and chemometric tools. Scientific reports 2017, 7, 6546.

[10] Bártíková, H., Podlipná, R., Skálová, L.: Veterinary drugs in the environment and their toxicity to plants. Chemosphere 2016, 144, 2290–2301.

[11] Kolpin, D.W., Furlong, E.T., Meyer, M.T., Thurman, E.M., Zaugg, S.D., Barber, L.B., Buxton, H.T.: Pharmaceuticals, hormones, and other organic wastewater contaminants in US streams, 1999– 2000: A national reconnaissance. Environmental Science & Technology 2002, 36(6), 1202-1211.

[12] Olaleye, M.T., Rocha, B.J.: Acetaminophen-induced liver damage in mice: effects of some medicinal plants on the oxidative defense system. Experimental and Toxicologic Pathology 2008, 59(5), 319-327.

[13] Bessems, J.G., Vermeulen, N.P.: Paracetamol (acetaminophen)-induced toxicity: molecular and biochemical mechanisms, analogues and protective approaches. Critical Reviews in Toxicology 2001, 31(1), 55-138.

[14] Moreira, J.R.M.: Intoxicações por paracetamol: metabolismo, mecanismos de toxicidade e novas abordagens da terapêutica. 2016.

[15] Zezulka, Š., Kummerová, M., Babula, P., Hájková, M., Oravec, M.: Sensitivity of physiological and biochemical endpoints in early ontogenetic stages of crops under diclofenac and paracetamol treatments. Environmental Science and Pollution Research 2019, 26(4), 3965-3979.

[16] Trombini, C., Hampel, M., Blasco, J.: Evaluation of acute effects of four pharmaceuticals and their mixtures on the copepod *Tisbe battagliai*. Chemosphere 2016, 155, 319-328.

[17] Nunes, B., Antunes, S.C., Santos, J., Martins, L., Castro, B.B.: Toxic potential of paracetamol to freshwater organisms: a headache to environmental regulators? Ecotoxicology and Environmental Safety 2014, 107, 178–185.

[18] Nunes, B., Pinto, G., Martins, L., Gonçalves, F., Antunes, S.C.: Biochemical and standard toxic effects of acetaminophen on the macrophyte species *Lemna minor* and *Lemna gibba*. Environmental Science and Pollution Research 2014, 21, 10815–10822.

[19] Avelino, A.P., de Oliveira, D.F.A., da Silva, H.A., de Macêdo, C.E.C., Voigt, E.L.: Regulation of reserve mobilisation in sunflower during late seedling establishment in continuous darkness. Plant Biology 2017, 19, 335–344.

[20] Agathokleous, E., Kitao, M., Calabrese, E.J.: Human and veterinary antibiotics induce hormesis in plants: Scientific and regulatory issues and an environmental perspective. Environment International 2018, 120, 489-495.

[21] Calabrese, E.J.: Paradigm lost, paradigm found: the re-emergence of hormesis as a fundamental dose response model in the toxicological sciences. Environmental Pollution 2005, 138(3), 378-411.

[22] Pan, M., Chu, L.M.: Phytotoxicity of veterinary antibiotics to seed germination and root elongation of crops. Ecotoxicology and Environmental Safety 2016, 126, 228-237.

[23] Barros, A.L.D.: Estudos de degradação de fármacos em meio aquoso por processos oxidativos avançados. 99 f. Tese (doutorado em química) - Universidade Federal do Ceará, Fortaleza-CE. 2014.

[24] ILO - International Labour Organization: Paracetamol. 2008. Available at: http://www.ilo.org/dyn/icsc/showcard.display?p_lang=en&p_card_id=1330&p_version=2; Accessed: 25 August 2019.

[25] Christou, A., Maratheftis, G., Eliadou, E., Michael, C., Hapeshi, E., Fatta-Kassinos, D.: Impact assessment of the reuse of two discrete treated wastewaters for the irrigation of tomato crop on the soil geochemical properties, fruit safety and crop productivity. Agriculture, Ecosystems & Environment 2014, 192, 105-114.

[26] ATSM American Society for Testing, Materials: Standard guide for conducting terrestrial plant toxicity tests. ASTM International, West Conshohocken, PA, 2003, pp 1534–1554.

[27] Rusan, M.J., Albalasmeh, A.A., Zuraiqi, S., Bashabsheh, M.: Evaluation of phytotoxicity effect of olive mill wastewater treated by different technologies on seed germination of barley (*Hordeum vulgare* L.). Environmental Science and Pollution Research 2015, 22(12), 9127-9135.

[28] Wang, Y.R., Yu, L., Nan, Z.B., Liu, Y.L.: Vigor tests used to rank seed lot quality and predict field emergence in four forage species. Crop Science 2004, 44(2), 535–54.

[29] Corsato, J.M., Fortes, A.M.T., Santorum, M., Leszczynski, R.: Efeito alelopático do extrato aquoso de folhas de girassol sobre a germinação de soja e picão-preto. Semina: Ciências Agrárias 2010, 31(2), 353-360.

[30] Araújo, A.S.F., Monteiro, R.T.R.: Plant bioassays to assess toxicity of textile sludge compost. Scientia Agricola 2005, 62(3), 286-290.

[31] Schmidt, W., Redshaw, C.H.: Evaluation of biological endpoints in crop plants after exposure to non-steroidal anti-inflammatory drugs (NSAIDs): Implications for phytotoxicological assessment of novel contaminants. Ecotoxicology and Environmental Safety 2015, 112, 212-222.

[32] D'Abrosca, B., Fiorentino, A., Izzo, A., Cefarelli, G., Pascarella, M.T., Uzzo, P., Monaco, P.: Phytotoxicity evaluation of five pharmaceutical pollutants detected in surface water on germination and growth of cultivated and spontaneous plants. Journal of Environmental Science and Health, Part A, 2008, 43(3), 285-294.

ASSESSMENT OF A WASTEWATER TREATMENT PLANT PERFORMANCE AND SUITABILITY OF TREATED WATER FOR IRRIGATION PURPOSE IN ADJOUGBA AREA: CASE OF THE BREWERY BB LOME

L. Tampo[1], A.R. Bouari[2], G. Boguido[1], N. Gnofam[2], M. Ayah[1], L.M. Bawa[1], G. Djaneye-Boundjou[1]

[1]*Laboratory of Applied Hydrology and Environment, Faculty of Sciences, Université de Lomé, BP 1515, charlestampo@gmail.com*

[2]*Brewery BB Lomé, SA society-Lomé, Togo*

Keywords: Brewery, Irrigation, Nutrients, Wastewater Treatment Plant

Abstract

One of the options for coping with water scarcity problems is not only the improvement of wastewater treatment technologies, but also opportunities for reuse of treated wastewater. This study was conducted to evaluate the performance of the wastewater treatment plant (WWTP) of a brewery located at Lomé and to present the opportunities of water reuse taking this brewery as an example. The results showed that the COD removal efficiencies ranged from 74% to 96%. Microbiological parameters showed high values for total coliforms (120,000/100 mL), but compatible with agricultural use because of the absence of *Escherichia coli* in the final effluent. Regarding the potential in irrigated agriculture, it was found that the final effluent is more suitable for irrigated agriculture with a Sodium Absorption Ratio (SAR) under 10 in comparison with groundwater. The final effluent provided a supply of nutrients (13.2 mg N/L and 7.9 mg P/L) and a potential of fertilizing elements in irrigated agriculture for crops growth. This paper includes effluent and freshwater characterization, treatment scheme and performance of a brewery effluent treatment plant. In addition to it, the problems associated with this water's suitability for irrigation purpose was discussed and suitable recommendations were made.

1 Introduction

Freshwater scarcity is a growing problem in the world and freshwater resources are becoming insufficient to satisfy the demand. Water scarcity is in most cases a climate-bound regional problem and is present all over the world, e.g., in North Africa, the Middle East, southern Europe, Australia, and the southern states of the USA [1]. If a global temperature increase of 3–4 °C is reached, changes in runoff patterns and glacial melt could force an additional 1.8 billion people to live in a water scarce environment by 2080 [2]. The United Nations estimate that globally a third of the world's population has a serious water shortage problem and this number could grow to two-thirds by 2025, if no corrective measures are taken [3]. The Food and Agriculture Organization [4] notes that millions of people are seriously affected by drought in several countries in the near East and South Asia. There is an increasing demand for freshwater resources, which is creating a situation that is unsustainable, with projections that 90% of all available freshwater could be allocated to agricultural practices by the year 2025 [5, 6]. For these reasons, over the past few decades, there has been a growing interest in the development of alternative sources of water, including used urban water and desalinized brackish water and seawater [7]. According to UNESCO-WWAP [8], more than 70% of the water that is withdrawn all over the world is used for agricultural irrigation. Therefore, there is a big potential for the application of treated wastewater in irrigation [9], despite the fact that the amount of wastewater satisfies just a fraction of the required amount of irrigation water. Notably, water reuse can be combined with nutrient reuse, particularly N and P, which are of obvious importance for agricultural production. The use of treated wastewater in agriculture has been practiced for centuries [10], often compared with effluent disposal in cities such as Berlin, London, Milan and Paris [11]. Indeed, the practice of irrigation with wastewater is old and worldwide. Already in the late 1800 and early 1900, there existed sewage farms throughout Europe, Australia, Latin America and the USA. In the outskirts of Paris, 5,300 ha of land were sewage-irrigated land, and in Berlin 17,200 ha. The interest during that era was mainly due to the ambition to keep the rivers free from fecal contamination. With technical development, better treatment systems and an increasing awareness of the importance of microbes in disease transmission, wastewater irrigation fell out of fashion. After World War II, the practice once again gained attention, not only to prevent river pollution, but also as a way to come to terms with the worlds increasing water demand [12]. Nowadays, the use of treated wastewater in agriculture is usually planned and intended to be without environmental or human risks. Much effort is made to make use of wastewater and it is looked upon as a valuable resource. Therefore, wastewater reuse, when appropriately applied, is considered as an example of Environmental Sound Technology [13] and has various benefits. First, recycled wastewater can serve as a more dependable water source, containing useful substances such as organic carbon and nutrients for some applications. The use of nutrient-rich water for agriculture and landscaping may lead to a reduction or elimination of application of chemical fertilizers. Second, wastewater reuse leads to reduced treatment needs, which results in a cost savings. Finally, by reusing treated wastewater, more freshwater can be allocated for uses that require higher quality, thereby contributing to more sustainable resources utilization [14].

About two-thirds of urban wastewater generated in the world receives no treatment before discharge to a receiving water body, and this can be a particular problem in developing countries [15]. The costs of providing conventional (up to tertiary) wastewater treatment cannot be supported by many cities, particularly in the developing world. An alternative to the disposal of untreated wastewater in surface water is to reuse the treated wastewater for agriculture. In this way, wastewater may be seen as a resource, which provides an opportunity for increasing food security in rapidly growing urban areas. Furthermore, Wichelns et al. [16] noted that in developing countries reuse of untreated wastewater by small farmers is frequently done informally (and even unintentionally), which increases health risks to both the farmer and the consumer. An organized and regulated system of water treatment and reuse could result in economic and environmental benefits and concurrently reduce health risks. Clearly, then, treated wastewater may be considered as "new" water resource, which can add to the general water balance of a region. This "new" source can be an effective substitute for freshwater used in irrigation and should be considered as an integral component of local water resources [17].

Regarding the beer brewing, very large quantities of water are spent during the process. It is estimated that for the production of 1 L of beer, 3–10 L of waste effluent is generated, depending on the production process and the specific water usage [18-20]. As beer is the fifth most consumed beverage in the world behind tea, carbonates, milk and coffee [19], a formal and regulated wastewater reuse can put brewery treated effluent at the scale of a "new" water resource or "renewable" water resource for the supply of freshwater used in irrigation. The economic appraisal of case studies in Spain and Mexico shows that there is potential for "win–win" arrangements among cities, farmers and the environment involving the use of reclaimed water [21].

But formal and regulated wastewater reuse has not received much attention in Togo, and wastewater is used indirectly in agriculture by few farmers in Zio River basin [22] because of the lack of information about the potential for wastewater reuse in agriculture as a climate change adaptation measure, as fertilizer supply for farmers and as a new water resource [6, 15]. Indeed, treated wastewater provides farmers renewable nutrients, reduces pollution of the environment and eases demand for freshwater [23, 24]. It is demonstrated that in Mexico, farmers save on artificial fertilizers; this saving is around US$ 135/ha/year, which is a significant amount of money for farmer's subsistence [15, 23]. The experiments at the Cuu Long Rice Research Institute [25] found that irrigation with wastewater from catfish farming increased rice yield by 1 t/ha and large amounts of nitrogen and phosphorus fertilizers were saved in Vietnam.

The most important concerns in the agricultural use of treated urban waters are related to the human and environmental health aspects, in other words, the quality and safety of the produced food [10] and the health concerns of agricultural workers. Although the potential health and environmental risks related to wastewater reuse are well known in general terms [26-28], and the economic costs and benefits are analyzed and acknowledged in various studies [21, 29-32], no effort has been made to integrate these key aspects into an analytical framework, which provides useful information to policy makers in countries like Togo, faced with the challenge of developing climate change adaptation strategies. In addition, there are no studies and information about wastewater treatment plant (WWTP), performance of few existing WWTPs, quality of treated wastewater and suitability for irrigated agriculture in Togo. Accordingly, the objective of this study was to measure the performance of a brewery effluent treatment plant and to characterize the suitability of the treated wastewater for irrigation purpose as climate change adaptation strategy.

2 Material and Methods

2.1 Wastewater Treatment Plant (WWTP)

The Brewery of Lomé is a Share Company located in Lomé town in the south of Togo. It is found in the plain altitude at 30 m, 6°16'55,9''N ; 1°12'34''E.

All effluents coming from the different production units of the brewery are collected in a collector tank. The raw wastewater collected follows pretreatment, primary and secondary treatment processes. The purpose of preliminary treatment is the removal of coarse solids and other large materials, found in raw wastewater. It includes screening, separation/flotation and equalization processes. The screening operation is used to eliminate the gross solids present in raw wastewater. Screening includes coarse and fine screening, mechanically operated to intercept floating and suspended debris. Separation/flotation is applied to separate the particles in suspension, such as oil, fat and other light substances dissolved or emulsified. The equalization unit is composed of a tank set downstream the preliminary treatments. This enables to attenuate the peaks in terms of flow rate, polluting load, pH and temperature, thus ensuring the supply of a constant flow rate to the subsequent process and improving the performance of the treatment. The primary treatment is performed only by sedimentation of suspended materials in an intermediate tank, incorporated in equalization unit, where they settle before the secondary treatment system. The secondary treatment is performed in an anaerobic tank (reactor/digester) and finally in an aeration tank to avoid or to reduce odor of reactor effluents. The aeration tank effluent (final effluent) is conducted by pipe line and discharged in Zio River (Figure 1).

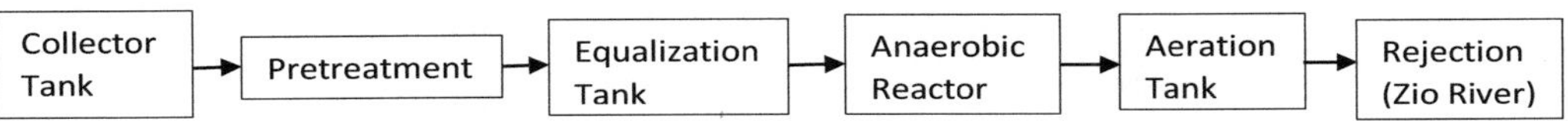

Figure 1: Wastewater Treatment Plant of the Brewery

2.2 Sample collection

For WWTP performance evaluation, samples from the equalization tank effluent and the reactor effluent were analyzed daily from January 1st to August 1st, 2019. The parameters used to evaluate WWTP performance were Chemical Oxygen Demand (COD), Total Suspended Solids (TSS), Volatile Fatty Acids (VFA), Total Nitrogen (TN) and Total Phosphorus (TP).

The samples for the characterization of treated wastewater for irrigation purpose concerned only the final effluent and were analyzed monthly during the same period. For this purpose, samples were taken at outlet point without mixing with surface water (E_0), and physicochemical and bacteriological analyses were performed in the Laboratory of Water Chemistry. Data from samples of surface water and groundwater at the vicinities of the outlet in Adjougba area in Zio River basin were also collected from 2014 to 2019. These data concerned only physicochemical and some bacteriological parameters of borehole and surface waters of the Zio River basin (Figure 2).

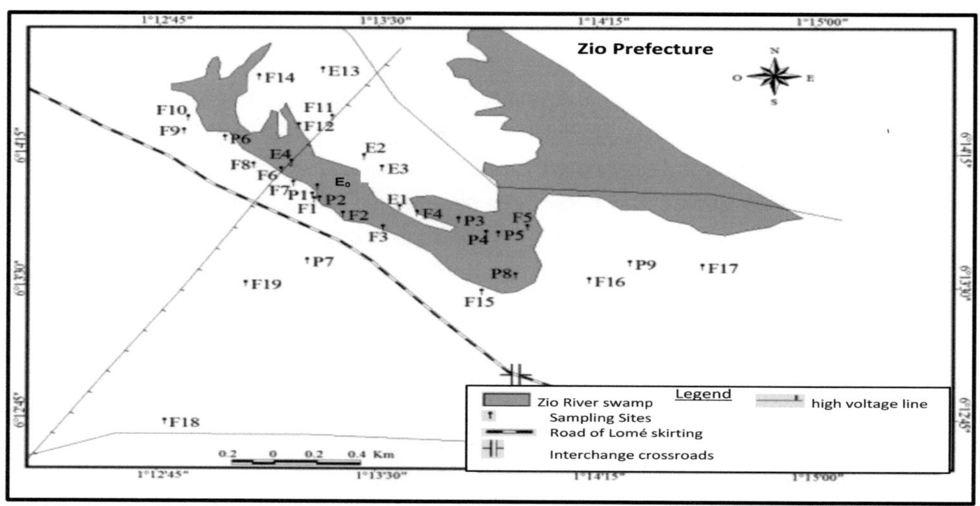

Figure 2: Localization of sampling sites of surface water and groundwater

2.3 Physicochemical analyses

In the field, parameters such as pH, Electrical Conductivity (EC), and temperature were measured using a multiparameter apparatus. For the other parameters, water was sampled in a plastic bottle for further analyses in the laboratory. Samples for COD were preserved by adding 2 mL of concentrated nitric acid to prevent growth of algae. In the laboratory, HCO_3^-, Mg^{2+}, Ca^{2+} and Cl^- were determined by titration methods using NTF 97 of AFNOR methods. K^+ and Na^+ were determined by Flame Emission Spectrophotometer. TSS was determined by gravimetric method (dried at 105 °C), TN was determined titrimetrically following the Kjeldahl's method (digestion, distillation and titration) as described by AFNOR methods. BOD was determined by the 5 d BOD test, while COD was determined in the laboratory following method described by AFNOR using potassium dichromate. All parameters were measured in the Laboratory of Applied Hydrology and Environment of Université de Lomé (Togo) with an accuracy ranking from 1% to 2% according to the standard methods as prescribed by AFNOR [33].

2.4 Bacteriological analyses

The total germs were counted on standard agar for enumeration using the bulk seed technique. The total coliforms were also counted by the bulk seed technique. For the count of thermotolerant coliforms, the method is the same as for total coliforms, except that the incubation was at 44 °C. The Fecal Streptococci enumeration involves two steps: presumption and confirmation. The Sulfito-Reducing Bacteria (SRB) were enumerated by the bulk seeding technique using Tryptone Sulphite Neomycin (TSN) tube agar. The growth and enumeration process of all the germs were followed AFNOR [33].

2.5 WWTP Performance

As the main treatment process takes place in an anaerobic reactor, WWTP performance was measured firstly in this unit and secondly for the entire WWTP system by calculating the removal efficiency (RE) of pollutants load using formula (1).

$$RE = \frac{CiQi - CoQo}{CiQi} \times 100 \qquad (1)$$

Where C_iQ_i and C_oQ_o are the inflow and outflow of the same pollutant load for a given treatment unit. Ci and Co, respectively, are concentration of a given pollutant in effluent of a treatment unit at entry and exit.

2.6 Suitability for irrigation purpose

The main physicochemical and bacteriological parameters were considered to characterize the final effluent quality and to assess the suitability of these waters used for irrigation. To assess and to classify water quality for irrigation, there are most popular criteria, such as TSS, EC, Sodium Adsorption Ratio (SAR), Permeability Index (PI), concentration of elements like Na^+, Cl^- and bacteriological features among other criteria.

According to Richards [34] and Frenkel et al. [35], sodium adsorption ratio (SAR) is expressed as:

$$SAR = \frac{Na^+}{\sqrt{\frac{Ca^{2+} + Mg^{2+}}{2}}} \qquad (2)$$

Doneen [36], defined permeability index (PI) as:

$$PI = \frac{Na^+ + \sqrt{HCO_3^-}}{Ca^{2+} + Mg^{2+} + Na^+} \times 100 \qquad (3)$$

Another indicator that can be used to specify the Magnesium Hazard (MH) for irrigation water is Magnesium Adsorption Ratio (MAR) or Magnesium Hazard (MH) that was calculated according to Raghunath [37] as:

$$MAR = \frac{Mg^{2+}}{Ca^{2+} + Mg^{2+}} \times 100 \qquad (4)$$

with all ion concentrations given in meq/L.

3 Results and Discussion

3.1 Assessment of WWTP performance

Based on monitoring samples from January 1st to August 1st, 2019, performance of anaerobic reactor (Figure 3A) and whole WWTP (Figure 3B) using removal efficiency of pollutants were assessed. Figure 3 indicates that the removal efficiency in the anaerobic reactor is changeable during the monitoring time, but this removal efficiency of all pollutants is over 70% for 50% of monitoring samples (see median values of each pollutant). Except the TSS, the minimum removal efficiency of the other parameters in the anaerobic reactor is over 50%. The VFA are eliminated from 80% to 98% between the preliminary treatment and the secondary treatment (anaerobic reactor). This pollutant is eliminated at nearly the totality after anaerobic treatment. This result means that anaerobic effluent does not need other treatments for VFA removal. The removal of organic matters evaluated by COD showed removal efficiency between 80% and 95% with median over 90%. Nutrient removal ranged from 75% to 91% and from 55% to 85% for TN and TP, respectively. TSS are the less eliminated pollutants in the reactor with an efficiency ranging from 33% to 95%. This variability of efficiency removal in anaerobic reactor with some atypical and extreme values can be explained by few disturbance of anaerobic digester occurring some times. The range of pH variation (Table 1) during the monitoring period can induce this disturbance of anaerobic digestion [38]. This shows the importance of hourly or daily control and monitoring of wastewater treatment units using some key parameters. Indeed, the purpose of biological treatment is to remove dissolved organic matter from wastewater and needs a regular testing of water from different units of the plant, which is evidently important to ensure its optimal functioning and to prevent any local problems in the system [39, 40]. A regular regime of monitoring is essential, if good quality effluents are to be achieved and maintained.

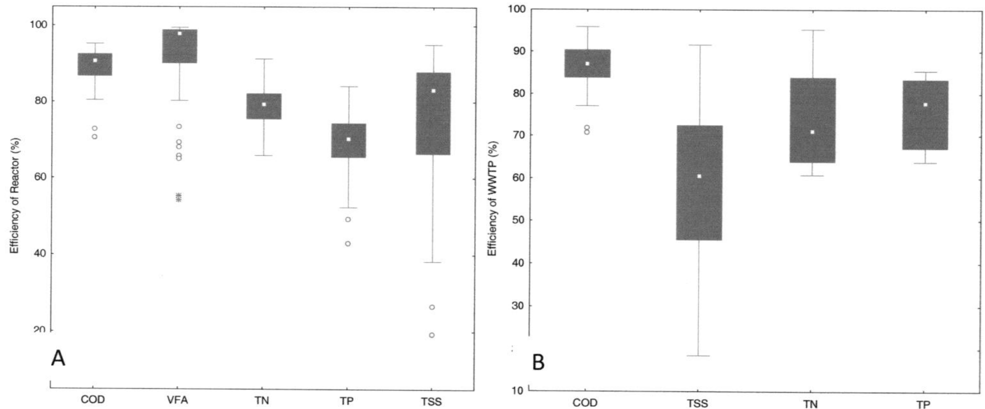

Figure 3: Removal efficiency of COD, VFA, TN, TP, TSS of (A) reactor effluent, (B) WWTP effluent

As the totality of VFA was nearly eliminated in the reactor, assessment of WWTP performance was only concerned of four components: COD, TN, TP and TSS. Figure 3 indicates the removal efficiency of the whole WWTP. For COD, the removal efficiency varied from 78% to 95% with few atypical values. TN and TP are removed over 60% with a median around 70%. TSS are the less removed in this wastewater treatment plant ranging from 18.5% to 91.4% with a median around 60%. The four components are removed over 60% in aeration tank effluent (final effluent), and these results indicate that aeration tank does not remove these components but contributes to reduce odor of effluent before rejection into the Zio River.

3.2 Quality of treated wastewater

3.2.1 Bacteriological quality

The bacteriological quality is often evaluated by some indicator organisms such as the Total Coliforms group, which belongs to the family of enterobacteriaceae and includes the aerobic and facultative anaerobic, gram-negative, non-spore-forming, rod-shaped bacteria that ferment lactose with gas production within 48 h at 35 °C [41, 42]. This group includes *Escherichia coli*, a Fecal Coliform which is one of the best indicators of treatment efficiency of the plant in water treatment plants. For example, this group has been found useful for assessing the safety of reclaimed wastewater in the Windhoek reclamation plant in Namibia [43]. Other indicator organisms are Fecal Streptococci, commonly inhabit the intestinal tract of humans and warm-blooded animals. They are used to detect fecal contamination in water and have been suggested as useful for indicating the presence of viruses, particularly in biosolids and seawater. The main anaerobic bacteria that have been considered as indicators are Sulfite Reducing Bacteria such as *Clostridium Perfringens*, mostly opportunistic pathogens, but are also implicated in human diseases. The hardy spores make this bacterium too resistant to be useful as an indicator organism. Many regulations consider more specifically *C. perfringens* as the indicator of choice, because they are indicators of past pollution and as a tracer to follow the fate of pathogens and suitable indicators for viruses and protozoan cysts in water treatment plants [44]. According to Table 1, the bacteriological quality of the effluent is affected by the presence of Total coliforms and Fecal coliforms. But the absence of Fecal Streptococci, *Escherichia coli* and sulfite reducing bacteria such as *Clostridium perfringens* indicates that the effluent is suitable for irrigation purpose and can be used for irrigated agriculture without pathogenic contamination of crops.

3.2.2 Chemical quality

Regarding biodegradability or natural purification of an effluent, a ratio of COD/BOD ranging from 1.5 to 2 often indicates an easy biodegradability. As the ratio of the final effluent range from 1.29 to 2.1, these results suggest that the most remaining organic materials can be easily degraded by biological processes or natural purifications. Thus, the rejection of these effluents cannot disturb the natural purification capacity of Zio River in the normal conditions.

Nitrogen is a necessary primary macronutrient for plants that stimulates plant growth and is usually added as fertilizer but can also be found in wastewater as nitrate, ammonium, organic nitrogen, or nitrite. The most important factor for plants is the total amount of nitrogen (TN), regardless of whether it is in the form of nitrate, ammonium, or organic nitrogen. The concentration of nitrogen required varies according to the crop with more sensitive crops being affected by nitrogen concentrations above 5 mg/L, whilst most other crops are relatively unaffected until nitrogen exceeds 30 mg/L [45, 46]. In this study, TN of the final effluent ranged from 25 to 58.4 mg/L with a median of 35.5 mg/L (Table 1). These values are close to those found in mixed wastewater in some developing countries such as in India (28.5-73 mg/L), and some developed countries like USA (20-85 mg/L) and France (30-100 mg/L) [47].

Phosphorus is also a primary macronutrient that is essential for plant growth. Municipal wastewaters may contain between 4 and 16 mg P/L [48]. Wastewater with 6-20 mg P/L increases the productivity of crops, and when the concentration exceeds 20 mg P/L, the availability of copper, iron, and zinc is reduced in alkaline soils [49]. In this study, TP concentrations ranged from 7.13 to 24.2 mg P/L with a median of 16.1 mg P/L. These values of TP concentration are in the range of medium concentration according to classification of typical wastewater composition and suggest the suitability of these waters for irrigation purposes [15, 46]. The values in this study are close to characteristics of wastewater from many breweries according to review studies conducted by Rao et al. [50] and Geoffrey et al. [51]. During the monitoring period, the final effluent contained up to 58.4 mg N/L (13.2 mg N/L) and 24.2 mg P/L (7.9 mg P/L) of TN and TP, respectively. These concentrations represent a potential fertilizer supply for crops irrigated with this treated wastewater in developing countries such as Togo, where chemical fertilizers are often scarce.

Table 1: Summary of parameters and indices of irrigation suitability of treated wastewater

Parameter	Mean	Median	Min	Max	25 P	75 P	SD
TC (cfu/100 mL)	14.8	102	0.00	120,000	3.50	191	41.7
FC (cfu/100 mL)	1,204	61.5	0.00	96,000	1.00	100	33.9.
FS (cfu/100 mL)	0.00	0.00	0.00	0.00	0.00	0.00	-
E. c. (cfu/100 mL)	0.00	0.00	0.00	0.00	0.00	0.00	-
SRB (cfu/20 mL)	0.00	0.00	0.00	0.00	0.00	0.00	-
COD (mg/L)	130	131	78	210	104	155	37.9
BOD (mg/L)	92.1	98.8	46	148	65	107	33.3
COD/BOD	1.67	1.57	1.29	2.10	1.50	1.80	0.25
TN (mg/L)	38.7	35.5	25.0	58.4	29.4	49.2	11.8
TP (mg/L)	18.0	16.1	7.13	24.2	14.8	20.1	2.61
pH	8.10	8.28	7.15	8.8	7.65	8.5	0.55
SAR	3.57	3.71	1.99	5.10	2.62	4.64	1.15
PI (%)	101	102	80	115	97	104	10.4
MAR (%)	51.9	50	40	66	45	55	9.01
EC (µS/Cm)	1,851	1,874	1,670	2,025	1,730	1952	134
Cl (mg/L)	316	322	281	346	299	329	22.3

3.3 Suitability of treated wastewater for irrigation purpose
The values of pH varied from 7.15 to 8.8 with an average value of 8.10, which indicates that after the treatment process the nature of wastewater is alkaline but compatible with irrigation purpose, although few values are over FAO guidelines for irrigation, since the permissible pH range for irrigation waters is 6.5–8.4 [15, 45]. However, irrigation water with a pH outside the normal range may cause a nutritional imbalance or may contain toxic ions.

Electrical Conductivity (CE) is among the most relevant and recommended parameters that needs to be regularly checked during the irrigation season. EC values of monitoring samples ranged from 1,670 to 2,025 µS/cm with a median of 1,874 µS/cm, expressing a slight to moderate degree of restriction on the use of this wastewater in irrigation due to salt build-up in soils and its adverse effects on plant growth. For safe irrigation, the EC should be less than 750 µS/cm at 20 °C [15], but it is reported that irrigation water with EC in the range of 750–2,250 µS/cm is permissible for irrigation and widely used without significant effects on soil or plant growth. Wastewater with EC over 750 µS/cm can be used for irrigation purpose, when the concentration of sodium is about 180 mg/L [24]. Indeed, EC measures the abundancy of ions or salts in a given water, but the most effect of high EC is caused by sodium. In irrigation water, EC must be analyzed with SAR values before making a decision about the use of a given water for irrigation. For example, water with an EC ranging from 700 µS/cm to 3,000 µS/cm is suitable for irrigation purpose without any restriction, if the Sodium Adsorption Ratio (SAR) is < 10. Therefore, sodium hazard is usually expressed in terms of SAR and can be calculated from the ratio of sodium to calcium and magnesium.

SAR is an important parameter for the determination of the suitability of irrigation water, because it is responsible for the sodium hazard, since it is more closely related to exchangeable sodium percentages in the soil than the simpler sodium percentage. Sodium replacing adsorbed calcium and magnesium is a hazard, as it causes damage to the soil structure. In this study, SAR varied from 1.99 to 4.64 (Table 1), and these values indicates that treated wastewater from the WWTP is suitable for irrigation. When the Magnesium Hazard (MH), evaluated following the formula (5), is less than 50, the water is safe and suitable for irrigation. Magnesium Absorption Ratio in this study varied from 40% to 65%, and the treated wastewater can be classified with few exceptions as suitable for irrigation use. The Permeability Index in this study ranged from 80 to 115. According to this index, water with PI > 75 is suitable for irrigation. Therefore, the treated wastewater can be used without damage of soil permeability, if other phenomena do not affect soil structure. Regarding Cl^-, it is not adsorbed or held back by soils, therefore, it moves readily with the soil-water, is taken up by the crop, moves in the transpiration stream, and accumulates in the leaves. If the Cl^- concentration in the leaves exceeds the tolerance of the crop, injury symptoms develop such as leaf burn or drying of leaf tissue. The result about it (Table 1) suggests that this treated wastewater can be used in irrigation with a few restrictions for some crops, if this water is used for a long time in the same soil.

3.4 Suitability of treated wastewater for irrigation in comparison with surface and groundwater
The Figure 3A, 3B, 3C, and 3D show a summary of suitability indices for irrigation purpose in treated wastewater (TWW), surface water (SW) and groundwater (GW) (Well Water and borehole water) in the vicinity of the wastewater outlet. This suitability or water quality for irrigation purpose is often assessed in relation with soil structure and permeability, and is based mainly on EC and sodium content relative to the concentrations of the other major cations like calcium and magnesium, and on bicarbonate content in the water [24, 35]. For example, extremely high salinities in irrigation waters have several adverse effects on both the irrigation soil and the crops being irrigated [52]. Generally Ca^{2+} and Mg^{2+} maintain a state of equilibrium in most waters. During

equilibrium, more Mg^{2+} in water will adversely affect the soil quality, rendering it alkaline with a decrease of crop yield [53]. High concentrations of Ca^{2+} and Mg^{2+} ions in irrigation water can increase soil pH, resulting in reducing the availability of phosphorus [54]. Irrigation water containing Ca^{2+} and Mg^{2+} higher than 10 meq/L (200 mg/L) and with MAR over 50% cannot be used in agriculture [55]. According to this index, about 90% of surface water samples, 60% of groundwater and 70% of treated wastewater are suitable for irrigation in Adjougba area. Total salt concentration and probable sodium hazard of the irrigation water are the two major constituents for determining SAR. If water used for irrigation is high in Na^+ and low in Ca^{2+}, the ion-exchange complex may become saturated with Na^+, which destroys the soil structure due to the dispersion of clay particles [56] and reduces the plant growth. There is a close relationship between SAR values in irrigation water and the extent of Na^+, which is absorbed. According to SAR threshold values, all treated wastewater and surface water samples are suitable for irrigation. In ground-water, about 99% of samples are suitable for irrigation considering SAR despite of their high Na^+ content.

The permeability of soil is influenced by sodium, calcium, magnesium and bicarbonate contents. Doneen [36] has evolved a criterion for assessing the suitability of water for irrigation based on PI. The PI of all surface water and treated wastewater samples were over 100% except few samples, while the PI groundwater samples were under 80%. These results are near to those of [57] in deep aquifer in southwestern zone of Bangladesh.

EC, the most important parameter to demarcate salinity hazard and suitability of water for irrigation purpose, show low values with 80% of samples under 1000 µS/cm. EC of treated wastewater with a median value around 2000 µS/cm is lower than of groundwater. According to [58] classification, surface water and treated wastewater samples ranged from good to excellent classes of the most indices of suitability. Groundwater samples with high salinity linked to sea water intrusion [59] do not meet permissible conditions for some samples and some indices.

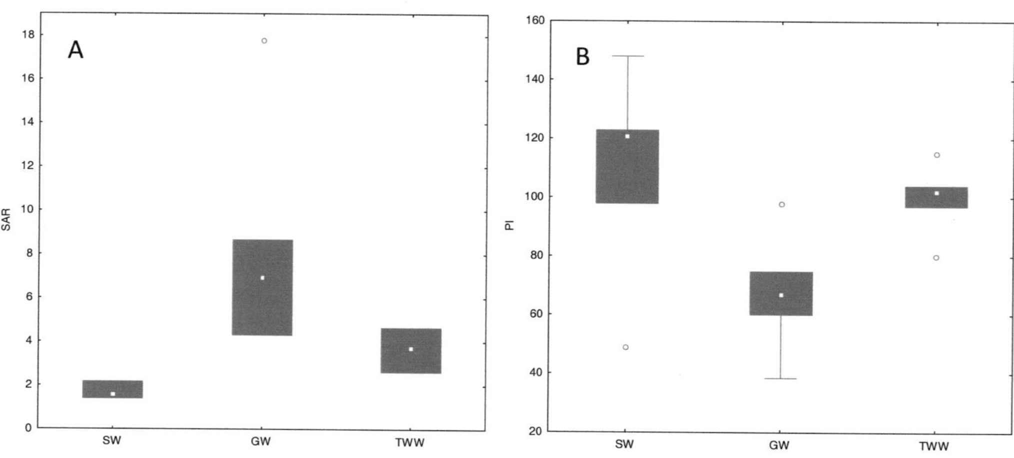

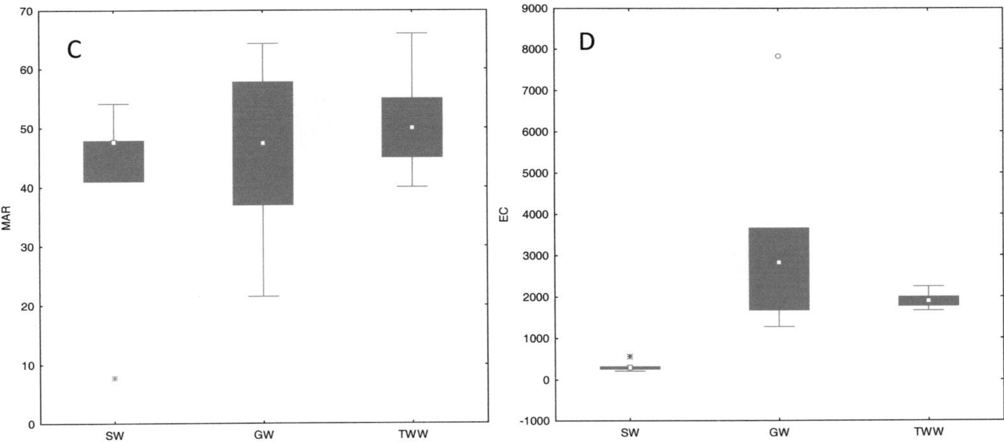

Figure 3: Variation of indices of suitability for irrigation in groundwater (GW) surface water (SW) and treated wastewater (TWW)

4 Conclusions and Perspectives

The results in this study showed that this WWTP removes VFA over 90%, COD over 80% and TN and TP over 60%. The absence of *E. coli* and *C. perfringens*, the ratio of COD/BOD indicating biodegradability of matters content in treated wastewater, and values of TN and TP in the final effluent indicate that the treated wastewater can be reused as supply of fertilizers. The indices of suitability for irrigation purpose showed that the final effluent is suitable for irrigated agriculture. But an additional treatment with an aeration process is recommended in order to improve the final effluent quality without restriction in irrigated agriculture. About surface water and groundwater of vicinity of the outlet, water quality indices for irrigation purpose indicate that surface water are suitable at 90% of samples without restriction, but groundwater samples are suitable at about 50% of samples because of sea water intrusion.

5 Acknowledgements

Authors thank the Exceed Swindon project and DAAD for taking charge of participation of the first author for experts workshop on "Water on Agricultural Practices: Training the Trainers" from 15 – 21 September 2019. The authors also thank the Engineers of Laboratory of Applied Hydrology and Environment, Université de Lomé for their help for the analyses of samples and field work.

6 References

[1]	Norton-Brandão, D., Scherrenberg, S.M., Van Lier, J.B.: Reclamation of used urban waters for irrigation purposes: A review of treatment technologies. Journal of Environmental Management 2013, 122, 85-98

[2]	UNDP: Fighting Climate Change: Human Solidarity in a Divided World. Human Development Report 2007/2008, 2007, New York.

[3]   McCarthy, M.: Water Scarcity Could Affect Billions: Is This the Biggest Crisis of All? Common Dreams News Center, September 2003; www. Arizonaenergy.org

[4]   FAO: Special Alert No 308: Millions of People Seriously Affected by Drought in Several Countries in the Near East and South Asia. FAO Global Information and Early Warning System on Food and Agriculture, 2000; www.fao.org

[5]   Salem, F.: Water sustainability – a national security issue for the Middle East and North Africa region. In: Paper presented at the 2nd International Water Conference in the Arab Countries. July 7–10, 2003

[6]   Trinh, L.T., Duong, C.C., Steen, P.V.D., Lens, P.N.L.: Exploring the potential for wastewater reuse in agriculture as a climate change adaptation measure for Can Tho City. Vietnam. Agricultural Water Management 2013, 128, 43–54

[7]   Bixio, D., Wintgens, T.: Water Reuse System Management Manual AQUAREC. Directorate-General for Research. European Commission Brussels 2006, Belgium, p 25, www.publications. europa.eu

[8]   UNESCO-WWAP: Water for People Water for Life: The United Nations World Water. Development Report. UNESCO and Berghahn Books 2003, New York, p. 145

[9]   Meda, A., Cornel, P.: Aerated biofilter with seasonally varied operation modes for the production of irrigation water. Water Science and Technology 2010, 61(5), 1173-1181

[10]   Pedrero, F., Kalavrouziotis, I., Alarcon, J.J., Koukoulakis, P., Asano, T.: Use of treated municipal wastewater in irrigated agriculture. A review of some practices in Spain and Greece. Agricultural Water Management 2010, 97(9), 1233-1241.

[11]   AATSE (Australian Academy of Technological Sciences and Engineering): Water Recycling in Australia. AATSE, 2004, Victoria, Australia, 205 p. www.applied.org.au

[12]   Shuval, H.I., Adin, A., Fattal, B., Rawitz, E., Yekutiel, P.: Wastewater Irrigation in Developing Countries: Health effects and technical solutions. Summary of World Band Technical Paper No. 51. Integrated Resource Recovery Series GLO/80/004 Number 6. The World Bank. Washington D.C. USA, 1986.

[13]   UNEP: Newsletter and Technical Publications, International Source Book on Environmentally Sound Technologies for Wastewater and Storm water Management, 2000; Available on: http://www.unep.or.jp/Ietc/Publications/TechPublications/TechPub-15/3-3AsiaPacific/3-4.asp *(accessed 2014).*

[14]   UNEP: Water and Wastewater Reuse – An Environmentally Sound Approach for Sustainable Urban Water Management, 2005; Available on: *(accessed 2014).* http://www.unep.or.jp/ietc/publications/water sanitation/wastewater reuse/index.asp

[15]   Mara, D.D.: Domestic Wastewater Treatment in Developing Countries. Earthscan Publ., 2004, London. England. ISBN 1844070190

[16] Wichelns, D., Owaygen, M., Redwood, M.: Developing country farmers need more than financial incentives to reduce the risks of wastewater irrigation. Water International 2011, 36(4), 467–475

[17] Scheierling, S.M., Bartone, C.R., Mara, D.D., Drechsel, P.: Towards an agenda for improving wastewater use in agriculture. Water International 2011, 36(4), 420–440

[18] Braeken, L., Van der Bruggen, B., Vandecasteele, C.: Regeneration of brewery waste water using nanofiltration. Water Research 2004, 38(13), 3075–3082.

[19] Fillaudeau, L., Blanpain-Avet, P., Daufin, G.: Water, wastewater and waste management in brewing industries. Journal of Cleaner Production 2006, 14, 463–471.

[20] Kanagachandran, K., Jayerantene, R.: Utilisation potential of brewery wastewater sludge as an organic fertilizer. Journal of the Institute of Brewing, 2006, 112(2), 92–96.

[21] Heinz, I., Salgot, M., Koo-Oshima, S.: Water reclamation and intersectoral water transfer between agriculture and cities. An FAO economic wastewater study. Water Science and Technology 2011, 63(5), 1067-1073.

[22] Tampo, L., Boguido, G., Ayah, M., Gnazou, D.T., M., Tchakala, I., Lazar, I.M., Djaneye-Boundjou, G., Bawa, L.M.: Assessment of Water Quality and its Suitability for Domestic and Irrigation Purposes: The case of Freshwater and Lomé Wastewater in Adjougba District (Togo), In: C. Kowenje. S. Paré, B. Asfaw (Eds.). Expert Workshop on Water Security. May 15th -20th, 2017. 2017, 105-122, Cuvillier Verlag, Göttingen, Germany.

[23] Archer, H.E., Mara, D.D.: Waste Stabilization Pond Developments in New Zealand. Water Science and Technology 2003, 48(2), 9–16

[24] Wakuma, B., Fita, T., Assaba, B.: Characterization of Brewery Wastewater for Irrigation Purpose (In case of Bedelle Brewery Share Company. Illubabor Zone. South-West Ethiopia). International Journal of Modern Chemistry and Applied Science 2017, 4(1), 485-491

[25] Cao, V.P., Nguyen, B.P., Hoang, T.K., Bell, R.W.: Using wastewater from catfish ponds for rice irrigation to reduce environmental pollution caused by catfish farming in the Mekong Delta. Technical Report (in Vietnamese). Cuu Long Rice Research Institute. O'Mon. Cantho Province, 2010, Vietnam, http://researchrepository.murdoch.edu.au

[26] WHO-FAO-UNEP: Guidelines for the Safe Use of Wastewater. Excreta and Greywater. Wastewater Use in Agriculture, Geneva, Switzerland, 2006, Vol. 2

[27] Asano, T.: The Role of Water Reuse in Water Resource Management. Workshop 4: Improved Water Use Efficiency through Recycling and Reuse. World Water Week. 2010, Stockholm.

[28] IWMI-IDRC: Wastewater Irrigation and Health: Assessing and Mitigating Risk in Low Income Countries. Earthscan, London, 2009.

[29] Danso, G., Drechsel, P., Wiafe-Antwi, T., Gyiele, L.: Income of farming systems around Kumasi. Ghana. Urban Agriculture Magazine 2002, 7, 5–6

[30] Asano, T., Burton, F.L., Leverenz, H.L., Tsuchihashi, R., Tchobanoglous, G.: Water Reuse. Issues. Technologies and Applications, McGraw Hill, New York, 2007, ISBN 9780071459273

[31] Hanseok, J., Hakkwan, K., Taeil, J.: Irrigation Water Quality Standards for Indirect Wastewater Reuse in Agriculture: A Contribution toward Sustainable Wastewater Reuse in South Korea. Water 2016, 8, 169

[32] Weldesilassie, A.B., Amerasinghe, P., Danso, G.: Assessing the empirical challenges of evaluating the benefits and risks of irrigating with wastewater. Water International 2011, 36(4), 441–454.

[33] AFNOR: Qualité de l'eau. Recueil des Normes Françaises Environnement. Tomes 1, 2, 3, 4. 1997, 1372 p.

[34] Richards, L.A.: Diagnosis and improvement of saline and alkaline soils. U.S. Department of Agriculture, Handbook, Washington D.C., U.S.A., 1954, Vol. 60.

[35] Frenkel, H., Goertzen, J.O., Rhoadesze, J.D.: Effects of Clay Type and Content. Exchangeable Sodium Percentage and Electrolyte Concentration on Clay Dispersion and Soil Hydraulic Conductivity. Soil Science Society, American Journal, 1978, 42, 32-39.

[36] Doneen, L.D.: The influence of crop and soil on percolating waters. Proceedings of 1961 Biennial Conference on Groundwater Recharge, California, USA, 1962, pp 156-163.

[37] Raghunath, H.M.: Groundwater. Wiley Eastern Ltd, New Delhi, India, 1987

[38] Kothiyal, M., Semwal, G.N.: Performance Evaluation of Brewery Biological Wastewater Treatment Plant. MOJ Ecology Environmental Science 2018, 3(1), 00058

[39] Li, L., He, Q., Wei, Y., He, Q., Peng, X.: Early warning indicators for monitoring the process failure of anaerobic digestion system of food waste. Bioresource Technology 2014, 171, 491-494.

[40] Alcaraz-Gonzalez, V., Fregoso-Sanchez, F., Seyer, J., Mendez-Acosta, H., Gonzalez-Alvarez, V., J.P., Garcia Sandoval: Exponential Regulation of Alkalinity and VFA in Continuous Anaerobic Digestion Processes under Uncertain Operational Conditions. WSEAS Transactions on Systems and Control 2015, 10, 453-460

[41] APHA: Standard Methods for the Examination of Water and Wastewater. 17th Ed. American Public Health Association, Washington D.C., 1989

[42] APHA: Standard Methods for the Examination of Water and Wastewater. 20th Ed. American Public Health Association, Washington D.C., 1999

[43] Grabow, W.O.K.: Microbiology of drinking water treatment: Reclaimed wastewater, In: Drinking Water Microbiology. G.A. McFeters (Ed.), Springer-Verlag, New York, 1990, 185–203.

[44] Payment, P., Franco, E.: *Clostridium perfringens* and somatic coliphages as indicators of the efficiency of drinking water treatment for viruses and protozoan cysts. Applied Environmental Microbiology, 1993, 59, 2418–2424.

[45] FAO: Wastewater Treatment and Use in Agriculture, Irrigation and Drainage Paper, International Water Management Institute, Colombo, Sri Lanka, 1992, p. 73.

[46] Ayres, R.S., Westcot, D.W.: Water Quality for Agriculture, FAO Irrigation and Drainage, Paper No. 29, Rev. 1, FAO, Rome, 1994, p. 59.

[47] Hussain, I., Raschid, M., Hanjra, M.A., Marikar, F., van der Hoek, W.: Wastewater use in agriculture: Review of impacts and methodological issues in valuing impacts, International Water Management Institute, Working Paper 37, Colombo, Sri Lanka. 2002

[48] Asano, T., Burto, F.L., Leveren, H.L., Suchihashi, R., Tchobanoglous, G.: Wastewater Engineering, Treatment and Reuse, 4th Edition, McGraw Hill, New York, 2006.

[49] WHO: Guidelines for the Safe Use of Wastewater, Excreta and Greywater, Geneva, Switzerland, 2008

[50] Rao, A.G., Reddy, T.S.K., Prakash, S.S., Vanajakshi, J., Joseph, J., Sarma, P.N.: pH regulation of alkaline wastewater with carbon dioxide: a case study of treatment of brewery wastewater in UASB reactor coupled with absorber. Bioresources Technologies, 2007, 98, 2131–2136.

[51] Geoffrey, S., Cluett, J., Iyuke, S.E., Musapatika, E.T., Ndlovu, S., Walubita, L.F., Alvarez, A.E.: The treatment of brewery wastewater for reuse: State of the art. Desalination 2011, 273, 235–247

[52] Yidana, S.M., Sakyi, P.A., Stamp, G.: Analysis of the Suitability of Surface Water for Irrigation Purposes: The Southwestern and Coastal River Systems in Ghana, Journal of Water Resource and Protection 2011, 3, 695-710

[53] [53] Kumar, M., Kumari, K., Ramanathan, A.L., Saxena, R.: A comparative evaluation of groundwater suitability for irrigation and drinking purposes in two intensively cultivated districts of Punjab, India. Environmental Geology 2007, 53, 553–574

[54] Al-shammiri, M., Al-saffar, A., Bohamad, S., Ahmed, M.: Waste water and reuse in irrigation in Kuwait using microfiltration technology in treatment. Desalination 2005, 185(3), 213- 225

[55] Khodapanah, L., Sulaiman, W.N.A., Khodapanah, D.N.: Groundwater Quality Assessment for Different purposes in Eshtehard District, Tehran, Iran. European Journal of Scientific Research 2009, 36(4), 543 - 553.

[56] Todd, D.K.: Groundwater hydrology. John Wiley and Sons Inc., New York, U.S.A., 1980

[57] Rahman, M.A., Tanvir, T.M., Syed, H.R., Ratan, K.M.: Groundwater quality for irrigation of deep aquifer in southwestern zone of Bangladesh. Songklanakarin Journal of Science and Technology 2012, 34, 345-352.

[58]   WHO: Water Quality Assessments: A Guide to Use of Biota, Sediments and Water in Environmental Monitoring, 2nd Edition, 1996

[59]   Tampo, L., Gnazou, D.T.M., Kodom, T., Oueda, A., Bawa, L.M.: Suitability of groundwater and surface water for drinking and irrigation propose in Zio River basin (Togo). Journal de Recherches Scientifiques, 2015, 17(3), 35-51

PRODUCTION OF FORAGE CACTUS FERTILIZED AND IRRIGATED WITH REUSE WATER IN THE SERTÃO OF CRATEÚS, BRAZIL

Marcus Roberto Góes Ferreira Costa, Joaquim Batista de Oliveira Neto, José Lopes Viana Neto, Gabriela Duarte Freitas

Federal Institute of Education, Science and Technology of Ceará, Animal Science Department, Crato, Ceará, Brazil; marcusgoes@ifce.edu.br

Keywords: *Opuntia ficus indica*, semiarid region, wastewater

Abstract

The objective of this research was to evaluate the number of sprouts, to measure the area of the cladodes, to determine the dry matter (DM), organic matter (OM) and mineral matter (MM), and to evaluate the morphometric measurements of fertilized forage palm and those irrigated with reuse water. Forage palm seedlings were planted at the Água Branca-Junco Experimental Didactic Unit of the IFCE, Crateús campus. Five treatments were used, three combinations of chemical fertilization, one organic fertilization and one control that did not receive fertilization. The chemical fertilizations were carried out in three combinations of nitrogen, phosphate and potassium fertilization: 130/70/30 (T1); 130/130/50 (T2); (N), P_2O_5 and K_2O, respectively, an organic fertilizer (T 4) and a control treatment (T 5). The irrigation method adopted was of the localized type consisting of motor pump assembly; control head, consisting of disc filter, pressure and record; pipes, main line, tapping line, in this one contained an easel with hydrometer; self-compensating type drippers, katif model, with a flow rate of 3.75 L/h at a service pressure of 100 kPa. Statistical analyses were performed using the ASSISTAT Version 7.7 Beta computational package. The results show that the application of chemical fertilizers together with the irrigation of reuse water provide higher cladodes production, greater development and higher morphometric measurements in forage palm plants, since in the dry matter (DM) contents it did not provide change, but it can change

1 Introduction

The Brazilian semi-arid region is characterized by low rainfall and is subject to climate dependence for agricultural and livestock production. Alternatives are sought that enable sustainable coexistence with rainfall irregularities in this region, one of these alternatives is the cultivation of forage palm (*Opuntia ficus indica*). It is of the *Cactaceae* family and has the necessary requirements to withstand the rigors of climate and the physicochemical characteristics of soils in semi-arid zones [5]. Forage palm is an important resource for ruminant feeding, especially in dry seasons.

Another important passive technology of application in the semi-arid regions is the reuse of water containing treated sewage effluents, which enables the use of a water source unfit for human and animal consumption in the production of roughage. With the use of reuse water, drip irrigation application allows tight control of the amount of water supplied to plants, water saving, lower labor, fertilizer optimization, salinity water and good uniformity of water application [4], characteristics that make this system a good option for sewage irrigation.

This research was carried out with the objective of evaluating the production of forage palm fertilized and irrigated with reuse water in the "Sertão of Crateús", Ceará State, Brazil, analyzing sprouting number, measuring the cladodes area, determining the DM, organic matter (OM) and mineral matter (MM) and evaluate the morphometric measurements.

2 Materials and Methods

The experiments were conducted at IFCE's farm, in the municipality of Crateús, Ceará, Brazil. The farm is located in the "Sertões of Crateús" region on the "Poti" River. The geographic location of the experimental area is 5º08'42 "south latitude, 40º41'04" west longitude and 271 m elevation relative to sea level, with annual average rainfall 697 mm and the relative humidity 77%. Five treatments were evaluated in a randomized block design with five replications.

The treatments were three chemical fertilizers, an organic fertilizer and a control treatment, which received no fertilizer. The chemical fertilizers were determined based on preliminary results of research conducted by the Center for Teaching and Studies in Foraging of the Federal University of Ceará. The combinations 130/70/30; 130/130/50; 190/130/70 pounds per hectare of nitrogen, diphosphorus pentoxide (P_2O_5) and potassium oxide (K_2O), respectively, were used as fertilizers.

As source of nitrogen (N) agricultural urea was used with approximately 46.6% N, for phosphate fertilization simple superphosphate with approximately 18% of P_2O_5, and potassium fertilizer was applied potassium chloride (60% K_2O). For organic fertilization, tanned manure at a rate of 5 kg per linear meter was used. Phosphate, potassium and organic fertilizers were applied in a single dose at the time of planting. The application of urea was divided into three applications: 40% of the recommended dose during planting, 30% 45 days after planting and the remaining amount after 90 days.

The irrigation method adopted was of the localized type, consisting of pump set; control head, consisting of disc filter, pressure tap and registers; pipelines, main line, branch line, this contained an easel with hydrometer; self-compensating drippers, katif model, with a flow rate of 3.75 L/h at a working pressure of 100 kPa. The water for irrigation came from the Poti River, which borders a strip of approximately 570 m from the experimental farm. Because it is located downstream of the Sewage Treatment Station of Ceará Water and Sewage Company, the stretch of river that runs along the farm's far end is perpetuated by the constant discharge of water containing effluent from the primary sewage treatment.. The experimental area is approximately 2,176 meters from the domestic Sewage Treatment Station following the natural riverbed. Table 1 shows the quality of water used for irrigation.

Table 1: Average values of water quality parameters at five points of the Poti River in Crateús - CE

Parameters	Units	Values
Total Alkalinity	mg/L $CaCO_3$	559
Alkalinity Carbonates	mg/L $CaCO_3$	294
Alkalinity Baking	mg/L $CaCO_3$	364
Cloreto	mg/L Cl^-	544
Acidity	mg/L $CaCO_3$	0.6
Total hardness	mg/L $CaCO_3$	305
Calcium Hardness	mg/L $CaCO_3$	109
Magnesium Hardness	mg/L $CaCO_3$	194
Color	uH	70.6
Conductivity at 25 °C	µS/cm	2,134
pH at 25 °C		8.93
Total Dissolved Solids	mg/L	1,451
Turbidity	UNT[1]	57.3
Dissolved oxygen	mg/L	2.33

Ten months after planting, the forage palm plants were cut at the base of the primary cladode. Then the cladodes were weighed separately according to the order in the plant. To estimate the production per plant and productivity, the weight of cladodes per plant was added and then the average production of each treatment was multiplied by the estimated number of plants per hectare according to the spacing used. For morphometric measurement of cladode, width and length, using a ruler, then the cladode area was determined by the mathematical formula of the ellipse area, where the area equals the product of half the width, half the length multiplied by π.

The samples for chemical analyses were separated the according to cladodium order on plant. The samples were then taken to the Chemistry and Environment laboratory of IFCE campus Crateús, where they were weighed, fractionated into approximately 4 cm pieces and placed in identified paper bags, and placed in a forced ventilation oven at 55 ± 5 °C. After a period of 76 h, the material was reweighed for pre-drying determination, and then the samples were milled in a Willey knife mill. After milling, the dry matter (DM) contents were determined by conditioning samples of approximately 2 g in an oven at 100 °C for 24 h. For the determination of mineral matter (MM), a muffle furnace heated to 600 °C was used to incinerate all organic matter (OM) from a pre-dried sample of approximately 2 g. For the determination of the OM content, the MM content was subtracted from 100% of the natural matter.

Statistical analyzes were performed using the ASSISTAT Version 7.7 Beta computer package.

3 Production of Forage Palm

Based on the results obtained with this research, Table 2 shows the cladode weight in plant order and estimated yield of green forage mass of forage palm. For the treatments with chemical fertilization, the total weights of the cladodes were superior to the treatments with organic fertilization and without fertilization. This behavior is directly reflected in the estimated yield of green forage mass, which was also superior for the treatments with chemical fertilization. The highest value for green forage mass yield is lower than the values found in the literature, as can be observed in [2], where the average yield for the same forage palm variety was 241.7 t/ha. The poor performance presented in this study can be attributed to the low salinity tolerance presented by the forage palm.

Soil salinity may promote inhibitory effects of forage palm development by up to 30%, reflecting the induction of lateral root abscission and inhibiting cell expansion in the root elongation zone [6]. This can directly compromise the absorption capacity of nutrients by the plant, providing a significant reduction in its development.

Regarding the number of cladodes per plant (represented in Table 3), the maximum death value in this research was below the maximum value obtained by [4], which was 15.64. This difference is explained by the age of the plants, as those authors worked with plants at the age of 18 months, while in this research the plants were evaluated at 10 months.

Table 2: Cladode weight in plant order and estimated yield of green forage mass of forage palm fertilized and irrigated with reused water in Sertão of Crateús

Treatment	Primary Cladodes	Secondary Cladodes	Tertiary Cladodes	Total weight	Estimated Productivity (t/ha)
T1	1.33^a	0.81^a	0.09^a	2.23^a	49.6^a
T2	1.25^a	0.85^a	0.07^a	2.16^a	48.1^a
T3	1.10^a	0.66^{ab}	0.08^a	1.85^{ab}	41.1^{ab}
T4	1.01^a	0.38^{ab}	0^a	1.39^b	30.8^b
T5	0.95^a	0.29^b	0	1.24^b	27.6^b
CV*(%)	27.0	43.5	32.2	32.1	32.1
P-Value	0.2894	0.0105	0.1019	0.047	0.0464

Coefficient of variation, T1-130/70/30 kg per hectare of N, P_2O_5 and K_2O; T2-130/130/50 pounds per hectare of N, P_2O_5; T3-190 / 130/70 kg per hectare of N, P_2O_5 and K_2O, T4-Organic fertilization and T5-control without fertilization. Means followed by the same lowercase letter in the column within each parameter do not differ by Tukey's test at the 5% probability level.

Table 3: Total average number of cladodes and average number of primary, secondary and tertiary cladodes of forage palm fertilized and irrigated with reused water in the Sertão of Crateús

Treatment	Total of Cladodes	Cladodes Primary	Cladodes Secondary	Cladodes tertiary
1	7.61^{abc}	3.40^{ab}	3.34^{a}	0.85^{a}
2	8.27^{ab}	3.66^{ab}	4.00^{a}	0.61^{a}
3	8.80^{a}	4.11^{a}	4.01^{a}	0.68^{a}
4	4.54^{c}	2.85^{b}	1.67^{a}	0.01^{a}
5	5.05^{bc}	3.52^{ab}	1.52^{a}	0.01^{a}
CV(%)*	28.1	17.7	44.9	46.3
P-Value	0.0079	0.0692	0.0168	0.0719

*Coefficient of variation, T1-130/70/30 kg per hectare of N, P_2O_5 and K_2O; T2-130/130/50 pounds per hectare of N, P_2O_5; T3-190/130/70 kg per hectare of N, P_2O_5 and K_2O, T4-Organic fertilization and T5-control without fertilization.

The highest results obtained for primary and secondary cladodes were similar to those reported by [7], which present a value 4.1 and 4.8 for primary and secondary cladodes, respectively. For the values of the third order cladodes, the results in this research were below those reported by the mentioned authors, where they cite the value of 3.6 tertiary cladodes per plant.

The results obtained for the cladode area are presented in Table 4, where it can be observed that for the primary cladodes the treatments, in which chemical fertilizers were applied, did not differ from each other (P > 0.05), being also the same as the organic fertilizer treatment. The control treatment showed statistical similarity to the treatment with organic fertilization, being them the lowest values observed. For the area values of the secondary cladodes, no significant difference (P > 0.05) was observed between them. When compared with primary cladodes only for treatment 2, there was a statistical difference (P < 0.05), where the primary cladodes presented larger cladode area.

When comparing the results obtained in this research with some scientific studies, it is observed that the values in Table 4 are lower than cited in [8] as 455 days after planting, forage palm plants presented a cladode area variation of 334.9 to 407.8 cm^2. This difference may be related to the difference in the age of the evaluated plants, being the plants of the work in [8] 145 days older than this research.

The variation in cladode area over time depends on edaphoclimatic conditions, cultivar and population density, among other factors. Generally, the cladode area increases to a maximum, decreasing after some time [2]. Evaluating the evolution of the cladode area of eight forage palm varieties observed that the varieties of the genus Opuntia presented larger areas of cladodes, when compared with those of the genus Nopalea. Of the varieties studied by this author, the one with the largest cladode area was IPA-20 (385.7 cm^2) followed by Italian (379.0 cm^2).

Even comparing with plants of near age, the values in Table 4 were also lower, as observed in the results of [9], where plants had nine months after planting an average area of 811.1 cm². This lower than expected result for forage palm cladodes area may reflect the high concentration of salts in the reused water used for irrigation, mainly bicarbonates.

Table 4: Area of cladodes (in cm²) in order of forage palm plant fertilized and irrigated with reused water

Order of the cladode	T1	T2[2]	T3	T4	T5	Mean	P-value	CV(%)
Primary[1]	240.50^a	245.33^{aA}	244.66^a	222.00^{ab}	170.63^b	224.62	0.01	14.98
Secondary	185.97	187.14^B	212.48	177.25	128.94	178.35	0.289	33.11
Mean	213.23	213.23	228.57	199.63	149.78	-	-	-
P-value	0.31	0.027	0.181	0.075	0.121			
CV(%)	37.90	15.79	15.18	17.36	25.39	-	-	-

[1]*Means followed by the same letter in the line direction do not differ (P> 0.05) by the Tukey test.*[2]*Means followed by the same letter in the column direction do not differ (P> 0.05) by Tukey's test.CV = coefficient of variation.*

The results of the morphometric measurements obtained in this research are presented in Table 5. It can be observed that the application of chemical fertilizer provided the largest measures of length and thickness for the primary cladodes. This effect was not as pronounced for the secondary cladodes, where no significant difference (P > 0.05) was observed for the length measurements. Making a joint evaluation between the treatments used, the plants submitted to treatment 2 presented superiorities over the other treatments in most of the measured variables.

Table 5: Morphometric measurements of the cladodes of forage palm fertilized and irrigated with reused water in the Sertão of Crateús

Variables	Treatments					CV(%)*	P-value
	1	2	3	4	5		
Primary length	23.47^{AB}	24.32^A	23.72^A	22.57^{AB}	17.98^B	12.79	0.0199
Secondary length	17.98^A	19.73^A	19.84^A	17.60^A	14.16^A	29.17	0.448
Tertiary length	9.72^B	12.19^{AB}	13.68^A	3.60^C	1.40^C	48.99	0.1847
Primary width	13.04^A	12.82^A	13.12^A	12.53^A	12.02^A	6.55	0.2658
Secondary width	12.79^A	12.06^A	13.76^A	10.50^{AB}	9.58^B	18.75	0.0524
Tertiary width	5.84^B	8.23^A	8.11^A	2.10^{BC}	0.90^C	48.86	0.0815
Primary thickness	17.33^A	18.92^A	13.69^B	14.23^B	14.93^B	15.53	0.0463
Secondary thickness	11.16^A	11.25^A	9.41^{AB}	9.57^{AB}	8.63^B	17.05	0.1032
Tertiary thickness	4.53^B	7.44^A	6.21^A	2.19^{BC}	0.91^C	47.57	0.1187

** Averages followed by the same uppercase letter in the column and the same lowercase letter in the row within each parameter do not differ by Tukey's test at the 5% probability level. *Coefficient of variation.*

When evaluating the forage potential of forage palm varieties in Cariri of Paraíba State, verified for the giant palm averages of 27 and 16 cm for the length and width of the cladodes [7], similar results to those are obtained in this research.

4 Chemical Composition of Forage Palm

The contents of dry matter, organic matter and mineral matter are presented in Table 6. Regarding the dry matter content, there was no statistical difference (P > 0.05) between the evaluated treatments, presenting averages of 9.80 and 10.39% for primary and secondary cladodes, respectively. The average values for dry matter content are close to those observed in the literature, such as that reported by [11], who reported the value of DM equal to 9.10%, being also close to the value presented by [1], which was 9.39%.

Comparing the DM content according to cladode order only for T1, a significant difference (P < 0.05) was observed, where the secondary cladode presented a higher MS content than the primary cladode. Despite being below the values commonly found for dry matter content of forage palm cladodes, one can also observe values similar to that obtained for the primary T1 cladodes, as reported by [10], presenting the value of 7.83% of DM.

For the values of organic matter and mineral matter, a significant difference (P < 0.05) was observed for the treatments studied for the primary cladodes. The combination of chemical fertilizers provided a greater accumulation of organic matter was used in T2. The values obtained corroborate those observed in the literature, where there is a variation of 83.70 to 88.25% of OM in MS for the works of [12] and [2] and 12.57 to 16.30 MM in MS ([12] and [11]).

Table 6: Contents of dry matter, organic matter and mineral matter of forage palm fertilized and irrigated with reused water

Order of the cladode	T1[2]	T2	T3	T4	T5	Mean	P-valoe	CV(%)
	Dry Matter (%)							
Primary	7.92[B]	10.46	10.51	10.08	10.03	9.80	0.0573	14.63
Secondary	9.25[A]	11.17	10.76	11.25	9.50	10.39	0.252	16.87
Mean	-	10.81	10.64	10.67	9.76	-	-	-
P-value	0.042	0.660	0.956	0.317	0.364	-	-	-
CV(%)	10.13	22.77	13.22	16.42	9.01	-	-	-
	Organic Matter (%Dry Matter)							
Primary	81.36[b]	86.81[a]	84.39[ab]	85.43[a]	83.79[ab]	-	0.009	2.53
Secondary	84.04	86.86	85.39	86.05	85.12	85.49	0.214	2.18
Mean	82.70	86.84	84.89	85.74	84.45	-	-	-
P-value	0.132	0.969	0.354	0.693	0.272	-	-	-
CV(%)	2.37	2.29	1.92	2.79	2.11	-	-	-

	Mineral Matter (%Dry Matter)							
Primary	18.64[a]	13.19[b]	15.61[ab]	14.57[b]	16.21[ab]	-	0.009	13.62
Secondary	15.96	13.14	14.61	13.95	14.88	14.51	0.214	12.86
Mean	17.30	13.16	15.11	14.26	15.55	-	-	-
P-value	0.083	0.718	0.359	0.693	0.272	-	-	-
CV(%)	12.32	13.67	10.77	16.76	11.47	-	-	-

[1]*Means followed by the same letter in the line direction do not differ (P> 0.05) by Tukey's test.* [2]*Means followed by the same letter in the column direction do not differ (P> 0.05) by Tukey's test. CV = coefficient of variation.*

4 Conclusions

The use of chemical fertilization together with irrigation with reused water provided the production of heavier cladodes, consequently also presented higher green mass yield of forage palm plants. It also increased the cladode area in forage palm plants, which may reflect a higher forage yield and provided higher morphometric measurements on the plants. The use of wastewater in forage palm cultivation may be a limiting factor for plant development due to the excess of dissolved salts in irrigation water, but a sustainable way to reuse water.

5 Acknowledgments

Acknowledgment is dedicated to PRPI-PIBIC-IFCE and PROINFRA-IFCE for granting scholarship funds and resources for this study and to FUNCAP for granting funds for this research. The author also likes to thank to Exceed Swindon project for supporting his participation at the workshop in Rio de Janeiro, Brazil, September 2019.

6 References

[1]   S.B. Bispo, M.A. Ferreira, A.S.C. Véras, E.C. Modesto, A.V. Guimarães, R.A.S. Pessoa: Comportamento ingestivo de vacas em lactação e de ovinos alimentados com dietas contendo palma forrageira, Revista Brasileira de Zootecnia 2010, 39(9), 2024-2031.

[2]   L.M. Silva, J.L. Fagundes, P.A.V. Almeida, E.A.R.N. Muniz, J.H. Moreira, A.L. Luciane, A.A Backes: Produtividade da palma forrageira cultivada em diferentes densidades de plantio, Ciência Rural 2014, 44(11) 2064-2071.

[3]   M.L.L.V. Leite, D.S. Silva, A.P. Andrade, A.T. Sales, B.L. Viana, A.C. Araújo, N.M. Souza: Eficiência de uso da chuva em Opuntia ficus-indica em função da adubação, no semi-árido paraibano, In: Congresso Nordestino de Produção Animal, 5, 2008, Aracaju, Anais. .Aracaju: SNPA, 2008.

[4]   E.B. Lopes, L.B. Costa, A.F. Cordeiro Jr., C.H. Brito: Rendimento e aspectos fenológicos de espécie de palma forrageira em relação ao cultivo com dois tipos de cladódios, Tecnologia & Ciência Agropecuária, João Pessoa 2013,7, 59-61.

[5]   E.C. Mantovani, S. Bernardo, L.F. Palaretti: Irrigação: princípios e métodos, Ed. UFV, 2006, 318p.

[6] P.S. Nobel: Biologia ambiental. In: Agroecologia, cultivo e usos da palma forrageira, Traduzido por SEBRAE/PB, João Pessoa, SEBRAE/PB 2001, 36-48.

[7] S. Oliveira Jr., M. Barreiro Neto, J.P.F. Ramos, M.L.M.V. Leite, E.A. Brito, J.P. Nascimento: Crescimento vegetativo da palma forrageira (Opuntia ficus-indica) em função do espaçamento no Semiárido paraibano, Tecnologia & Ciência Agropecuária, João Pessoa 2009, 3(1),7-12.

[8] A.T. Sales: Potencial de adaptação de variedades de palma forrageira (Opuntia ficus-indica e Nopalea cochenilifera) no Cariri paraibano, In: Congresso Nordestino De Produção Animal, 4. 2006, Petrolina, Anais, Petrolina: SNPA 2006.

[9] J.P.F. Ramos, M.L.M.V. Leite, S. Oliveira Jr.S.: Crescimento vegetativo de Opuntia ficus-indica em diferentes espaçamentos de plantio, Revista Caatinga, Mossoró 2011, 24(3), 41-48.

[10] M.M., Teles, M.V.F. Santos, J.C.B Dubeux Jr.: Efeitos da adubação e de nematicida no crescimento e na produção da palma forrageira (Opuntia ficus indica Mill) cv. Gigante, Revista Brasileira de Zootecnia 2002, 31(1), 52-60.

[11] M.S.L. Tosto, G.G.L. Araújo, R.L. Oliveira: Composição química e estimativa de energia da palma forrageira e do resíduo desidratado de vitivinícolas, Revista Brasileira de Saúde e Produção Animal 2007, 8(3), 239-249

[12] W.L. Wanderley, M.A. Ferreira, A.M.V. Batista: Consumo, digestibilidade e parâmetros ruminais em ovinos recebendo silagens e fenos em associação à palma forrageira, Revista Brasileira de Saúde e Produção Animal 2012, 13(2), 444-456.

EVALUATION OF FLY ASH PELLETS FOR PHOSPHORUS REMOVAL IN A LABORATORY SCALE DENITRIFYING BIOREACTOR

Shiyang Li[1], Xiangfeng Huang[1], Jia Liu[1], Lijun Lu[1], Kaiming Peng[1], Rabin Bhattarai[2]

[1]*College of Environmental Science and Engineering, State Key Laboratory of Pollution Control and Resource Reuse, Ministry of Education Key Laboratory of Yangtze River Water Environment, Shanghai Institute of Pollution Control and Ecological Security, Tongji University, Shanghai 200092, People's Republic of China; hxf@tongji.edu.cn*

[2]*Department of Agricultural and Biological Engineering, University of Illinois at Urbana Champaign, 1304 W Pennsylvania Ave, Urbana IL 61801, USA*

Keywords: Denitrification bioreactor, non-point pollution, subsurface drainage

Abstract

Nitrate and orthophosphate from agricultural activities contribute to the main nutrient loading in surface water bodies. The objective of this study was to evaluate the efficacy of using woodchips and fly ash pellets (FAP) in flow-through tests for their abilities to remove NO_3-N and soluble P from the agricultural runoff. A lab scale bi-column system with two sections was designed to target the nitrate and orthophosphate removal, respectively. The front section was a 3 m long, 0.152 m diameter PVC pipe, filled with woodchips. The subsequent section was 1 m long, filled with fly ash pellets facilitated by an adjustable height outlet pipe. The influent nutrient concentrations were 12 mg/L and 5 mg/L for nitrate and orthophosphate, respectively. Three tests were conducted with different flow rates in this study. The results showed, the woodchip bioreactor section demonstrated average nitrate removal efficiencies of 49.2–85% and hydraulic retention times (HRTs) of 0.67–4.04 h. The fly ash pellet showed a very stable removal efficiency of 68.4-74.8% and HRTs of 0.67–4.04 h. The woodchips had a nitrate removal rate of 40.2-49.2 mg N/m^3.d. The fly ash pellets effectively removed phosphate in the bioreactor effluent, and the total phosphate adsorption was 0.059-0.114 mg P/g, which is far less than the saturated capacity (1.69 mg/g). The effluent phosphate concentrations increased with decreasing HRTs. The fly ash pellet section reduced a certain level of nitrate in the bioreactor effluent, but was not significant. Overall, the results of this study suggest that woodchip denitrification followed by fly ash pellet filtration can be an effective treatment technology for nitrate and phosphate removal in subsurface drainages.

1 Introduction

The quality of any body of surface or ground water is a function of both natural influences and/or human influences. The increasing amount of agricultural production activities produces large amounts of contaminants, including nutrients, herbicides, pesticides and solids that greatly threaten watercourses. The fertilization practice in agricultural activities surely increased the burden of self-purification of water by bringing in more nutrients. Although this has helped increase crop yields over the years, fertilizers are often applied in quantities in excess of crop uptake. This increase in runoff nutrient concentrations, along with a variety of other factors (e.g., tillage practices, landscape, timing of fertilizer application, weather patterns, soil type, and other environmental factors), has led to long-term adverse effects on the receiving waters of agricultural watersheds. Especially in the Midwestern United States, the pervasive utilization of subsurface drainage systems critically reduces the time that nutrient flux reaches the receiving water body. A more recent simulation study, conducted in the Mississippi River Basin, estimated that agricultural watersheds (highest subsurface drainage density area) accounted for 70% of the total N and P delivered to receiving waters [1]. The nutrients transport though the Mississippi River is the main cause of the hypoxia in the Gulf of Mexico [2].

There are many on-farm and edge-of-field conservation practices to control the nutrient flux from the agricultural area [3-5]. The edge-of-field approach has been identified as an effective method to help controlling the subsurface drainage nutrient loss [6]. This method is designed and maintained by a filter media (biological or chemical) chamber, before the water is allowed to enter the receiving water body. It is usually located at the outlet of the subsurface drainage tile. Well-designed systems are installed parallel to receiving streams and take little to no agricultural land out of production. With NO_3-N being the primary contaminant transported through artificial subsurface drainage, the early focus of this approach was to promote denitrification through the introduction of an additional carbon source. Although no standard filtration media exists yet, woodchips and other carbon containing substrates have shown great promise for promoting denitrification and greatly reduced NO_3-N loads in several studies [1, 7-11].

Multiple studies on the woodchip bioreactor have been conducted to determine factors influencing nitrate removal efficiency and removal rate. The retention time is an important parameter, which will strongly influence the nitrate removal by limiting the water contact time with denitrifying bacteria and a release for the source of carbon [12, 13]. Additionally, the bioreactor operating temperature is another vital factor for the denitrification rate. Higher temperatures can boost the denitrification rate of bacteria by increasing the cell membrane permeability and accelerate the carbon source release speed [14]. At the same time, the level of dissolved oxygen and pH of the water also affects the performance of the bioreactor [15]. Meanwhile, the rising concern of removing the soluble P with N simultaneously was strongly advocated, because the considerable loss of P element through the tile system was addressed. In many studies, multiple materials had been reported for their ability to remove soluble P from solution [16-20].

Fly ash, which is an industrial waste product, was suggested as an ideal sorbent, because it contains large percentages of natural minerals, such as calcium (Ca), aluminum (Al), and iron (Fe) in various forms [21, 22]. Li et al. [23] compared the P absorption efficiency of fly ash and red mud. The result showed that the fly ash provided an average 52% P removal ability with the pH at 7 and at room temperature. Wang et al. [24] developed a fly ash based lanthanum oxide hybrid material to capture the P run into lakes. Waters from five lakes were compared and this material maintained the absorption rate under the pH varied situation. Li et al. [21] used fly ash as the base materials to create a pellet form sorbent, which was proved to have more than 90% P absorption efficiency in a batch test. This pellet form sorbent is ideal for the chamber structure and stable enough for the hydraulic status in tile system.

While existing research has begun to address efficient and cost-effective approaches for controlling contaminant release from tiled drained agricultural lands, current methods and materials are far from refined. This allows for a large increase in the removal efficiency of both nutrients, before the methods behind this approach could be considered ideal. Bioreactors are effective in removing N, but the current design needs to be modified to make it effective for P removal as well. The goals of this study was to conduct lab scale column experiments to test the efficacy of using woodchips and fly ash pellets (FAP) developed in an earlier study of the authors [21] in flow-through tests for their abilities to remove NO_3-N and soluble P from agricultural drainage.

2 Materials and Methods

2.1 Research Site and Bioreactor

Two horizontal-flow laboratory-scale column reactors were constructed to test FAP for its capacity to remove soluble phosphorus from tile drainage water, while simultaneously examining the effect of woodchip media used for nitrate removal (Figure 1, A and B). To allow for simultaneous experimental repetitions, each reactor consisted of two identically constructed horizontal PVC columns (0.1524 m diameter), and each configuration consisted of an upstream 3 m section filled with woodchips, and a downstream 1 m long FAP-filled section. The material was secured in the columns by PVC plates with drilled holes (12× ϕ1 cm) covered by a non-reactive mesh at each end. Flow through the columns was regulated using a controlled drainage structure connected to a manifold, which diverted the flow equally and served as the inlet for each of the two columns. The outlet of each column consisted of a 5.08 cm diameter PVC pipe. The outlet pipe was rotated to any angle to achieve head differences ranging from 0 to 1.5 m below the inlet water level to induce flow.

The configurations were calibrated by measuring volumetric flow rates (in triplicate) from the paired column outlets, while the configurations were operating at three outlet placements. Effective hydraulic conductivity (K_e) for each column was calculated using Darcy's equation (1).

$$K_e = \frac{Q}{A} \cdot \frac{L}{\Delta H} \tag{1}$$

Where are K_e = effective hydraulic conductivity [L/T], Q = Flow [L^3/T], A = Area [L^2], L = column length [L], ΔH = head difference [L].

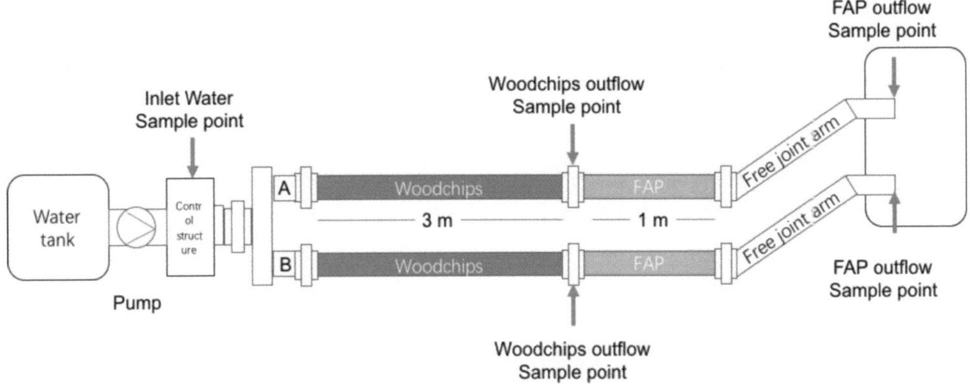

Figure 1: Illustration of two replicated paired-column configurations using (A) woodchips and (B) fly ash pellets (FAP) fed by a control structure

2.2 Bioreactor Media

The coal fly ash pelleting method used in this study was done just similar to as described by Li et al. [21]. All dry materials (fly ash 60 %, lime 30%, and clay 10% by weight) were mixed uniformly with a blender. Deionized water (15% by weight) was added to the mixture and blended again to prepare the slurry. The slurry mix was then sealed, left to stabilize for 24 h at room temperature, then converted into pellets using a commercial pelletizer (Colorado Mill Equipment-ECO-10, USA) equipped with a 10 HP, 3-phase motor. The pellets were then baked in a high-temperature furnace (Thermolyne BOX furnace, MA, USA) for a total of 7 h, raising the temperature 200 °C each hour for 5 h, then keeping it at 1000 °C for another 2 h. Once the pellets were removed from the oven, they were cooled for 6 h and then rinsed with distilled water [21].

The woodchips used in the study were collected from an existing field-scale bioreactor at a university research farm at the University of Illinois (Urbana, IL). This bioreactor was established in 2015, and previous research showed a promising result for Nitrate-N removal [25]. The woodchips were collected from the surface to a depth of approximately 0.61 m. Previous study at this site showed, this depth was often/always fully submerged, thus a selection of aerobic and anaerobic-exposed woodchips was included. The woodchips were collected in plastic bins and packed into the PVC columns. The columns sections were vertically compacted with a long tamping rod at 2.54 cm increments during loading and filled the columns to a height of 290 cm to achieve an approximately uniform density throughout all of the columns.

2.3. Characterization of Materials

Gravimetric measurement of drainable porosity was undertaken by draining the columns from the bottom over a 1-h period, measuring the weight of the drained water, and then extracting the media from the columns and measuring the difference in wet and dry media weights after air drying in a fume hood for 3 d. Primary porosity was calculated as the ratio of the drainage water volume to the open bed volume of the columns, and secondary porosity (porosity internal to the wood particles) was calculated as the ratio of the water loss upon media drying to the open bed volume. Total porosity was then calculated as the sum of the primary and secondary porosities. Bulk density was measured using Oertling YP4 balance.

The size of materials, based on observation, ranged from approximately 0.25 cm to 10 cm for the woodchips, and 0.5 cm to 1.5 cm for the FAP (Figure 2). The woodchips were of varying thickness and length, while the FAP tended to be more uniform and of pellet shape.

Figure 2: The fly ash pellets (left) and woodchips (right) used in this study

2.4 Experimental Method
The column reactor was used for a series of three experiments, each using a different flow rate that tested for both soluble N and P removal efficiency of woodchips and FAP. The flow rates were 2.80 mL/s, 5.68 mL/s and 1.98 mL/s for test 1, test 2, and test 3, respectively. Subsequently, the retention times for the woodchip sections of the paired configurations were 2.9, 1.5, and 4.2 h, respectively; the retention times for FAP sections were 1.36, 0.67 and 1.92 h, respectively. This low-high-low flow rate sequence was set to avoid the flow rate bias on the bacterial growth on woodchips.

Prior to the experiment, the systems were initialized by flushing them with pond water for 3 d at a flow rate of 4.6 mL/s. The flushing water was spiked with KH_2PO_4 and $KHNO_3$ to produce an approximate 5 mg P/L and 12 mg N/L solution. This solution was pumped by an electric pump into the controlled drainage structure (Agri-Drain, Adair, Iowa) that served as a constant head device, and then into the manifold connecting the configurations (Figure 1). The overflow from the constant head device spilled back into the holding tank, continuously mixing the solution.

3 Results and Discussion

3.1 Materials characterization
The effective hydraulic conductivities of the two columns were not statically significantly different from each other (Column A and B: 3.10 ± 0.18 and 3.13 ± 0.10 cm/sec, respectively; p = 0.85). The bulk density of woodchips in the packed columns was 200 kg/m^3 (5.96 kg, 0.03 m^3) with a

drainable porosity of 54.5% (Table 1). The bulk density of fly ash pellets was 1.037 g/cm^3, and the drainable porosity was 75.3% in the column. Based on the bulk density and drainable porosity of the fly ash pellets, the effective volume and weight in FAP section was calculated, which was 0.014 m^3 and 14.2 kg.

Table 1: Bulk density and drainable porosity of fly-ash pellets and woodchip used in this paired column study

	Bulk Density (dry basis)	Drainable Porosity	Effective volume
Pellet	1,037 kg/m^3	75.3%	0.014 m^3
Woodchips	200 kg/m^3	54.5%	0.03 m^3

3.2 Orthophosphate Removal

Significant reductions in orthophosphate concentration were observed as the result of passing the solution through both the woodchips and the fly ash pellets (Figure 3 A, B, C). The main orthophosphate removal in this system was achieved through the FAP section. The FAP section exhibited an average orthophosphate removal of 71.4%, and this was not statistically significantly different between the three tests, even though the retention time varied (Test 1, 2, and 3: P removal: 74.7%, 68.3% and 71.2%, respectively). This result indicated that the FAP had a stable performance of trapping orthophosphate under the different flow rates and even under higher inflow P concentrations when an increase in concentration was observed exiting the woodchips (Test 3 in Figure 3C). Phosphorus adsorption is an ion exchange reaction, which takes place on the surface of calcium, iron, and aluminum hydroxides [21]. Binuclear or bridging complexes are formed between HPO$_4^{2-}$ ions and metal oxide surfaces, and H$_2$O and OH$^-$ are displaced [26]. Previous work demonstrated the P sorption ability of FAP, and this flow-through study further shows that FAP, which can be rich in calcium, iron or aluminum oxide, can be a proper filter material for P absorption in a tile drainage system. The performance of FAP in removing the orthophosphate in this study was better than other industry by-products, which were tested in similar experimental settings.

In all three tests, the woodchip contributed to orthophosphate removal, despite their primary purpose being for nitrate removal, with average concentration reductions ranging from 10 to13%. There was no significant difference in the woodchip outflow average orthophosphate concentration between the three tests (Test 1, 2, and 3: 3.99, 4.04, and 4.21 mg P/L; p > 0.05). At the beginning of test 3, the first two sample's concentrations (5.5 and 5.2 mg P/L) were higher than the inlet concentration (average 4.5 mg/L), which may have been due to a release of absorbed orthophosphate from woodchips during the anaerobic time between the tests (Figure 3C, box). Two potential pathways in which woodchip could remove orthophosphate, include physical attachment/sorption or bacterial assimilation. Both the change in flow orthophosphate concentration and bacteria metabolism could be the cause of the release of the orthophosphate.

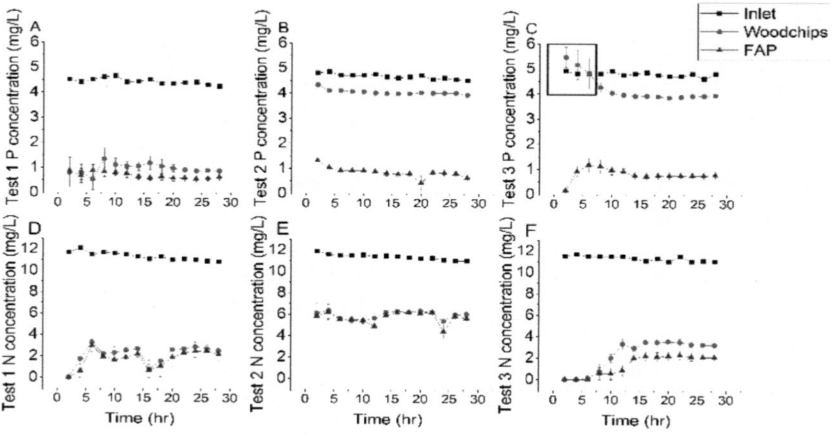

Figure 3: Phosphate concentrations for the inlet, effluent from the woodchips, and effluent from the fly ash pellet (FAP) sections for three tests operated at hydraulic retention times of 2.9, 1.5, and 4.2 h, which correspond to panel A, B and C. Nitrate concentrations for the inlet, effluent from the woodchips, and effluent from the fly ash pellet (FAP) sections for three tests operated at hydraulic retention times of 2.9, 1.5, and 4.2 h, which correspond to panel D, E and F.

3.3 Nitrate Removal

All three tests resulted in a significant reduction in nitrate concentration after water passed through the woodchip section (Figure 3 D, E, F). The general nitrate removal rate at the woodchip section was stable during the whole test period in all three tests. In test 1, there was an average nitrate reduction of 70% observed after the woodchip section and an additional 2% average reduction after the FAP section. Lower nitrate removal efficiency in test 2 (average 47.5%) versus test 1 was due to the decrease in retention time (test 1 and 2 woodchip section retention time: 2.9 vs. 1.45 h). In test 3, the retention time was 4.1 h. Nitrate was not detected in the first 2 samples, because the water leaving in the woodchip section was fully denitrified. At 10 h during test 3, the nitrate concentration stabilized at 2.3 mg N/L resulting in a nitrate removal efficiency of 80.8%.

The notably lower woodchip effluent concentration at the beginning occurred, because the sampling interval was shorter than the retention time (2.0 vs. 2.9 h, respectively), and it is likely the initial water had already been sufficiently denitrified. Other low effluent concentration events (at 16 h during test 1 and at 12 and 24 h during test 2) were due to refilling the water supply tank, which extended the water retention time (3 h for pumping water from farm and transport it back into tank).

There was a small-scale reduction in concentration of nitrate after the FAP section in all three test results, especially when the retention time was raised. In test 3, the nitrate concentration difference averaged 0.8 mg/L before and after the FAP section. It is likely this was due to the active composition in FAP, which also possibly absorbed nitrate. In fly ash pellets, the aluminum and magnesium oxide, which had an active composition, could have a physical and chemical sorption reaction with nitrate, when there is enough contact time [27]. The chemical absorption reaction of aluminum and magnesium oxide with nitrate follows the equation below (Eq. 2):

$$[Mg_{1-x}{}^{II}Al_x{}^{III}O]^x \cdot yH_2O + xNO_3^- \rightarrow [Mg_{1-x}{}^{II}Al_x{}^{III}(OH)_2]^{x+} [NO_3^-]_x \cdot yH_2O \qquad (2)$$

This reaction involves the breaking up of electrostatic interactions as well as the hydrogen bonds between the hydroxide layers and the outgoing anion, and the reformation of these bonds with the incoming NO_3^-. The only factors that could have impacted these reactions were the contact time and the pH levels in the FAP section in this test. Further results and discussion about pH are shown in section 3.3. A similar denitrification efficiency was observed by Goodwin et al. [28], which reported that the general denitrification efficiency was 88.8% for a 12 mg/L inlet nitrate concentration with average 4.9 h retention time. Hua et al.[29] also found a similar denitrification efficiency (84.4% removal efficiency) in a vertical tube test for a six-month period, the general retention time was 6-24 h and inlet concentration was 20 mg/L. Additionally, both Lepine et al. [12] and Hoover et al. [11] reported 55-65% removal efficiencies in a lab scale woodchip bioreactor, and the influent concentration and retention time were 20-80 mg/L and 6.5-55 h, and 10-50 mg/L and 1.7-21.2 h, respectively.

3.5 The Effect of Retention Time on Nutrient Removal

Total nitrate removal rates were 46, 55 and 55 g/m³·d for test 1, test 2 and test 3, respectively. Correspondingly, nitrate removal rates for only the woodchip section were 40, 49 and 41 g/m³·d. This result indicated the retention time was negatively related with the nitrate removal rate. The reason for this result is that the higher flow rates (lower RTs) have more mass loading into the bioreactor. On the contrary, the nitrate removal rates contributed by FAP for all three tests were impacted by the retention time positively.

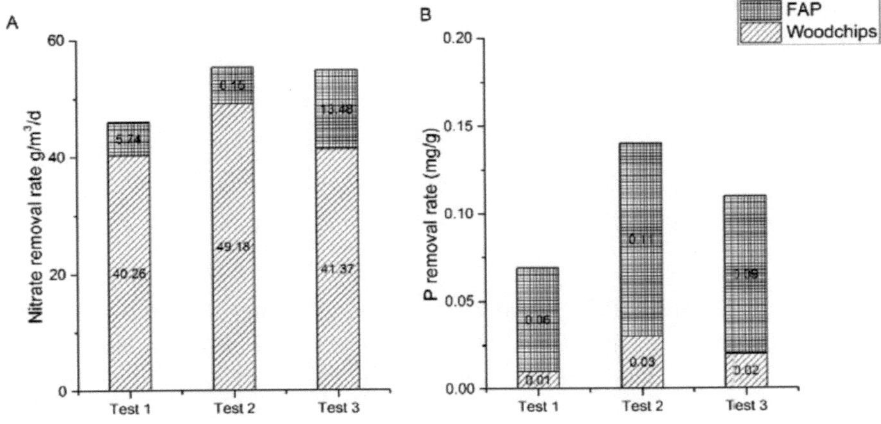

Figure 4: Nitrate and orthophosphate accumulation/removal rates of both sections for all tests

Like nitrate, the P removal rate also increased, when retention time decreased. The P removal rate for each test in the FAP section was 0.06, 0.11 and 0.09 mg/g, which was far below the saturation absorption capacity of FAP (1.98 mg/g) [11]. Therefore, the P removal rate in FAP section was negatively related to retention time. Both results in Figure 4 A and B showed, those two materials could provide stable nutrient removal under the normal flow rate situations of tile drainage and could be applied in the field for further tests.

4 Conclusions

In this study, a bi-column test of both woodchips and FAP has been applied as bioreactor and filter for nutrient retention from a tile drainage system. The results showed that the woodchips section's nitrate removal efficiency was strongly influenced by the retention time, and the shorter retention time had less efficiency. The removal efficiency range was 47.8-80.8%, but the N removal rate was higher, when woodchips run under higher flow rate conditions. The FAP section also contributed to the nitrate removal by chemisorption, of which the removal rate raised along with the retention time, but this contribution was not significant. The total nitrate removal rates of this system were 46.0, 55.3, and 54.8 mg/m^3·d for test 1, test 2 and test 3, respectively. The fly ash pellets effectively removed phosphate in the bioreactor effluent, and the total phosphate adsorption was 0.059-0.114 mg P/g, which was far less than the saturated capacity (1.69 mg/g). The total phosphate removal efficiency could be maintained at the range of 81.7-84.5% under different flow rate situations. Both the nitrate and orthophosphate in this effluent were effectively removed by the woodchips bioreactor-FAP filter system.

Overall, the results of this study suggest that FAP can be applied as effective adsorption materials for phosphate removal in subsurface drainage. Also, this in-stream two-stage nutrient removal system is a highly promising technique and could be applied in the field. The follow-up research on FAP for phosphorus retention should be focused on the hydraulic parameters combined with the analysis of how to avoid the possible field application failures. Likewise, more attention should be given to investigating obvious constraints such as poor saturation time, high pH of outflow, or possible desorption conditions in the field.

5 Acknowledgements

This work was supported by DAAD and Exceed Swindon project.

6 References

[1]   García, A.M., Alexander, R.B., Arnold, J.G., Norfleet, L., White, M.J., Robertson, D.M., Schwarz, G.: Regional Effects of Agricultural Conservation Practices on Nutrient Transport in the Upper Mississippi River Basin. Environ. Sci. Technol. 2016, 50, 6991-7000.

[2]   Rabalais, N.N., Turner, R.E., Wiseman, W.J.: Hypoxia in the Gulf of Mexico. J. Environ. Qual. 2001, 30, 320-329.

[3]   Haas, M.B., Guse, B., Fohrer, N.: Assessing the impacts of Best Management Practices on nitrate pollution in an agricultural dominated lowland catchment considering environmental protection versus economic development. J. Environ. Manage. 2017, 196, 347-364.

[4]   Ritesh, K., Mary Love, M.T., Joel, O.P., Juan, D.P.-G.: Assessment of On-Farm Water Storage (OFWS) Systems as a BMP for Sustainable Irrigation and Nutrient Loading Control in Mississippi. 2015 ASABE Annual International Meeting. (2015)

[5]   Rudolph, D.L., Devlin, J.F., Bekeris, L.: Challenges and a Strategy for Agricultural BMP Monitoring and Remediation of Nitrate Contamination in Unconsolidated Aquifers. Ground Water Monit. Rem. 2015, 35, 97-109.

[6]   Rittenburg, R.A., Squires, A.L., Boll, J., Brooks, E.S., Easton, Z.M., Steenhuis, T.S.: Agricultural BMP Effectiveness and Dominant Hydrological Flow Paths: Concepts and a Review. J. Am. Water Resour. Assoc. 2015, 51, 305-329.

[7]   Jaynes, D.B., Kaspar, T.C., Moorman, T.B., Parkin, T.B.: In Situ Bioreactors and Deep Drain-Pipe Installation to Reduce Nitrate Losses in Artificially Drained Fields. J. Environ. Qual. 2008, 37, 429-436.

[8]   Li, R., Feng, C., Hu, W., Xi, B., Chen, N., Zhao, B., Liu, Y., Hao, C., Pu, J.: Woodchip-sulfur based heterotrophic and autotrophic denitrification (WSHAD) process for nitrate contaminated water remediation. Water Res. 2016, 89, 171-179.

[9]   Robertson, W.D., Merkley, L.C.: In-Stream Bioreactor for Agricultural Nitrate Treatment. J. Environ. Qual. 2009, 38, 230-237.

[10]   Schipper, L.A., Robertson, W.D., Gold, A.J., Jaynes, D.B., Cameron, S.C.: Denitrifying bioreactors—An approach for reducing nitrate loads to receiving waters. Ecol. Eng. 2010, 36, 1532-1543.

[11]   Hoover, N.L., Bhandari, A., Soupir, M.L., Moorman, T.B.: Woodchip Denitrification Bioreactors: Impact of Temperature and Hydraulic Retention Time on Nitrate Removal. J. Environ. Manage. 2016, 45, 803-812.

[12]   Lepine, C., Christianson, L., Sharrer, K., Summerfelt, S.: Optimizing Hydraulic Retention Times in Denitrifying Woodchip Bioreactors Treating Recirculating Aquaculture System Wastewater. J. Environ. Qual. 2016, 45, 813-821.

[13]   Nojiri, M., Koteishi, H., Nakagami, T., Kobayashi, K., Inoue, T., Yamaguchi, K., Suzuki, S.: Structural basis of inter-protein electron transfer for nitrite reduction in denitrification. Nature 2009, 462, 117-U132.

[14]   Thomas, K.L., Lloyd, D., Boddy, L.: Effects of oxygen, pH and nitrate concentration on denitrification by pseudomonas species. FEMS Microbiol. Lett. 1994, 118, 181-186.

[15]   Boujelben, N., Bouzid, J., Elouear, Z., Feki, M., Jamoussi, F., Montiel, A.: Phosphorus removal from aqueous solution using iron coated natural and engineered sorbents. J. Hazard. Mater. 2008, 151, 103-110.

[16]   Dobbie, K.E., Heal, K.V., Aumônier, J., Smith, K.A., Johnston, A., Younger, P.L.: Evaluation of iron ochre from mine drainage treatment for removal of phosphorus from wastewater. Chemosphere 2009, 75, 795-800.

[17]   Jayarathna, L., Bandara, A., Ng, W.J., Weerasooriya, R.: Fluoride adsorption on gamma - Fe_2O_3 nanoparticles. J. Environ. Health Sci. Eng. 2015, 13, 10.

[18]   Kunaschk, M., Schmalz, V., Dietrich, N., Dittmar, T., Worch, E.: Novel regeneration method for phosphate loaded granular ferric (hydr)oxide A contribution to phosphorus recycling. Water Res. 2015, 71, 219-226.

[19]   Xiong, J., He, Z., Mahmood, Q., Liu, D., Yang, X., Islam, E.: Phosphate removal from solution using steel slag through magnetic separation. J. Hazard. Mater. 2008, 152, 211-215.

[20]   Li, S., Cooke, R.A., Wang, L., Ma, F., Bhattarai, R.: Characterization of fly ash ceramic pellet for phosphorus removal. J. Environ. Manage. 2017, 189, 67-74.

[21]   Xie, J., Lai, L., Lin, L., Wu, D., Zhang, Z., Kong, H.: Phosphate removal from water by a novel zeolite/lanthanum hydroxide hybrid material prepared from coal fly ash. J. Environ. Sci. Health., Part A 2015, 50, 1298-1305.

[22]   Li, Y., Liu, C., Luan, Z., Peng, X., Zhu, C., Chen, Z., Zhang, Z., Fan, J., Jia, Z.: Phosphate removal from aqueous solutions using raw and activated red mud and fly ash. J. Hazard. Mater. 2006, 137, 374-383.

[23]   Wang, Z., Fan, Y., Li, Y., Qu, F., Wu, D., Kong, H.: Synthesis of zeolite/hydrous lanthanum oxide composite from coal fly ash for efficient phosphate removal from lake water. Microporous Mesoporous Mater. 2016, 222, 226-234.

[24]   Rendall, T.J.: Effect of passive and active heating on the performance of denitrifying bioreactors. Master thesis, University of Illinois at Urbana-Champaign, 2015

[25]   Vohla, C., Kõiv, M., Bavor, H.J., Chazarenc, F., Mander, Ü.: Filter materials for phosphorus removal from wastewater in treatment wetlands—A review. Ecol. Eng. 2011, 37, 70-89.

[26]   Christianson, L.E., Lepine, C., Sibrell, P.L., Penn, C., Summerfelt, S.T.: Denitrifying woodchip bioreactor and phosphorus filter pairing to minimize pollution swapping. Water Res. 2017, 121, 129-139.

[27]   Islam, M., Patel, R.: Nitrate sorption by thermally activated Mg/Al chloride hydrotalcite-like compound. J. Hazard. Mater. 2009, 169, 524-531.

[28]   Goodwin, G.E., Bhattarai, R., Cooke, R.: Synergism in nitrate and orthophosphate removal in subsurface bioreactors. Ecol. Eng. 2015, 84, 559-568.

[29]   Hua, G., Salo, M.W., Schmit, C.G., Hay, C.H.: Nitrate and phosphate removal from agricultural subsurface drainage using laboratory woodchip bioreactors and recycled steel byproduct filters. Water Res. 2016, 102, 180-189.

WASTEWATER REUSE IN IRRIGATED AGRICULTURE

S. El Hajjaji[1], J. Mabrouki[1], C. Bakkouche[1], A. Dahchour[2]

[1]*Laboratory of Spectroscopy, Molecular Modeling, Materials, Nanomaterials, Water and Environment, CERNE2D, Mohammed V University in Rabat, Faculty of Science, AV Ibn Battouta, BP1014, Agdal, Rabat, Morocco; hajjajisouad@yahoo.fr*

[2]*Agronomic and Veterinary Institute Hassan II, Rabat, Morocco*

Keywords: Irrigation, Wastewater Reuse, Agriculture, Morocco

Abstract

Morocco's interest in the reuse of wastewater for irrigation purposes comes from its water resources, which are limited and unequally distributed over space and time, particularly in some cities located in regions characterized by an arid or semi-arid climate with a large water deficit and high agricultural activities. The agricultural reuse of wastewater has had variable applications and has become an unavoidable practice for the majority of farmers already weakened by lack of water. Direct irrigation of crops from raw sewage is practiced despite its prohibition. This practice is not without dangerous consequences causing damage to human health and the environment. As a result, Morocco has been called upon to develop its regulatory and legal aspects of wastewater development through integrated approaches to water resources management, while taking into account technical, socio-economic and environmental factors, to preserve the quality of its bodies of water and to obtain safe alternative supplies.

1 Introduction

The increase of the Moroccan population, the unequal distribution of water resources (Figure 1), increasing of social and economic activities, climate change and intensive agricu ture has increased the demand for water supply [1]. These combined factors have had a serious impact on the water resources potential in Morocco. The ratio per capita is expected to be below the alarming level of 550 CM/cap by 2020 instead of 2000 CM/cap in 1970. This called up different water resource management programs and strategies including improving water storage capacity, water use governance and the use of unconventional water resources.

In Morocco, a national liquid sanitation program has been launched since 2005 to ensure the connection of sewers in urban and rural areas, improving the quality of water before it is discharged into rivers or streams and its utilization in the agriculture. Wastewater potential was estimated at 600 MCM (million cubic meter) in 2005 and is expected to reach 900 MCM/yr by 2020 (Figure 2). Rural areas produce 500,000 m^3 WW/d, which are directly rejected on the ground. The quality of wastewater (WW) depends on the size of the urban center and the pollution load (pollutants and pathogenic microorganisms).

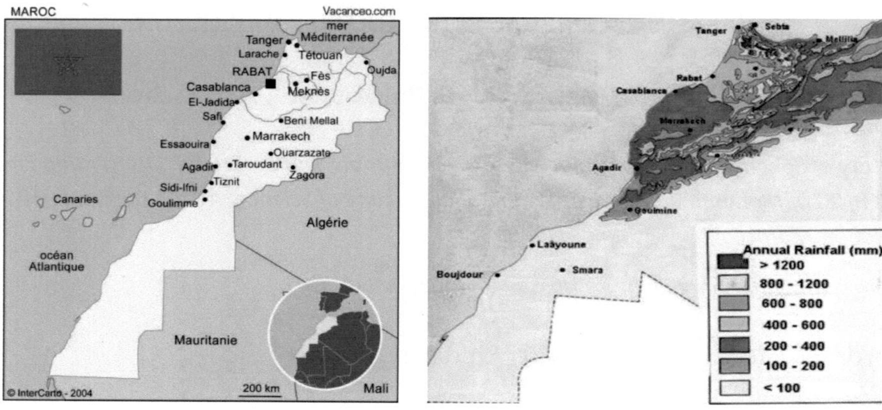

Figure 1: *(left)* Main Rivers in Morocco, *(right)* Rainfall Distribution [1]

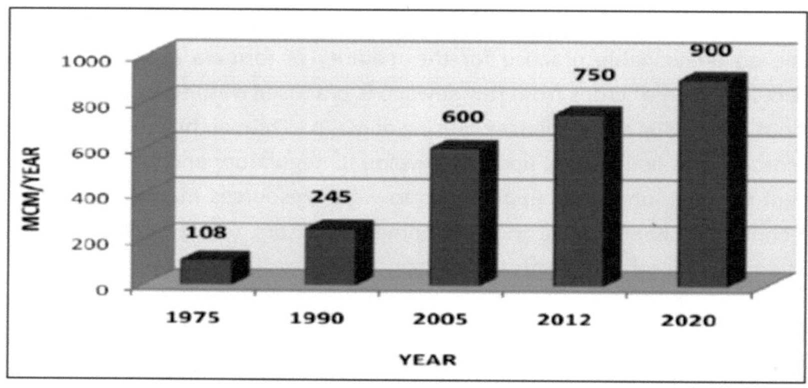

Figure 2: Temporal evolution of the total volume of wastewater discharged in Morocco [2]

The present work aims to discuss the current situation in Morocco regarding the reuse of raw and treated wastewater in agriculture, and its positive and negative impacts on humans and the environment, while giving a general overview of the guidelines adopted by the country for better and safe reuse.

2 Wastewater Reuse Standards and Regulations

At their national level, some developed countries have guidelines for agricultural irrigation with wastewater, while internationally, the most widespread guidelines are those established by the United Nations (UN), specifically those of the World Health Organization (WHO) and of the United Nations Food and Agriculture Organization (FAO). Their major objectives are to ensure the protection of the environment, to improve public health and to facilitate the rational use of wastewater in agriculture [3].

2.1 WHO standards

Since the 1970s, WHO has begun to take into account the environmental and socio-economic benefits of agricultural wastewater reuse and its importance. However, following the results of an epidemiological study that was conducted, it was concluded by WHO experts that the consumption of food from crops irrigated with wastewater might have adverse effects on health from pathogens, such as protozoa, viruses, bacteria and helminths [4]. Therefore, in order to reduce the danger associated with the reuse of wastewater on public health, WHO has developed a directive called *"Directive on the microbiological quality of wastewater used in agriculture"*, which focuses on four measures, namely: water treatment, crop limitation, control of the use of wastewater, exposure control with improvement, and the purpose of which is to promote the adequate reuse of wastewater in agriculture [5].

Table 1: Water quality recommended by WHO (2006) [5]

Parameter		Units	Degree of restriction on use		
			None	Slight to moderate	Severe
Salinity EC_w^a		dS/m	< 0.7	0.7 – 3.0	> 3.0
TDS		mg/L	< 450	450- 2000	>2000
TSS		mg/L	< 50	50 - 100	>100
SAR^b	0 - 3	meq/L	> 0.7 EC_w	0.7 – 0.2 EC_w	< 0.2 EC_w
SAR	3 - 6	meq/L	> 1.2 EC_w	1.2 – 0.3 EC_w	< 0.3 EC_w
SAR	6 - 12	meq/L	> 1.9 EC_w	1.2 – 0.5 EC_w	< 0.5 EC_w
SAR	12 - 20	meq/L	> 2.9 EC_w	2.9 – 1.3 EC_w	< 1.3 EC_w
SAR	20 - 40	meq/L	> 5.0 EC_w	5.0 – 2.9 EC_w	< 2.9 EC_w
Sodium (Na^+)	Sprinkler irrigation	meq/L	< 3	>3	
Sodium (Na^+)	Surface irrigation	meq/L	<3	3 - 9	>9
Chloride (Cl^-)	Sprinkler irrigation	meq/L	< 3	>3	
Chloride (Cl^-)	Surface irrigation	meq/L	<4	4 - 10	> 10
Chlorine (Cl_2)	Total residual	mg/L	< 1	1 – 5	> 5
Bicarbonate (HCO_3^-)		mg/L	< 90	90 - 500	> 500
Boron (B)		mg/L	< 0.7	0.7 – 3.0	> 3.0
Hydrogen sulfide (H_2S)		mg/L	< 0.5	0.5 – 2.0	> 2.0
Iron (Fe)	Drip irrigation	mg/L	< 0.1	0.1 – 1.5	> 1.5
Manganese (Mn)	Drip irrigation	mg/L	< 0.1	0.1 – 1.5	> 1.5
Total nitrogen (TN)		mg/L	< 5	5 - 30	> 30
pH			Normal range 3.5 - 8		

2.2 United Nations Guidelines for Food and Agriculture (FAO)

From its part, FAO has similarly formed numerous guidelines on the use of wastewater in agriculture. They established the relationship between the degree of restriction of water use and the parameters of salinity, infiltration and toxicity of specific ions [6]. FAO has also suggested guidelines for "*Agricultural reuse of treated water and treatment requirements*", in which the type of agricultural reuse was classified according to the type of irrigated crop (Table 2) [7].

Table 2: Types of irrigated crop [7]

Type of Agricultural Reuse	Agricultural reuse in crops that are consumed and not processed commercially	Agricultural reuse in crops that are consumed and not processed commercially	Agricultural reuse in crops that are not consumed
Type of Treatment	Secondary: Filtration-Disinfection	Secondary-Disinfection	Secondary-Disinfection
Quality Criterion	pH = 6.5 – 8.4 BOD < 10 mg / L < 2 UNT <14 NMP *E. coli* /100 ml < 1 Egg /L	pH = 6.5 – 8.4 BOD < 30 mg /L SS < 30 mg /L <200 NMP *E. coli* / 100 mL	pH = 6.5 – 8.4 BOD < 30 mg /L SS < 30 mg /L <200 NMP *E. coli* / 100 mL

2.3 Moroccan regulatory framework

Morocco is a country, whose economy is based mainly on agriculture. However, agriculture is the most consumer sector in terms of water resources around the world. These resources have varied characteristics according to their regions and their eventual contact with pollution sources. In addition, given the reduction in water supplies over the last decades, and in order to mitigate the impacts on groundwater, policy makers are encouraging the reuse of treated wastewater as a substitute for agricultural practices [8].

In this context, usage of WW is adopted illegally by an increasing number of farmers in Morocco, mainly in regions under freshwater scarcity. To prevent negative impact on health and environment, water quality standards for irrigation have been established to protect [9]:

- The public and farm workers;
- Consumers of agricultural products;
- Surface and groundwater resources and soils;
- Irrigation equipment.

In this sense, the State Secretary of Ministry of Energy in charge of water and the environment has established its quality standards for wastewater for irrigation. They are based on general legislation (law 36-15 of water) for the protection and integrated management of water resources. More specifically, the Article 84 of Law 36-15 on water in relation to wastewater is stating that the usage of wastewater for agricultural purposes is forbidden, if these waters do not meet national standards [9, 10].

2.4 Water quality standards for irrigation

Moroccan standards for water quality for irrigation of the Secretariat of State at the Ministry of Energy, Mines, Water and Environment, in charge of Water and the environment [9, 10] are not applied only to irrigation with treated wastewater, but to all types of irrigated water. The quality of wastewater designed for irrigation, which specifies bacteriological, parasitic and physicochemical parameters, is given in Table 3.

Table 3: Bacteriological, parasitic and physicochemical parameters [10]

	PARAMETERS	LIMIT VALUES
	BACTERIOLOGICAL	
1	Fecal Coliform	1000/100 mL *
2	Salmonella	Absence in 5 L
3	*Cholera Vibrio*	Absence in 450 mL
	PARASITOLOGICAL	
4	Pathogen	Absence
5	Eggs, Parasite cysts	Absence
6	Ancylostoma larva	Absence
7	Flurococercaires of *Schistomosa hoematobium*	Absence
	TOXIC PARAMETERS	
8	Mercury (Hg) mg/L	0.001
9	Cadmium (Cd) mg/L	0.01
10	Arsenic (As) mg/L	0.1
11	Chromium (Cr) mg/L	1
12	Lead (Pb) mg/L	5
13	Copper (Cu) mg/L	2
14	Zinc (Zn) mg/L	2
15	Selenium (Se) mg/L	0.02
16	Fluorine (F) mg/L	1

17	Cyanide (CN) mg/L	1
18	Phenol mg/L	3
19	Aluminum (Al) mg/L	5
20	Beryllium (Be) mg/L	0.1
21	Cobalt (Co) mg/L	0.5
22	Iron (Fe) mg/L	5
23	Lithium (Li) mg/L	2.5
24	Manganese (Mn) mg/L	0.2
25	Molybdenum (Mo) mg/L	0.01
26	Nickel (Ni) mg/L	2
27	Vanadium (V) mg/L	0.1
PHYSICOCHEMICAL		
	Salinity	
28	Total salinity (TSD) mg/L	7680
	Electrical Conductivity (EC) mS/cm at 25 °C	12
29	Infiltration	
	SAR*** = 0 - 3 & EC =	<0.2
	SAR*** = 3 - 6 & EC =	<0.3
	SAR*** = 6 - 12 & EC =	<0.5
	SAR*** = 12 - 20 & EC =	<1.3
	SAR*** = 20 - 40 & EC =	<3
TOXIC IONS (AFFECTING SENSITIVE CROPS)		
30	Sodium Na mg/L	
	Surface irrigation (SAR***)	69
	Sprinkler irrigation	9
31	Chloride Cl⁻ mg/L	
	Surface irrigation	350
	Sprinkler irrigation	15

32	Boron (B) mg/L	3
	Various effects (affecting sensitive crops)	
33	Temperature °C	35
34	pH	6.5 – 8.4
35	Suspended Solids mg/L	
	Gravity irrigation	200
	Localized sprinkler irrigation	100
36	Nitrate nitrogen (N-NO$_3^-$) mg/L	30
37	Bicarbonate (HCO$_3^-$) [sprinkler irrigation] mg/L	518
38	Sulfates (SO$_4^{2-}$) mg/L	250

Indeed, treated wastewater to be in accordance with irrigation regulations, must respect these three criteria [10]:

- *Sampling*
  The collection of treated wastewater samples must take place at the end of the treatment plants. The minimum number of samples is fixed as follows:
  - For heavy metals analysis: 4 per year (1 per quarter)
  - For the analysis of physicochemical, bacteriological and parasitological parameters: 24 per year (1 every 15 days)

- *Analyses*
  Indicator parameters for water quality for irrigation should be measured using standard methods.

- *Conformity assessment*
  Any water, whose characteristics meet the limit values in the table of water quality standards for irrigation, is valid for irrigation.

3 Impact of Wastewater Use in Agriculture

At present, millions of m^3 of raw sewage are discharged into the environment due to the lack of treatment in several cities. Most of this water is discharged into the sea (about 60%). The continuation of these discharges could unleash a serious degradation of the surface and underground water resources of the different regions of the country. At the same time, continued reuse of raw wastewater can have serious consequences for the environment and the health of the local population, who will be exposed to waterborne diseases [11]. However, they are widely used in agriculture for their socio-economic implications, supply of fertilizers and sustainable availability [8 -19].

3.1 Impacts on health

One of the most recognized benefits of reusing treated wastewater in irrigation is the associated reduction in pressure on freshwater sources [12] and fertilizers, increasing crop yields, and the protection of consumer health and the environment. The reuse of raw wastewater is likely to endanger human health. As a result, wastewater can carry many pathogens (parasites, bacteria, viruses and fungi) that are highly resistant to the environment and can cause harm to humans.

As a result, wastewater is, therefore, considered to be an important transmitter of biological and chemical agents resulting from human and/or industrial activities. Irrigated agriculture with wastewater concentrates many infectious and toxic agents. In addition, it represents an environment conducive to the proliferation of certain pathogens emitted in human or animal waste. In addition, sewage resulting from discharge from hospital or other infected environments may be a source of dangerous contamination.

Due to the complexity of pathogen biology and their mode of transmission, mankind is likely to be contaminated by various means to know [11]:

- *Direct contact with wastewater*: This risk is associated with farmers, who cultivate the soil without taking the necessary precautions to protect themselves from the risk of contamination, which can take place by simple direct contact with wastewater containing parasites of human or animal origin; these are able to cross the skin barrier and enter the body.
- *The direct consumption of crops irrigated with waste water or animal meat fed with fodder irrigated by wastewater:* This is a kind of frequent human contamination. This risk is not limited to farmers, who consume their own agricultural products, but it also concerns the population that buys contaminated products.

3.2 Environmental impacts

The reuse of wastewater makes it possible to exploit the nutrients it contains in agriculture or green spaces. This will contribute to the recycling of nutrients in plant biomass and the reduction of their pollution capacity, especially when they seep into groundwater or are found as particles or in solution in surface water, while increasing the agricultural production in regions in crisis, thus contributing to the food security of their inhabitants [13]. However, it is necessary to point out that the reuse of treated wastewater must be done according to good irrigation practices: types of irrigation, choice of agricultural practice, mode of rotation, etc. Nevertheless, the risk of infection depends on the pathogens and the circumstances that favor their transmission [11].

3.3 Health risks related to reuse of agricultural wastewater

Depending on health and socio-economic conditions of a particular community, variations in levels and types of pathogen and chemical pollution in wastewater are observed from one region to another [12]. Indeed, in developing countries, the concentration of viruses, parasitic protozoa and helminths in wastewater can be 10 to 1000 times higher than in developed countries [15]. Table 4 gives the types of major enteric pathogens and substances of health interest that can be found in wastewater used for agricultural irrigation [5, 12-15].

Table 4: Risks related to the reuse of agricultural wastewater

Type of Risk		Pathogens
Chemical	Pesticides (1)	Aldrin, DDT
(Substances of sanitary	Heavy Metals (2)	As, Cd, Hg
interest)	Hydrocarbons (2)	PCBs, Dioxins, Furans,
Biological	Bacteria (1)	*E. coli, Vibrio cholerae, Salmonella spp., Shigella spp.*
	Helminths (1)	*Ancylostoma, Ascaris, Tenia spp.*
	Protozoans (1)	Intestinal *Giardia, Crysptospridium, Entamoeba spp.*
	Virus (1)	Hepatitis A&E, Adenovirus, Rotavirus, Norovirus
	Schistosoma (2)	Blood-flukes

(1) Contact and/or consumption; (2) consumption

4 Moroccan Strategies for Reuse of Agricultural Wastewater

Being in water deficit and having the will to control and to limit the old practices of reuse of raw WW, Morocco has launched since tens of years many multidisciplinary projects, carrying on the reuse of treated wastewater. The main purpose of which was to address the country's agronomic, health and environmental concerns, but their applications remain limited [16]. For instance, to faces the challenge of the scarcity of water resources, the Green Morocco Plan and the National Water Strategy consider the management of water demand and the valorization of water as a priority axis of strategic importance for the sectors water and agriculture.

4.1 Green Morocco Plan

In 2008, the Green Morocco Plan highlighted the structural water shortages in most of Morocco's major agricultural production basins, while considering that the scarcity of water resources is a major constraint to the development of agriculture in the different regions of the country [17].

4.2 The National Water Strategy

In 2010, the Moroccan public authorities approved the National Water Strategy, a strategy that believes that wastewater reuse is an important unconventional water resource, and its valorization must be part of the integrated management of water resources at national level [17]. To develop a master plan for the reuse of treated wastewater in irrigation, the Ministry of Agriculture launched in 2011 a study, whose objective is to remove water constraints with a view of promoting the reuse of wastewater in agriculture and taking into account the economic, social, environmental and health issues of reuse of treated wastewater and the context of growing scarcity of water resources [17].

4.3 Water use statistics in agriculture

Because of the increasing water deficit, the use of raw wastewater effluents in Morocco is a very old practice. Generally, these effluents are used, where they are most valuable. Farmers use this practice in small areas surrounding the treatment networks as well as on the periphery of few large cities, where farmland is located downstream of effluent discharges [11]. However, irrigation of vegetable crops with raw sewage is prohibited in Morocco, but this is not respected. This poses the consumer of agricultural products and the farmer risks of bacteria or parasitic diseases [11].

4.4 Pilot projects in Morocco

To ensure safe usage of WW in agriculture, four pilot projects were conducted with WW treated at different locations (Ouarzazate, Bensergao, Bensliman and Drarga). The systems of WW treatment are different: Ouarzazat (lagoon), Bensergao (filtration, percolation), Benslimane (aerated lagoon) and Drarga (filtration, percolation). The objectives were the assessment of the usefulness of such projects for the population in terms of economic benefits, health and environmental protection, defining the most appropriate techniques and reinforcing national capacities in reuse of treated WW (Table 5).

Results from tests in Benslimane were positive in terms of reduction of nuisance and preservation of groundwater used for irrigation of golf courses. The level of treatment is very high in the majority of the treatment plants tested, mainly in terms of pathogenic microorganisms. The project in Drarga was designed for a maximum reuse of WW treated and all derivatives from the plant. Thus, WW treated is sold to farmers, while sludge is dried and composted. Under environmental aspects, treatment of WW allows the reduction of pollution in terms of BOD5 from 270,000 tons to 27,000 tons per year. Nitrogen pollution will be reduced from 54,000 tons to 25,000 tons. For instance, results obtained in Drarga pilot plant showed a reduction of nitrogen compounds from 12.6 tons to 1.6 tons.

Nutritive materials could be then recycled in the vegetable biomass reducing their pollution capacities to the environment [19]. Comparison of production costs in the four tested plants revealed that the lowest price was recorded for Bensergao plant, functioning under filtration percolation system (0.12-0.15 USD/m^3). However, wastewater is rich in pathogenic microorganisms such as viruses, bacteria and parasites that could cause diseases and impacts on human health. Protozoa and helminth eggs are most virulent, and they are most difficult to remove by treatment processes.

Table 5: Efficiencies of Wastewater Treatment Plants (WWTP) [18]

WWTP	Oua		B S	D	BSL	Mar	BZK
System used	Lag	Lag AHR	Oer		Ae Lag	Lag fac	Lag
Resident Time (days)	25	21.9	-	-	30 to 40	30	--
BOD_5 (mg/L	81.7	65.3	98	98.5	78	97	75
COD (mg/L)	72	56.4	92	96	79	76	71
MES (mg/L)	281	-	100	96.6	-	69	76
NTK (mg/L)	31.5	48	85	96.8	75	71	14
P_{tot} (mg/L)	48.5	54	36	95.9	41	85	-
CF/100 ml	99.9	99.9	99.9	99.9	100	99.4	99.9
O. he/L	100	100	100	100	100	100	100

Oua: Ouarzazat, BS: Bensergaou, D: Drarga, BSL: Benslimane, Mar: Marrakech, BZK: Bouznika, Lag: laggon, AHR: at higher activity, fac: facultative

5 Conclusions

On a global scale, agriculture is a big consumer of water. The search for alternative sources of irrigation water is considered essential for supplying food security and preserving natural water bodies. The safe use of wastewater as an alternative source of irrigation is a recognized strategy for the effective use and prevention of water pollution, which is gaining relevance around the world, particularly in countries facing water shortages, especially in Morocco. However, some risks associated with this type of reuse must be assessed against a local framework, while taking into account the receiving environment and ensuring that pollution will not be transferred from one environment to another.

6 Acknowledgements

The authors would like to thank EXCEED Swindon project and DAAD (German Academic Exchange Service) for support to participate at the *"International Expert Workshop on Water in Agricultural Practices: Training the Trainers"*, 15 – 21 Sep, 2019, Rio de Janeiro, Brazil.

7 References

[1]   Ouhssain, M. : La gestion sociale de l'eau au Maroc de AZERF à la loi sur l'eau. Revue HTE, 141, December 2008.

[2]   Mandi, L.: Review of wastewater situation in Morocco. In: I. Chorus, U. Ringelband, G. Schlag and O. Schmoll (Eds.): Water, Sanitation and Health. World Health Organisation Water Series, IWA Publishing, London, UK, 2000, 383-387.

[3]   Drechsel, P.: International Water Management Institute and International Development Research Centre. Wastewater Irrigation and Health: Assessing and Mitigating Risk in Low-Income Countries. 2010.

[4]   WHO: Guidelines for the safe use of wastewater and excreta in agriculture and aquaculture. Geneva, 1989. ISBN 9241 54248 9

[5]   WHO: Good irrigation practice in: OMS. Guidelines for the safe use of wastewater, excreta and greywater, Volume 2, Wastewater use in agriculture. 2006, vol. 2, ISBN 924154683 2.

[6]   Ayers. R, Wescott, D.: Water Quality for Agriculture; FAO, 1985, p. 174.

[7]   FAO: Wastewater Treatment and Use in Agriculture. 1992, 47. ISBN 92-5-103135-5

[8]   Malki, M., Bouchaou. L, Mansir, I., Benlouali, H., Nghira, A, Choukr-Allah, R.: Wastewater treatment and reuse for irrigation as alternative resource for water safeguarding in Souss-Massa region, Morocco, Eur. Water, 2017, 59, 365-371.

[9]   Secrétariat d'Etat auprès du Ministère de l'Energie, des Mines, de l'Eau et de l'Environnement, chargé de l'Eau et de l'Environnement : Normes de qualité des eaux destinées à l'irrigation. 2007.

[10]   Dadi, E.M. : l'évaluation de la possibilité de réutiliser en agriculture, l'effluent traité de la commune de Drarga. Centre universitaire de formation en environnement, Université de Sherbrooke. Sherbrooke, Québec, Canada, 2010.

[11]   Jemali, A., Kefati, A.: Ministry of Agriculture, Rural Development and Waters and Forests, Rural Engineering Administration Development and Irrigation Management Directorate. Water Demand Management Forum « Wastewater Reuse in Morocco », 2002.

[12]   Jaramillo, M.F., Restrepo, I.: Wastewater Reuse in Agriculture: A review about its limitations and benefits. Sustainability, 2017, 9, 1734.

[13]   Corcoran, H., Nellemann, E., Baker, C., Bos, E., Osborn, R., Savelli, D.: Sick Water? The Central Role of Wastewater Management in Sustainable Development, A Rapid Response Assessment. United Nations Environment Program, UN-HABITAT, GRID-Arendal. 2010, ISBN: 978-82-7701-075-5. www.grida.no.

[14] Gerba, C.P., Rose, J.B.: International guidelines for water recycling: Microbiological considerations. Water Sci. Technol., 2003, 3, 311–316.

[15] Jimenez, B., Mara, D., Carr, R, Brissaud, F.: Wastewater treatment for pathogen removal and nutrient conservation: Suitable systems for use in developing countries. 2010. IWMI Books, Reports H042608, International Water Management Institute.

[16] Condom, N., Declecq, R.: Rapport ecofilae. Réutilisation des eaux usées pour l'irrigation agricole en zone péri-urbaine de pays en développement. Pratiques, défis et solutions opérationnelles. Agriculture dans le Delta du Nil, 2015, 1, 10-16.

[17] Ministère de l'Agriculture et de la Pêche Maritime. Secrétariat d'Etat auprès du Ministère de l'Energie, des Mines, de l'Eau et de l'Environnement chargé de l'Eau et de l'Environnement : Projet de Renforcement des Capacités sur l'Utilisation sans danger, 2011.

[18] Choukr-Allah, R., Hamdy, A.: Wastewater treatment and reuse as potential water resource for irrigation. In: Atef Hamdy (ed.): The use of nonconventional water resources. Proceedings of the international Workshop Alger, Algeria, 12-14 June 2005. Options Méditerranéennes, Séries A, 2005, (66), 101-124. ISBN: 2-85352-318-7.

[19] Choukr-Allah, R.: Wastewater Treatment and Reuse in Morocco: Situation and Perspectives. In: Hamdy, A., El Gamal, F., Lamaddalena, N., Bogliotti, C., Guelloubi, R., (eds.): Non-conventionnel Water Use - WASAMED Project, CIHEAM. 2005, pp 104-114.

CHARACTERIZATION OF EFFLUENTS FROM FECAL SLUDGE TREATMENT PLANT IN CAMEROON AND ASSESSMENT OF THEIR POTENTIAL USE FOR AGRICULTURAL IRRIGATION

L.S.T. Tchianzeu, W.A.L. Nzouebet, C. Wanda, V.G.D. Wafo, E.S. Kengne, G. Liégui, J.P. Noutadié, A.P. Agendia, W.K. Arnold, I.M.K. Noumsi

University of Yaounde I, Faculty of Science, Wastewater Research Unit, Po. Box.812 Yaounde, Cameroon; tchianzeus@gmail.com

Keywords: Agriculture, fecal sludge, purification plant, tropical urban area

Abstract

The present research work is a contribution to the characterization of wastewater treatment plants' discharges in the city of Bafoussam with a view of their use for agricultural irrigation. From July 2017 to February 2018, fecal sludge treated effluents from the treatment plant of Kuekong, Bafoussam were analyzed in order to determine their physicochemical and bacteriological characteristics following standard protocols. Physicochemical parameters recorded at the outlet of the treatment plant are at the order of 25.6 mg/L for TSS, 791 µS/cm for EC, 2.93 mg/L for NO_3^-, 5.46 mg/L for NH_4^+, and 3.31 mg/L for PO_4^{3-}. The average values recorded were at the order of 227 mg/L for COD and 70 mg/L for BOD_5, respectively. As for the values recorded at the exit of the station for bacteriological parameters, they were at the order of 130 CFU/100 mL for fecal coliforms and 64 CFU/100 mL for fecal streptococci. In general, the system allows a reduction of fecal sludge pollutants with discharge values in accordance with the discharge standards for treatment plants as prescribed by MINEP DED and WHO, except for COD, BOD_5 and TSS that are above the standards. In view of the parameters analyzed and their comparison with standards, this shows that the latter being in conformity with national and international standards and can be used in agriculture without having considerable health and environmental impacts. However, fecal sludge also contains trace metals that can have an impact on health and the environment. So, these waters from sludge treatment must be analyzed before reuse.

1 Introduction

Because of its geographical location, Cameroon has a great diversity of soils and climates that allow the country to be divided into five large agro-ecological zones. With a growth rate of about 3%, Cameroon's population is fast growing, thereby increasing the food demand. Agriculture is and remains the predominant sector of the economy accounting for more than 44% of GDP (Gross domestic product) and occupying about 56% of the labor force in Cameroon [1]. In the Western Highland region, the dominant cash crop is coffee. The associated food crops are maize, groundnuts, plantain, yams, beans, bananas, onions, soybeans, sweet potatoes, and potatoes [2].

To increase agricultural yields, farmers sometimes use chemical fertilizers. These fertilizers are of bad impact on the environment. Indeed, they are known to have a significant negative impact on the quality of the water bodies [2, 3]. They are often associated with eutrophication, because they stimulate algal growth and, as a result, increase the potential for cyanobacteria outbreaks [2]. A recent report has highlighted many other negative impacts of excess nitrogen: soil acidification, leading to a decrease of its fertility and an increased risk of leaching heavy metals; loss of biodiversity, especially in the grasslands, contribution to global warming by emissions of nitrous oxide (N_2O), which contributes 300 times more to the greenhouse effects than carbon dioxide at equal quantity [4]. Natural fertilizers are a solution for this problem. In order to reduce the public health and environmental hazard of poor fecal sludge management while benefiting from the by-products of sanitation, a treatment plant has been constructed in Bafoussam, a city of about 200,000 inhabitants, situated in the western region of Cameroon. Indeed, as shown by Bassan et al. [5] and Soh et al. [6], fecal sludge is rich in organic and mineral matter and can be used for agricultural irrigation, provided that their treatment generates by-products (dried fecal sludge or bio-solids and urine), whose quality complies with the standards for reuse in agriculture. With this regard, the present work focusses on the characterization of the effluents from the Bafoussam fecal sludge treatment plant for reuse in agriculture.

2 Material and Methods

2.1 Study site

The fecal sludge treatment plant of Bafoussam is located 20 km from the city center of Bafoussam, Western Region of Cameroon. Its geographical coordinates are 5°28' North Latitude and 10°33' East Longitude, and the average altitude is 1,450 m. This sludge treatment plant is under tropical climate with an average annual temperature of 20.0 °C and an average annual rainfall of 1,871 mm. The climate is mild and cool with a dry season from mid-November to mid-March and a rainy season from mid-March to mid-November [7]. Its population was estimated at 347,517 in 2008 [8]. The treatment plant has a total surface area of 1,200 m² and consists of an assemblage of 3 anaerobic bed in series combined to 10 lagoons in series (Figure 1).

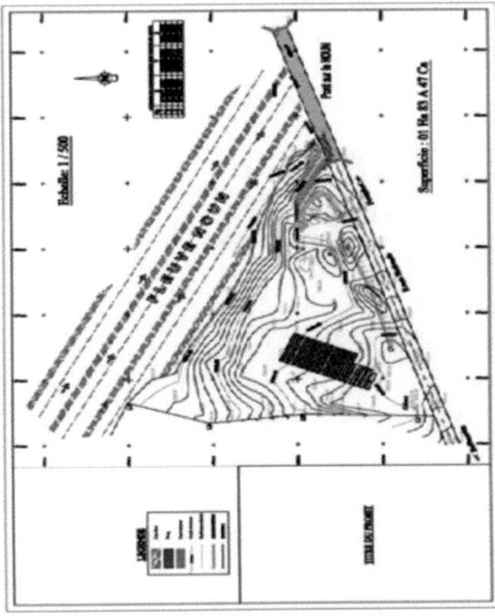

Figure 1:   Overview of the Bafoussam
Fecal Sludge Treatment Plant

2.2 Wastewater sampling and analysis

The methodology consisted of sampling of fecal sludge samples and water samples (using sterile polyethylene bottles) at the outlet of the treatment system. These samples were then returned to the laboratory in a refrigerated cooler for physicochemical, bacteriological and biological analyses according to the standard protocols [9]. The main parameters considered for analysis were: pH, temperature, electrical conductivity (EC), total dissolved solids (TDS), suspended solids (TSS), biochemical oxygen demand (BOD_5), chemical oxygen demand (COD), phosphate (PO_4^{3-}), nitrate (NO_3^-) and ammonium (NH_4^+).

3 Results and Discussion

3.1 Results

Purification efficiency of the station

The evolution of the purification performance of the lagooning system downstream of the bio-digester follows a decreasing trend (Table 1). The removal efficiency rates for EC, TSS, NO_3^-, NH_4^+, PO_4^{3-}, COD, BOD_5, fecal coliforms (FC), and fecal streptococci (FS) are given in this table. It can be seen from Table 1 that the treatment performance is overall above 85%, except for electrical conductivity (59.8%). There is a strong reduction of fecal coliforms and fecal streptococci.

Table 1: Rate of reduction of physicochemical and bacteriological parameters of the station

Parameters	Concentration		Removal efficiency (%)
	Entry	Exit	
pH	7.07 ± 0.3	7.52 ± 0.1	/
Temperature (°C)	26.87 ± 1.2		/
TSS (mg/L)	$8{,}739 \pm 1{,}182$	25.7 ± 13.4	99.7
TDS (mg/L)	966 ± 486	$3{,}061 \pm 4{,}934$	/
EC (µS/cm)	$1{,}969 \pm 978$	791 ± 611	59.8
NO_3^- (mg/L)	436 ± 129	2.93 ± 1.62	99.3
NH_4^+ (mg/L)	120 ± 147	$5{,}46 \pm 7.12$	95.4
PO_4^{3-} (mg/L)	28.6 ± 6.44	3.31 ± 2.16	88.4
COD (mg/L)	$10{,}545 \pm 3{,}998$	227 ± 58.1	97.8
BOD_5 (mg/L)	$1{,}523 \pm 140$	70 ± 10	95.4
FC (CFU/100 mL)	$(70 \pm 36) \times 10^5$	130 ± 52.9	99.9
FS (CFU/100 mL)	$(53 \pm 46) \times 10^5$	64 ± 36.5	99.9

Effluent quality

The effluent quality is summarized in Table 2. The concentrations of the parameters T, pH, TSS, TDS, EC, NH_4^+, NO_3^-, PO_4^{3-}, COD and BOD_5 at the outlet of the treatment plant are listed there. Fecal coliform and fecal streptococci concentrations are 130 ± 52.9 CFU/100 mL and 64 ± 36.5 CFU/100 mL, respectively. This Table shows that the concentrations of NH_4^+, NO_3^-, PO_4^{3-} of the effluents are in compliance with the release standards.

Table 2: Comparison of physicochemical and bacteriological parameters of the releases of the Bafoussam treatment plant with the national release standards of Cameroonian [10] and international ones of WHO [11].

Parameters	Effluent	Cameroonian Norms	WHO Norms
Temperature (°C)	25.6 ± 0.81	30	35
Ph	7.52 ± 0.11	6 - 8,5	6.5 - 8.4
TSS (mg/L)	25.7 ±13.4	50	50 – 100
TDS (mg/L)	3,061 ± 4,934	/	450 – 2000
EC (µS/cm)	791 ± 611	< 200/	900
NH_4^+ (mg/L)	5.46 ± 7.12	/	/
NO_3^- (mg/L)	2.93 ± 1.62	30	/
PO_4^{3-} (mg/L)	3.31 ± 2.16	/	0
COD (mg/L)	227 ± 58.1	200	100
BOD_5 (mg/L)	70 ± 10	50	02
FC (CFU/100 mL)	130 ± 52.9	2000	2000
FS (CFU/100 mL)	64 ± 36.5	1000	1000

3.2 Discussion

The levels of physicochemical and bacteriological parameters of the fecal sludge at the entrance of the Bafoussam treatment plant are similar to those in the literature [5, 6, 12, 13]. Indeed, those authors working on the characterization of raw sludge of sewage treatment plants in Sub-Saharan Africa have found that the sludge has high levels of organic substances, minerals, particulate and bacteriological matter. This variability of the sludge obtained at the entrance of the treatment plant is due to the factors influencing the quality of the fecal sludge. Indeed, [14] identified several factors such as temperature, groundwater intrusion, home performance of the structure, sludge storage time and presence of solid waste.

The evolution of physicochemical and bacteriological parameters along the station follows a regression. During this study, the values of the physicochemical parameters recorded at the exit of the treatment plant and their respective removal efficiency were at the order of 25.6 mg/L (99.7%) for TSS, 791 µS/cm (59.8 %) for EC, 2.93 mg/L (99.3%) for NO_3^-, 5.46 mg/L (95.5%) for NH_4^+, and 3.31 mg/L (88.4%) for PO_4^{3-}, respectively. The elimination of organic pollution averaged 227 mg/L (97.8%) for COD and 70 mg/L (95.4%) for DBO_5. Bacteria concentration and removal averaged 130 CFU/100 mL (99.9%) for fecal coliforms and 64 CFU/100 mL (99.9%) for fecal streptococci, respectively. Except for COD and BOD_5, these results are in agreement with those of [15], who worked on the purification of domestic wastewater in Cameroon. While comparing the purification performances of macrophyte and microphyte lagooning, they observed good physicochemical performances in their work. Indeed, these authors obtained 96.6% reduction for TSS, 89.5% for COD, 89.0% for BOD_5, 99.8% for fecal coliforms and 99.9% for fecal streptococci, respectively. The strong reduction in physicochemical and bacteriological parameters along the treatment profile is related to the predominance of physicochemical pollutant removal processes, including sedimentation, reduction, adsorption, predation and volatilization, and the effect of UV irradiation [16, 5].

At the exit of the treatment plant, the pH values vary from 7.4 to 7.6 with an average of 7.5. pH is a synthetic expression of the physicochemical conditions that partly govern the structure of the soil, the activity microorganisms and the availability of nutrients [17]. Average temperatures approach 25.6 °C at the treatment system. These results are in agreement with those of [15], who observed that pH increases slightly between the 2nd and the 8th lagoon and the temperatures are constantly high (27.2 °C in average). These values are consistent with the rejection standards of WHO and Cameroon.

As for the nitrogenous forms (NH_4^+, NO_3^-), the treatment allows a significant decrease. In fact, the concentration of nitrate decreases from 436 mg/L at the inlet to 2.93 mg/L at the outlet of the station with a reduction rate of 99.3%. This rejection conforms to the Cameroonian rejection standards. Ammonium decreases from 120 mg/L at the inlet to 5.46 mg/L with a reduction rate of 95.5%. The elimination of these parameters would be mainly the result of oxidation-reduction processes, volatilization, and absorption by algae [18]. This can also be explained by the fact that nitrate (NO_3^-) is a plant nutrient and highly soluble in water, therefore, directly assimilated by plants [19].

Nitrogen is a macronutrient necessary for plants found in waters and is used in the form of nitrate, ammonium, and nitrite. The sum of nitrogen concentrations in all various forms is called total nitrogen. Most plants absorb only nitrate, but usually other forms of nitrogen are transformed into nitrate in soil [20]. Municipal wastewater containing 20-85 mg/L of total nitrogen does not pose problems of acidification unlike synthetic fertilizers and increases the productivity on a quantitative and qualitative basis depending on their concentration in the soil and type of culture.

The system reduced phosphate (PO_4^{3-}) from 28.6 mg/L to 3.31 mg/L with a reduction rate of 88.4%. These values are consistent with the rejection standards of Cameroon. These results are similar to those of [15], who observed that these parameters decrease during treatment. Indeed, these authors obtained good reduction rates for phosphate and ammonium. At the outlet, the strong removal of phosphate would be partly due to the sedimentation of dead algae. Indeed, algae are able to metabolize and accumulate high levels of phosphorus in form of polyphosphates beyond the necessary needs for their growth and energy transfer [21]. Municipal wastewater containing 6-20 mg/L of phosphate does not affect soil properties, but increases agricultural productivity. Phosphate is a plant macronutrient, rarely present in soils under a bioavailable form for plants, and so it is almost always necessary to fertilize it with fertilizers. This element is relatively stable in soils, in which it can accumulate, especially on the surface or near the surface. These ions are necessary for the growth and development of plants; they can be valued in agriculture. Wastewater normally contains small amounts of phosphate, which makes its use for irrigation beneficial and does not have a negative impact on the environment [22]. This impact remains non-detrimental, even when sewage water containing high levels of phosphate is used for extended periods of time [23]. It is recognized that wastewater has a direct effect on the chemical properties of the soil. Several studies have shown an increase of soil water holding capacity following sludge application [24].

4 Conclusions

The purpose of this study was to characterize the discharges from the Fecal Sludge Treatment Plant in the city of Bafoussam for reuse in agriculture. This work allowed at the same time to evaluate the effectiveness of the treatment system. It appears that the Bafoussam station has a good purification performance for the elimination of physicochemical and especially bacteriological pollution. In general, the system allows a reduction of sludge pollutants with discharge values consistent with the discharge standards, as prescribed by the MINEPDED and the WHO, except for COD, BOD_5 and TSS, which are higher than standards. In these discharges, the nitrate, phosphate and ammonium are compliant and can be used in agriculture in the area. The system allows for better elimination of physicochemical and bacteriological pollution, since the concentrations obtained in the discharge waters are compatible with the national (Cameroon) and international (WHO) discharge standards.

In view of these results, one can safely recommend the reuse of this wastewater for irrigation in the agriculture due to its richness of nitrogen and phosphorus and also its low level of pathogenic germs. The parameters analyzed and their comparison with standards (Cameroonian and WHO) show that the values are in conformation with these standards, and the treated wastewater can be used in agriculture without having a considerable health and environmental impact. However, fecal sludge contains also trace amounts of heavy metals that can have an impact on health and the environment [25]. So, these waters from sludge treatment must be analyzed before reuse.

5 Acknowledgements

This study was conducted at the laboratory of the Wastewater Research Unit of the Faculty of Science, University of Yaounde I. The author likes to thank DAAD and Exceed Swindon project for supporting him to participate at the workshop in Rio de Janeiro, September 2019.

6 References

[1]   Anonymus: Institut national de la statistique Annuaire statistique du Cameroun. 2015.

[2]   Rice, P., Horgan, B.: Nutrient Loss in Runoff from Turf: Effect on Surface Water Quality. Turfgrass and Environmental Research Online, 2010, 9(1), 1-10.

[3]   Anonymus: Informations actualisées sur le Partenariat mondial sur les sols, y compris le Code de conduite international sur l'utilisation et la gestion des engrais. 2018.

[4]   Anonymus: Terres cultivées @ Pays du Monde Les engrais azotés chimiques, une catastrophe écologique. 2012.

[5]   Bassan, M., Tchonda, T., Yiougo, L., Zoelling, H., Mahamane, M., Mbérguéré, M., Strande, L.: Characterization of faecal sludge during dry and rainy season in Ouagadougou, Burkina Faso. *In*: 36 th WDC International conference, Nakuru, Kenya. 2013.

[6]   Soh, K.E., Kengne, I.M., Nguetsop, V.F., Ida, S.F., Amougou, A., Linda S.: Algal diversity and distribution in Waste Stabilization Ponds treating faecal sludge leachate from drying vegetated beds. Int. J. Biol. Chem. Sci. 2014, 8(3), 946-955.

[7]   Mpakam, H.G. : Étude de la vulnérabilité à la pollution des ressources en eau dans la ville de Bafoussam (Ouest Cameroun) et indices socio-économiques et sanitaires: modalités d'assainissement. Thèse de doctorat/PhD, Université de Yaoundé I, 2008, 272 p.

[8]   Anonymus: Results of the general population census in Cameroon. Technical Report of BUCREP. 2012. Available from: http:/ www.bucrep.cm,p 36.

[9]   Rodier, J.: L'analyse de l'eau naturelle, eaux résiduaires, eaux de mer, 8ème Edition DUNOD technique, Paris, 2005, 1008-1043.

[10]   Anonymus: Normes environnementales et procédure d'inspection des installations industrielles et commerciales au Cameroun. MINEPDED, 2008, 128 p.

[11]   Anonymus: L'utilisation sans risque des eaux usées, des excreta et des eaux ménagères OMS: volume ii utilisation des eaux usées en agriculture. 2012.

[12]   Kengne, I.M., Amougou, A., Soh, E.H., Tsama, V., Ngoutane, M.M., Dodane, P.H., Kone D.: Effects of feacal sludge on growth characteristics and chemical composition of *Echinochloa pyramidalis* (Lam). Hitch. And Chase and *Cyperus papyrus* L. *Ecol. Eng.*, 2008, 34(3), 233-242.

[13]   Seck, A., Gold, M., Nyang, S., Mbérguéré, M., Diop, C., Strande, L.: Faecal sluge drying beds: Increasing drying rates for fuel resource recovery in sub-saharan Africa. *J. Wat . San . Hyg. D.* 2015, 5, 72-80.

[14]   Heinss, U., Larmie, S.A., Strauss, M.: Solid separation and pond systems for the treatment seepage and public toilet sludges in tropical climate. Lessons learnt and Recommendations for preliminary design. EAWAG/SANDEC. 1998.

[15]   Nya, J., Brissaud, F., Kengne, I.M., Drakides, C., Amougou, A., Atangana, E.R., Fonkou, T., Agendia P.L.: Traitement des eaux usées domestiques au Cameroun: Performances épuratoires comparées du lagunage a macrophytes et du lagunage a microphytes. *In*: Proceedings of International Symposium on Environmental Pollution Control and Waste Management: cvcTunis: 2002, 726-736.

[16]   Nya J.: Peuplement phytoplanctonique et performances épuratoires de la station de lagunage à microphytes de Biyem-Assi (Yaoundé). Thèse doctorat 3e Cycle, Université de Yaoundé, 2001, 148 p.

[17]   Genot, V., Colinet, G., Bock, L.: Fertilité des sols agricoles et forestiers en région Wallonne. Dossier scientifique, rapport analytique 2006-2007 sur l'état de l'environnement Wallonne. Faculté universitaire agronomiques de Gembloux, 2007, 13 p.

[18]   Vinnerås, B., Palmquist, H., Balmér, P., Weglin, J., Jensen, A., Andersson, Å., Jönsson, H.: The characteristics of household wastewater and biodegradable waste – a proposal for new Swedish norms, Urban Water, 2006, 3, 3–11.

[19]   Bouziani, M.: L'eau (de la pénurie aux maladies). Ed., Ibn Khaldoun, 2000, 247 p.

[20]   National Research Council : Use of reclaimed water and sludge in food crop production. Washington, DC, National Academy Press, 1996, pp. 64–65. and National Research Council: Issues in potable reuse: the viability of augmenting drinking water supplies with reclaimed water. Washington, DC, National Academy Press. 1998.

[21]   Reddy, K.R., Kadlec, R.H., Flaig, E., Gale P.M.: Phosphorus retention in streams and wetlands. A review. CRC Press LLC. Critical review. Env. Sci. Technol., 1999, 29, 83-146.

[22]   Girovich, M.J. (ed.) : Biosolids treatment and management : processes for beneficial use. New York, Marcel Dekker, Inc. (Environmental Science and Pollution Control 18). 1996.

[23]   Degens, B.P., Schipper, L.A., Claydon, J.J., Russell, J.M., Yeates, G.W.: Irrigation of an allophanic soil with dairy factory effluent for 22 years: Responses of nutrient storage and soil biota. Aust. J. Soil Res., 2000, 38, 25-36.

[24]   Chang, A.C., Pan, G., Page, A.L., Asano, T.: Developing human health-related chemical guidelines for reclaimed water and sewage sludge applications in agriculture. WHO, Geneva, 2002, *(unpublished document)*

[25]   Letah, N.W.A., Kengne Noumsi, I.M., Rechenburg, A.: Prevalence and diversity of intestinal helminth eggs in pit latrine sludge of a tropical urban area. J. Water Sanit. Hyg. De., 2016, 74, 622-630.

THE RESULTS OF IMPLEMENTATION OF TREATED WASTEWATER AND BIO-SOLIDS GENERATED FROM WWTP ON THE PLANTATION OF FODDER AT MARGINAL AREAS AT AL-KARAK, SOUTH JORDAN

Tayel El-Hasan[1], Husam Hamaidah[2], Salah Al-Jbour[2], Nabeel Bani Hani[3], Faddel Ismail[3], Yahya Shakhatreh[3], Dua'a Al Majali[3]

[1]*Department of Chemistry, Faculty of Science, Mutah University, Al-Karak, Jordan;*
tayel@mutah.edu.jo

[2]*Faculty of Engineering, Mutah University, Al-Karak, Jordan;*

[3]*National Agricultural Research Center (NARC), Al-Baqa, Rajab Al-Selebi Street, opposite the Satellite Station, Building No. 2, 19381 Amman, Jordan;*

Keywords: Bio-Solids, Land application, Sorghum, Marginal areas, WWTP, Co-digestion

Abstract

Decentralized integrated management of sludge is considered a viable solution for solving the problem of sludge accumulated in wastewater treatment plants in Jordan. This research is part of a project that aims at reducing the carbon footprint of sludge handling and thus considers primarily the use of sludge as a source of energy and soil conditioner. The objective of this study is to assess the application of bio-solids (solid sludge and treated organic matter (TOM) that results from the co-digestion process) as soil amendments and its effect on plant and soil production, and water use efficiency (WUE). This assessment also includes analysis of the organic matter before application. The field application on sorghum plantation at marginal areas in south Jordan shows that bio-solids increases the dry and fresh yield of sorghum crops up to 40%. The effect on soil shows that heavy metals concentrations were slightly noticed by bio-solid addition, but is it still under the permissible limit and should be continuously monitored. Correlation analysis shows positive correlation between Zn, Mo and Co concentration in the soil and the added bio-solids. Moreover, it shows an increase of P but depletion of K and N as they are macronutrients for sorghum. Obviously, the use of bio-solids positively affected the water use efficiency by 2.4 and 3.2 times those of normal fertilizers. Therefore, sludge could be used as soil amendment in rangelands with < 200 mm/yr of rainfall. Nevertheless, bio-solid quality needs to be monitored before application. More research on land application measures, effect on food chain and effect on pollutants accumulation is needed.

1 Introduction

Jordan is one of the most water stressed countries in the world. Jordan shares significant transboundary surface and groundwater resources with Israel, Syria, Egypt, the West Bank and Saudi Arabia. The availability of adequate quantity and quality water in Jordan is deteriorating over time due to a number of factors such as rapid natural population growth, influx of high number of refugees due to regional political instability, rapid urbanization, limited water resources, unsustainable water use and degradation of water quality. Climate change is affecting Jordan severely in terms of lower precipitation trend and deteriorating the soil fertility, and increasing the desertification [1]. Unconventional mitigation measures are essential to overcome the water scarcity and high demands. In Jordan, there are 32 wastewater treatment plants (WWTPs) distributed in the urban centers of the country. Therefore, about 61% of the population has access to wastewater collection and treatment system. The quantity of effluent treated wastewater was about 75.4 MCM (million cubic meter); more than 90% of this water is used in the agriculture. The official plan is to increase the utilization of this water up to 250 MCM in the year 2025 [2].

In Jordan, rangelands are defined as lands, where the annual rainfall does not exceed 200 mm and no irrigation takes place. Rangelands have deteriorated considerably over the last 20 years, and thus, have been targeted for restoration activities since the early 2000s. Treated wastewater is used effectively in rangeland for the production of fodder plantation [3]. In 2009 and subsequently in 2011, instructions were issued banning the production and use of organic fertilizers from materials originating from WWTPs. Therefore, the national standard was modified in 2016 (JS: 1145/2016), where the reuse of Type I bio-solids as an agricultural fertilizer was banned. Nevertheless, it is worth mentioning that authorities are currently in the process of reviewing the technical regulations related to sludge management and the reuse of bio-solids.

Organic fraction of municipal solid waste (OFMSW) is a suitable co-substrate, since it is an easily degradable constituent [4]. Accordingly, anaerobic co-digestion of OFMSW and sewage sludge becomes one valuable alternative to the conventional management methods in terms of energy recovery and environment security. Compared with sewage sludge, OFMSW has a higher C/N ratio and lipid load that elevate the initial stage of anaerobic digestion, which may increase CH_4 production in case of co-digestion [5].

Most lands of Al-Karak province are characterized as marginal areas with a rainfall less than 250 mL/yr. As well, it suffers from severe climate change impacts in terms of low rainfall, drought and desertification. Thus, the demand for water is increasing, and authorities are looking to utilize the treated wastewater effluent from the two WWTPs in the province. In fact, they even started to use the untreated wastewater prior to building the sewage system in the city [3, 6]. The implementation of these waters is restricted to certain areas and crops, too. This current work is demonstrating the results of previous field experiments and laboratory analyses of implementing the treated wastewater and the bio-solids generated from the Mutah-Mazar WWTP in the irrigation and soil amendment for livestock production (Figure 1).

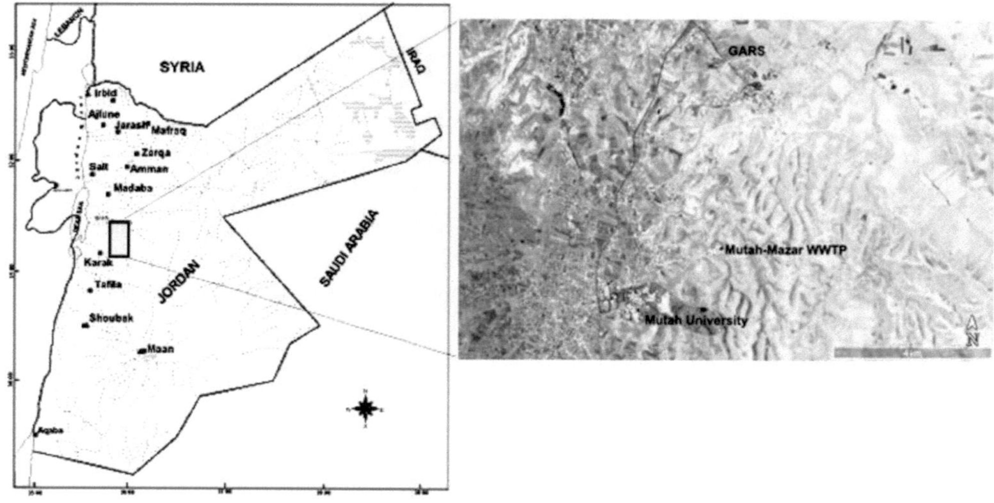

Figure 1: Location of Mutah and Al-Mazar WWTP, and Ghweir Agricultural Station

2 Materials and Methods

Mutah-Mazar WWTP is a mechanical computerized modern tertiary wastewater treatment plant located in al Al-Karak governorate, established 2013 and operated 2015 using aerobic and non-aerobic treatment for wastewater. Treated water is discharged to the side wades and the treated bio-solid planned to be dumped in an assigned place. The design capacity is 7,650 m^3/d and the actual operated capacity is about 1,200 m^3/d (16%). Therefore, the effluent wastewater is of relatively good quality (BOD < 30 ppm). The sludge is dewatered in drying beds after thickening with average production of 120 kg/d (90% total solid). The sewage sludge is considered as class I according to Jordanian standard (JS 1145:2016).

Representative food waste samples were obtained by multi-quartering standard procedure following the procedure of [7]. The final representative food waste sample was minced, and then stored in a refrigerator at 4 °C for subsequent use and characterization.

Anaerobic co-digestion process was implemented in a semi-continuous anaerobic reactor (Figure 2). Using the sludge from the Mutah-Mazar WWTP thickener and the food wastes collected from Mutah University campus, the experiments were done for series of food waste and sludge mixing ratios. The inoculum was brought from Al-Shlalah WWTP for building up a bacteria community. The generated biogas was collected, purified and measured using a gas gauge [9]. After 21 days, the treated organic material of the co-digestion was collected and dried for using in the agricultural experiments.

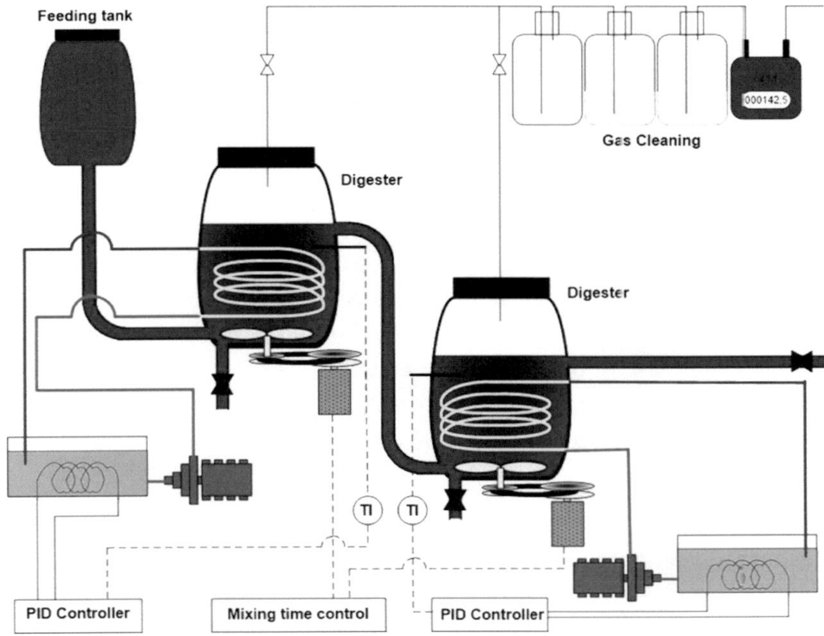

Figure 2: Semi continuous two-stage anaerobic co-digestion reactor

2.1 Agricultural Experimental Setup

Ghweir Agricultural Research Station (GARS) is located about 1.5 km from Ghweir village. The plantation experiments took place there. The station was established in 1971 with a total area of 50 ha at an elevation of 820 m. It has 3 shallow depth renewable ground water wells (156-332 m) with a total water abstraction of 0.5 MCM. These wells consider as an important source for drinking water for Al-Karak. Water salinity ranges from 1.5 to 2.5 dS/m. Ghweir soils lie within the xeric-aridic transitional moisture regime. Land use of Ghweir Catchment is field crops, mainly barley with total area about 6,200 ha (15.8%), followed by fruit trees 1,000 ha (2.5%), mainly olive trees, and a small irrigated area of about 38 ha (less than 0.1 %). Most of the soil is bare soil (69%). In the year 2016, wastewater was used in plantation of several fodder species (*Pennisetm glaucum, Medicago sativa, Atriplex hallimus,* and *Jatropha curcas*) at GARS, using drip irrigation system. The plants and soil were monitored and analyzed for one year [3].

Sorghum was planted under irrigation as summer crop in May 2017 in 0.75 m apart rows with 0.3 m between plants within the row. Drip irrigation system with emitter of 4 L/h discharge and spaced at 0.3 meter is used to irrigate the trial. The effect of two levels of bio-solids supplied by Mutah-Mazar WWTP, in addition, two controls are tested in the trial. Materials are mixed with surface soil layer (upper 15 cm) before planting. Due to lack of enough quantity of bio-solids, only two treatments C3 and C6 are conducted on a limited area within assigned plots (1 m^2) in three replicates, but C9 treatment could not be applied.

The experiments were designed as follows.

Compost-sludge experiment (sorghum-barley-sorghum sequence)
The compost-sludge experiment was conducted on sorghum-barley-sorghum sequence under irrigation and rain fed agricultural conditions.

First growing season, May to August 2017: Sorghum crop was planted under irrigation by fresh water using drip irrigation system with emitter of 4 L/h discharge and spaced at 0.3 m. Three levels of TOM and three levels of solid sludge were used. The three levels were 3, 6 and 9 t/ha, including two controls; one without any addition and the second one with a conventional fertilizer (in total eight treatments). Randomized complete block design (RCB) with three replications were used with 24 plots in total (Figure 3). The plot dimension was 3 x 3 m (treated TOM area was adjusted according to available TOM). Sorghum was planted in 0.75 m apart rows with 0.2 m between plants within row. Compost or solid sludge was mixed with surface soil layer (upper 15 cm) before planting.

Second growing season, November 2017 to May 2018: Barley crop was grown after sorghum with the same treatments mentioned above by new application to study the build-up effect of TOM and solid sludge for the same 24 plots by distribution of 8 treatments over three replicates in the same arrangement. Barley was grown under rain fed conditions at seeding rate of 100 kg/ha.

Third growing season, May to August 2018: The first growing season experiment of sorghum was repeated again (third growing season) with new application of TOM and sludge to the same previous 24 plots by distribution of 8 treatments over three replicates in the same arrangement.

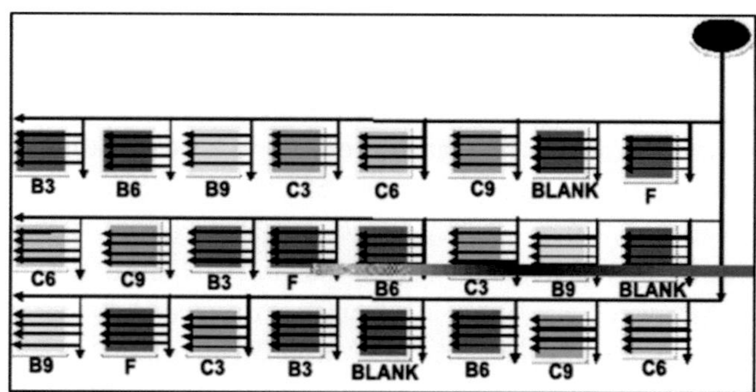

Symbol	Definitions	Symbol	Definitions
Blank	Without any application	F	DAP fertilizer 100 g/m^2
B3	Solid sludge 300 g/m^2	C3	TOM 300 g/m^2
B6	Solid sludge 600 g/m^2 (B6)	C6	TOM 600 g/m^2
B9	Solid sludge 900 g/m^2 (B9)	C9	TOM 900 g/m^2

Figure 3: Experimental layout for sorghum plantation using Randomized Complete Block design (RCB).

2.2 Analyses and field measurements

2.2.1 Crop data: Establishment, heading date, maturity date, and plant height, fresh and dry yields were measured and recorded.

2.2.2 Analyses

Initial composite soil samples from plots assigned for each treatment were collected from surface (0-15 cm) and subsurface (15-30 cm) for analysis (pH, EC, N, P, K, trace element and heavy element). At the end of each growing season composite plan, plant and soil samples (surface and subsurface) were collected from each treatment for analysis. The following analyses were conducted:

1.   Soil analyses (pH, EC, N, P, K, trace elements and heavy metals),
2.   Plant analyses (pH, EC, N, P, K, trace elements and heavy metals),
3.   Biological analyses of the plants: Total Coliform, Fecal Coliform, *E. coli* and Salmonella.

3 Results and Discussion

The treated organic matter (TOM) produced from the anaerobic co-digestion process was tested according to Jordanian standard (JS 1145:2016). Al-Jbour et al. [8] found that TOM is safe and contains low heavy metals and bacterial constituents. Besides, they were enriched with N and P (Table 1). The analysis of the total solid content in the final digested organic matter indicates a reduction in volume. The volume reduction is attributed to biogas generation and water evaporation. Anaerobic digestion of sewage sludge promoted 41.9% Total Volatile Solids (TVS) conversion. Mixing food waste with sewage sludge promoted 47.9% TVS conversion. The total phosphorus (TP) content in the final digested organic matter as a percentage increased slightly because of the reduction in the final volume of the digestate. Similar trend was noticed for the total nitrogen (TN) content in the final digested organic matter.

13 trace elements (TE) were analysed including some essential plant nutrients and some toxic elements that could affect plants and humans, such as Cd, As, Zn, Pb and U. The element contents (ppm) in TOM samples, and the Jordanian standards (JS1145; 2016) regarding the uses of treated sludge are shown in Table 1. It is noticeable that TOM with higher sludge contents (1 and 2) contains higher TE than other ratios with lower sludge content. This indicates that the major source of TE is the sludge, whereas the TE concentration in food waste is very low. TOM 2 possessed higher content of TE compared with other TOM samples. Nevertheless, these concentrations are lower than the permissible concentration of TE for treated sludge [9]. As for the Total Fecal Coliform Count (TFCC) in TOM, the counting results are shown in Table 1.

The results indicate that anaerobic conditions under normal pH are unable to destroy the TFCC. Thus, TOM needs to be sterilized before being used as soil amendment. The Jordanian standard for treated sludge indicates that the maximum permissible TFCC count is 1000 MPN (Most Probable Number). Therefore, TOM has much lower counts than the Jordanian standard.

Table 1: Total N and P (wt %); and Trace element concentration (ppm), beside the Total Coliform and Fecal Coliform bacteria (MPN count) in the different mixing ratios of TOM samples.

Parameter	TOM 1	TOM 2	TOM 3	TOM 4	TOM 5	JS 1145:2016
TP	0.16	0.17	0.13	0.12	0.11	
TN	0.44	0.45	0.28	0.47	0.53	
Ba	1,124	547	627	22	6	
Zn	803	1,288	260	121	88	2,800
Pb	13	14	9	2	1	300
V	56	69	31	16	10	
Cd	1	1	1	0	0	40
Ni	27	35	52	20	8	300
Cu	115	146	61	12	125	1,500
Co	3.5	3	4.1	1	0.4	
Cr	28	37	19	13	11	900
Ag	1	1	1	0	0	
Th	1	2	1	0	4	
As	27	26	39	7	6	41
U	0	1	0	0	0	
Total Coliforms	619	641	-	665	-	1,000
Fecal Coliforms	50	534	-	<1	-	

The investigation showed that the wastewater produced from Mutah-Mazar WWTP could be used for fodder production. *Pennisetm glaucum* and *Medicago sativa* are the best plant species to be planted as the yield was very high and these plants are known to contain high nutrient values for animal production. The use of wastewater in the investigation site can help for combating desertification. No organic or inorganic pollution thread was found in the wastewater, plant and soil of the investigation site.

This paper discusses the results of sorghum planting only. The biological assessment of sorghum plants (2017 and 2018) indicates that TOM and sludge treatments did not affect the presence of *E. coli* and Salmonella in sorghum and barley plants, and all plants (including those of blank and fertilizer treatments) were free from contamination. Fecal Coliforms were recorded only in sorghum 2017 after solid sludge treatment but not after TOM application. Sorghum (2018) plants were free from Fecal Coliforms (Table 2). Taking into consideration that total and fecal coliforms are just pathogen indicators, handling forages planted on dry sludge should receive more attention in order to reduce any unexpected or potential health risk in future.

Table 2: Fecal coliforms, *E. coli* and salmonella for sorghum crop under different levels of TOM and solid sludge at the Ghweir station, 2017 and 2018 (Symbols are explained in Figure 3)

Treatment	Fecal Coliforms MPN/g of plant		E. coli MPN/g of plant		Salmonella	
	2017	2018	2017	2018	2017	2018
Fertilizer	negative	negative	negative	negative	negative	negative
Blank	negative	negative	negative	negative	negative	negative
C3	negative	negative	negative	negative	negative	negative
C6	negative	negative	negative	negative	negative	negative
C9		negative		negative		negative
B3	240	negative	negative	negative	negative	negative
B6	240	negative	negative	negative	negative	negative
B9	240	negative	negative	negative	negative	negative

The added organic material (Table 3) affected sorghum yield and yield components in both growing seasons. The results of 2018 indicate cumulative effects of the organic materials on sorghum, and this is clearly seen by increase in yield and yield components in 2018 compared with 2017. The cumulative effect of TOM and solid sludge was more on dry weight than on fresh weight. The fresh yield of 2018 were higher using large dosages of TOM and sludge (i.e., C6, B6 and B9), and lower than 2017 using low dosages of TOM and sludge (i.e., C3 and B3) and fertilizers. However, dry yield of 2018 showed higher yields with all dosages than the 2017 yields. And fertilizers treatment shows lower yield than 2017 (Figure 4).

The percentage increase of dry yield after organic treatment compared with fertilizer treatment ranged from 76% to 289% in 2018 (Table 3). On the other hand, yield and yield components of blank and fertilizer treatments were less in 2018 compared with 2017. Although solid sludge treatments resulted in more yield compared with TOM, the effect of TOM levels on sorghum fresh and dry yield was comparable and non-significantly different from the effect of the equivalent levels of dry solid sludge (C3 and C6 compared with B3 and B6, respectively). Water use efficiency of TOM and solid sludge treatments was increased in 2018 compared with 2017 (Table 3), but it was reduced for blank and fertilizer treatments. As average, WUE after TOM and after dry solid sludge treatments were 2.4 and 3.2 times higher than after fertilizer treatment in 2018.

Based on Table 3, the percentage increase in fresh and dry yield of sorghum under different levels of TOM and solid sludge are compared with fertilizer at GARS. As for the application of TOM, the maximum increase recorded in C6 was 220% and 169% for dry and fresh yield, respectively. The next was C3, 129% and 93% for dry and fresh yield, respectively. Meanwhile, the least one was C9, where it was 76% for both dry and fresh yield. This means that application of 600 g/m^2 was the best for the yield, while, the application of bio-sludge shows that B9 (900 g/m^2) has the largest increase of percentage 289% and 236% for dry and fresh yield, respectively. The lowest was B3, where 133% and 96% for dry and fresh yield, respectively, were obtained.

Table 3: Plant height, fresh yield, dry yield and water use efficiency (WUE) of sorghum crop under different levels of TOM and solid sludge at Ghweir station, 2017 and 2018

Treatment	Plant Height (cm)		Fresh Yield (t/ha)		Dry Yield (t/ha)		WUE Dry (kg/m^3)	
	2017	2018	2017	2018	2017	2018	2017	2018
Fertilizer	132	179	58.1	22.6	18.7	9.7	3.74	1.76
Blank	90	167	48.0	17.4	14.4	8.2	2.88	1.49
C3	104	200	51.5	43.6	15.2	21.9	3.05	3.98
C6	129	210	53.0	60.6	18.1	30.7	3.63	5.58
C9*		176		39.8		16.9		3.07
B3	98	180	57.0	44.3	16.1	22.4	3.21	4.07
B6	116	183	58.2	63.8	17.6	33.3	3.51	6.05
B9	135	188	60.2	75.8	20.6	37.2	4.12	6.76

**C9 was not applied in 2017 season*

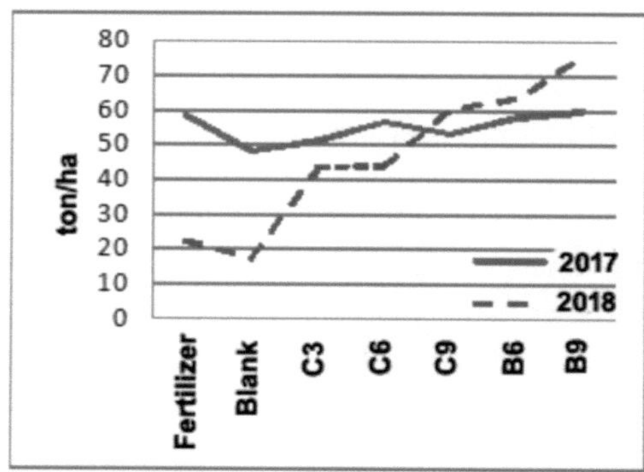

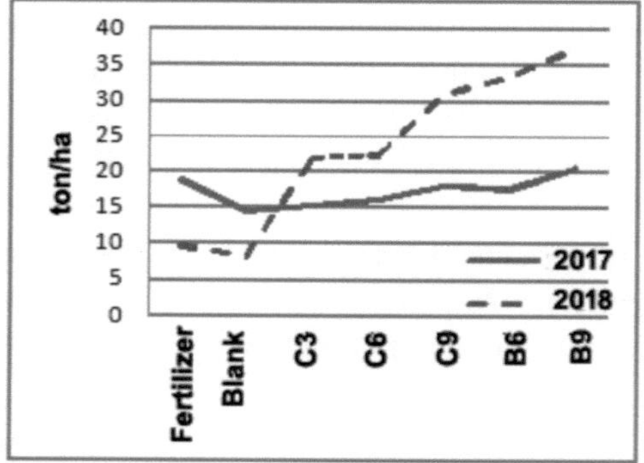

Figure 4: Fresh yield (left) and dry yield (right) of Sorghum crop under different levels of TOM and solid sludge at GARS, 2017 and 2018

Water use efficiency (WUE) for dry yield was improved by the applied treatments. WUE-dry amounts after TOM and after dry solid sludge treatments were 2.4 and 3.2 times higher than after fertilizer treatment in 2018.

Heavy metal contents of sorghum plants after TOM and dry solid sludge treatments were similar or lower than of fertilizer treatment for both seasons, except for Pb in 2018 (Table 4). The increase in Pb content was less after TOM treatments compared with solid sludge. Although Pb content for B9 treatment was the highest (3.25 ppm), it is still less than one tenth of the level considered to be toxic in animal feeds. Analysis of sorghum plants in both seasons showed similar and very low concentration of Ni, Cr and Cd after all treatments including blank treatment. Heavy metal contents of Sorghum plants after TOM and solid sludge were not detectable in most of the treatments, or very low in some treatments.

Table 4: Heavy metals concentration (ppm) in sorghum crop under different levels of TOM and solid sludge at GARS 2018

Parameter	Ni	Pb	Co	Cr	Cd
Fertilizer	<0.010	1.63	1.80	<0.006	<0.002
Blank	<0.010	1.33	1.10	<0.006	<0.002
C3	<0.010	2.50	0.60	<0.006	<0.002
C6	<0.010	2.75	1.10	<0.006	<0.002
C9	<0.010	2.38	1.00	<0.006	<0.002
B3	<0.010	2.25	0.50	<0.006	<0.002
B6	<0.010	3.06	0.70	<0.006	<0.002
B9	<0.010	3.25	1.40	<0.006	<0.002
FAO*	100	40	40	1000	10

**FAO, 1998. Animal feeding and food safety. Food and Nutrition Paper 69 [10]*

The concentration of all elements and salinity on the surface soil is higher than subsurface soil for all treatments. After three successive additions of TOM and solid sludge (soil after sorghum 2018), heavy metal concentrations in soil ranged from nil to very low concentration in soil and was not highly affected by addition of TOM and/or solid sludge as compared with blank, and the concentrations were much lower than the permissible range. Soil macronutrients K and N relatively not increased by TOM and solid sludge as compared with fertilizer, might be due to plant uptake and low dissolution from TOM and/or solid sludge, whereas P has slightly increased. The effect of TOM and solid sludge on trace element concentrations is not obvious. Trace element concentrations after fertilizer were similar or slightly lower than after solid sludge, but were similar or slightly higher than after TOM treatment. Zn, Mn, Co, and P concentrations in the soil amended by solid sludge were higher than in soil amended by TOM or fertilizer treated soil. Correlation analysis shows positive correlation between Zn, Mo and Co concentrations in the soil and the added organic materials quantities. This indicates the need for longer studies to evaluate the accumulation of these (and other) minerals in the soil. Organic matter at surface and subsurface soil increased as compared with previous seasons due to accumulation of TOM and/or solid sludge in soil and due to plant residue remains in the soil from previous growing season.

4 Conclusions

From these experiments it can be concluded that TOM and solid sludge clearly increases the dry and fresh yield of sorghum crops. Water use efficiency has been improved by the application of both solid sludge and TOM. Pollutants were not detected in sorghum plants, however, continuous monitoring for and testing is required. The accumulation of heavy metals in soil was slightly influenced by TOM and solid sludge addition and, therefore, should be continuously monitored.

5 Acknowledgements

This work was conducted in the context of a project financially supported by GIZ through the Decentralised Integrated Sludge Management project (DISM). The authors are thankful for the Prince Faisal Center for Dead Sea, Environment and Energy researches at Mutah University laboratories for providing the technical support for the co-digestion experiment. They also like to express their gratitude to Exceed Swindon Project and DAAD for supporting their participation at the Workshop in Rio de Janeiro, September 2019.

6 References

[1]   Matouq, M., El-Hasan, T., Al-Bilbisi, H., Abdelhadi, M., Hindiyeh, M., Eslamia, S., Duheisat, S.: The Climate change implication on Jordan: A case study using GIS and Artificial Neural Networks for weather forecasting. Journal of Taibah University for Science 2013, 7, 44-55.

[2]   El-Hasan, T.: Industrial Wastewater Resources In Jordan for the Period (1999-2006). In: Hamaideh et al. (Eds): Proceedings of the Expert Workshop Water Challenges in MENA region 2016-2050. April 24-29, Aqaba, Jordan, 2106, 36 - 46.

[3]   Jiries, A., Al-Nasir, F., El-Hasan, T.: Pasture land in a desert area at Al-Karak province – Jordan. In: Abou-Elnaga & Aydin (eds.): Proceedings of the regional workshop: Solutions to water challenges in MENA region. April 25-30, 2017, Cairo, Egypt, 2017, 10-17.

[4]   Cabbai, V., Ballico, M., Aneggi, E., Goi, D.: BMP tests of source selected OFMSW to evaluate anaerobic co-digestion with sewage sludge. Waste Management 2013, 33(7), 1626-1632.

[5]   Sakar, S., Yetilmezsoy, K., Kocak, E.: Anaerobic digestion technology in poultry and livestock waste treatment- a literature review. Waste management & Research 2009, 27(1), 3.

[6]   Jiries, A., El-Hasan, T., Al-Naqa, A., Al-Naser, F., Ismael. M., Al-Wahishi, S.: Wastewater Resources Management in Rural Areas in Jordan. UNESCO, Amman, Jordan, 2010, p. 42. *(Unpublished report)*

[7]   UNEP: 2006. *file:///H:/Methane%20Gas%20Yeild/Guidelines-WasteData_2%20(1).pdf*

[8]   Aljbour, S.H., El-Hasan, T., Al-Hamiedeh, H., Hayek, B., Abu-Samhadaneh, K.: Anaerobic co-digestion of domestic sewage sludge and food waste for biogas production: A decentralized integrated management of sludge in Jordan. Bioresources *(in press)*.

[9]   JISM. 2016. Water-Sludge-Uses of Related Sludge Disposal No. 1145/2006. Jordanian Institute for Standards and Meteorology, Amman, Jordan.

[10]   FAO: Animal feeding and food safety. Food and Nutrition Paper 69 (1998).

ENGAGING, ENABLING AND EMPOWERING BANGLADESHI YOUTH ON CLIMATE CHANGE PROJECTS THROUGH FACEBOOK USAGE TO ENSURE SUSTAINABLE DEVELOPMENT

Md. Abdur Razzak[1], Md. Serajul Islam[2], Jinzhang Jiang[3]

[1]School of Media and Communication, Shanghai Jiao Tong University, Shanghai-200240, China; abdurrazzak7545@gmail.com

[2]Bangladesh National Museum, Shahbag, Dhaka-1000, Bangladesh; sisun7@gmail.com

[3]School of Media and Communication, Shanghai Jiao Tong University, Shanghai-200240, China; jinzhangphd@163.com

Keywords: Bangladesh, Youth, Facebook, Climate Change, Sustainable Development

Abstract

The purpose of this study is to identify the involvement of Bangladeshi youth on climate change projects through Facebook usage to ensure sustainable development. The study seeks to investigate the combined roles of Facebook usage for engaging, enabling and empowering the youth force of Bangladesh accelerating sustainable development. The study used a sample of youth (N=250) and a closed ended survey questionnaire to probe the real youth Facebook users' contributing spirit on Bangladeshi climate change projects. The study found that youths are extremely vocal on Bangladeshi climate change issues, projects and sustainable development agenda in Facebook. Surprisingly, two special groups as youth Facebook users were found. The Facebook users were active to disseminate information promptly, had tendency to comment or to clear their opinions on viral issues. Finally, it was found that Facebook accelerates youth's contributing spirit on climate change issues that are really needed for the developing countries like Bangladesh.

1 Introduction

Climate Change is one of the established global problems from the last decade. It is a world-wide crisis. More extensively, it is multifarious, versatile, most horrible, and long standing against sustainable development. Head of the UK Government Economic Service and former World Bank Chief Economist Sir Nicholas Stern strongly emphasized rapid and international collaboration to tackle Climate Change risks [1]. Alongside, the jeopardy of Climate Change hazards are linked with global prosperity packages, world development pathway, sustainable development, country investment plan or Millennium Development Goals (MDG'). It is not only a cause of universal defenselessness toward development, but also troublemaker in various local, national, regional and global advances. The South Asian coastal region, which is highlighted as a large coastal populous area, is the witness of climate adverse [2].

There is no doubt that Climate Change is a vital in ecological, social and economic development [3]. It is also priorities issue in the policy agenda in developing countries. Sustainable development actually long term sustain development policies that are integrated with socio-economic and environmental development [4].

Sustainable development is also a futuristic strategy for sustained development. Being an uncertain road map to address the iconic sustainable commitment like Agenda 21, a joint integrated collaboration between governmental efforts, private sectors, civil society and stakeholders' participations are strongly required [5].

Who is the right choice to manage uncertainty of Climate Change and Sustainable Development? Even, who is the most powerful, trusted, accountable and technology friendly figure? There is no alternative to reply without mentioning the youth force. They are the real business dealers, who have right imagination to conduct small, medium or large business whatever necessary to address global environmental degradation [6]. In addition, global warming and climate change are worldwide issues, whereas youths are global ambassadors [7]. They are the future leaders to address the global adverse issues like Climate Change.

1.1 Clarification of Coined Concepts

Bangladeshi Climate Change Projects, youth, civic engagement, sustainable development and Facebook usage can be defined in different styles and angles. A clarification is needed, so that the audiences will understand the jurisdiction of used concepts in the study.

1.1.1 Climate and Existing Bangladeshi Climate Change Projects

Climate is plainly weather condition in a particular area. Actually, the term climate includes temperature, rainfall and wind pattern [8]. On the other hand, IPCC defines Climate Change as *"any change of climate over time to natural variability or as a result of human activity"* [9]. The study counted relevant Climate Change projects sponsored either by donor groups, agencies, countries or government revenue budget. The study has taken consideration of particular projects that planned, guided, implemented, funded or focused related to the national and global policy frameworks including National Adaption Program of Actions (NAPAs), Bangladesh Climate Change Trust Fund (BCCTF), the United Nations Framework Convention on Climate Change (UNFCCC), the Intergovernmental Panel on Climate Change (IPCC), Global Carbon Fund (GCF) and Millennium Development Goals (MDGs) [10].

1.1.2 Youth

According to the Bangladeshi national history, youth are the real heroes. They initiated all revolutionary steps and sacrificed their lives to the nation building march-past from the language movement in 1952 to the glorious independent of the country in 1971. There is no fix age bar for youth. The researchers are flexible to define and to fit for their purposeful work. Bangladeshi writer Samir Ranjan Nath defined the Bangladeshi youth as 15 to 35 years old in his research [11], whereas Middle East and North Africa (MENA) count their youth people from the ages of 14 to

24 years old [12]. This study figured the Bangladeshi youth as the ages of 15 to 24 years old. They are the powerhouse in every society. Scholar like Barry Goldson agreed to say youth, those have potentiality and adult responsibility [13]. Youth are the frontier force to build the community and the society. Surprisingly, they are change agent towards sustainable development [14]. Delli Carpini emphasized that the youth ages of 18 to 29 is the prime contributor on civic affairs engagement [15]. John Muncie framed the youth differently. He identified the youth age is exactly after child and before adulthood [16]. Taking all scholarly notes, the study considered the youth adults of ages 14 to 29 years old only.

1.1.3 Sustainable Development
Nowadays, it is a common slogan in governmental, non-governmental and private development strategies. Simultaneously, everybody concern about integral development concept of eco-friendly and long lasting approach [17]. In citizen science, Alan Irwin has outlined the sustainable development by fantastic wording. It is an outstanding policy framework that ensures the relationship between the environmental threat, it's scientific solution and the public [18]. Artur Pawlowski also described three elements of sustainable development. He leveled them as (a) ecological, (b) social and (c) economic [19].

1.1.4 Facebook Usage
Today, Facebook is pioneering social communication as a popular tool [20]. It is a world-wide social networking site that has enormous attraction power to the users without any race, religion and age limits. Audiences are drastic transforming from mass media to social media for digging news, opinion and civic information. Facebook is leading legend based on their requirement [21]. According to Facebook's CEO, Mark Zuckerberg, Facebook is playing facilitation of civic discourse, and user is amplifier their potential through it [22].

1.2 Research Question
Considering the above concepts, the study was centered to reveal the satisfactory answers from the following Research Questions (RQ):

1. What are the implications of Facebook usage to engage, to enable and to empower Bangladeshi youth in Climate Change Projects?

2. What are the specific roles of Facebook usage by youth in Climate Change Projects to ensure Sustainable Development (SD)?

1.3 Bangladesh Perspective
Being an agricultural nation, Bangladesh is facing significant threat from climatic dilapidation. Concurrently, high temperature is another remarkable impact on its food production process. The increasing warmth trends like 1.0 ^{O}C by 2030, 1.4 ^{O}C by 2050 and 2.4 ^{O}C by 2100 are considerable subject to project any development framework due to its detectable scientific pressure in Bangladesh [23]. The changing monsoon rainfall schedule is also affecting the South Asian and South East Asian livelihood, crops, property, bio-diversity and eco-system of countries

for continuing towards countless [24]. The book of *"Facing Up to Climate Change in South Asia"* published by International Institute for Environment and Development (IIED) highlighted four Climate Change effects in Bangladesh, Bhutan and Nepal as (1) changes in temperature, (2) distribution of rainfall, (3) rising of sea level, and (4) extreme weather events [25]. Finally, Bangladesh is an agricultural country. Its economy purely depends on agriculture. It is estimated that Climate Change related flood and drought will be reduced overall rice production 7.4 % every year that is a remarkable negative indicator to the national economic growth [26].

2 Methods

This study is based on Survey Method. It engaged both qualitative and quantitative approaches. A questionnaire is used for data collection. 270 questionnaire forms were distributed among the active Bangladeshi youth Facebook users, out of which 250 were analyzed, as these were completely filled up. The qualitative study is universal and a well established method among social scientists for doing large scale studies on the adoption of communication technologies [27]. Moreover, it involves the studied use and collection of various observed materials like interview, survey, life story, case study, personal experience, cultural, historical and visual texts [28].

3 Theoretical Framework

Youth are the core stakeholder with jam-packed power to re-built, to re-shape and to re-construct the society. They are the key participants to contribute in decision making process and social development [29]. Usually, youth starts after the childhood. Motivations, performances and activities are not quite similar between these two age groups. In addition, their ability to move, problem solving capacity, socialization communication patterns are different. Youths are growing up in a special society today. They are using technology and living in a digital society [30] . Youth psychological study is impactful to study the youth attitude, motivations and contribution to the societal development. Considering youth contribution feelings and thoughts for development, the study followed the "Role Taking" theory. It is established by Robert L. Selman of Boston University in 1976. Continued it discussed and developed by the experts time to time such as "Selman's theory of role taking and social awareness (1980, 2003) and exploring meaning psychological maturity (2003). This theory is more relevant and important to address civic engagement especially for youth empowerment. How the activity affect to others is the nucleus of the role taking theory. Role taking is described as the ability to understand someone's "thoughts", "feelings" and "point of view" [31]. These are quite similar and the study checked the basic four resources of role taking theory like (1) subjective role, (2) self reflective role, (3) mutual role and (4) societal role [32]. Deemed the role taking theory and the gist of Bangladeshi Climatic Projects, youth Facebook usage and Sustainable Developments, the graphical view of theoretical framework is as below (Figure 1).

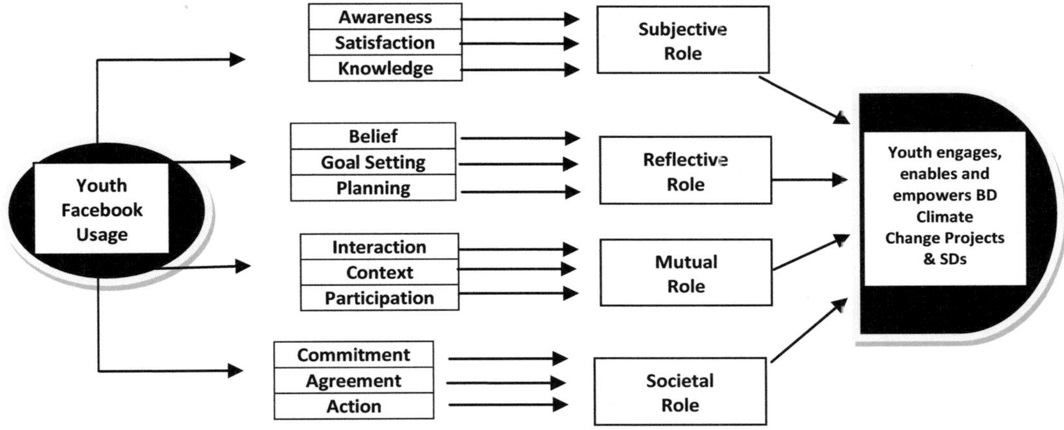

Figure 1: Theoretical framework of the study

4 Results and Discussion

The study gave the evidence that Facebook usage makes the youth engaged, enabled and empowered about the Climate Change Projects in regard of Sustainable Development (SD). The study gave the data that 171 male participants and 79 female participants were engaged in the survey, among which most of the participants were students (Table 1 and Figure 2). Moreover, the participants were service holders, business persons and others, which mean the differences in the youth force paving the way to interact with each other within a common context like climate change projects through Facebook usage and if Facebook properly used, it will ensure their participations in the dialogue of Sustainable Development. In this study, it is proved that the mutual roles (interaction, context, and participation) played among the Facebook users in Bangladesh.

As the youth learnt about Climate Change Projects which contributes to the achieve Sustainable Development, the youth has engaged themselves with the Climate Change Projects through Facebook usage. Moreover, Facebook usage affected the youth's belief, goal setting, and planning. The youth gets the platforms of interaction and participation in the context of Sustainable Development through Facebook usage. Furthermore, the youth has become committed, agreed and action-taking agents to Sustainable Development through the Facebook usage. This study gives clear evidence that Facebook usage has played subjective role, reflective role, mutual role and societal role over the youth to engage, to enable and to empower Climate Change Projects and Sustainable Development.

The 250 data-set each comprising 17 questions revealed different aspects of Engaging, Enabling and Empowering Bangladeshi Youths on Climate Change Projects through Facebook usage to ensure Sustainable Development. The survey revealed the following demographic information (Table 1).

Table 1: Participants' gender and age

			Age			
			14 to 18	19 to 23	24 to 29	Total
Sex	Male	Count	29	67	75	171
		% within participants' sex	17.0%	39.2%	43.9%	100%
	Female	Count	15	30	34	79
		% within participants' sex	19.0%	38.0%	43.0%	100%
Total		Count	44	97	109	250
		% within participants' sex	17.6%	38.8%	43.6%	100%

Table 1 shows that total 171 male and 79 female youth were surveyed in the study. Among them, male youth ranging the age of 14 to 18 were 17.0%, 19 to 23 were 39.2% and 24 to 29 were 43.9% whereas female youth ranging the age of 14 to 18 were 19.0%, 19 to 23 were 38.0% and 24 to 29 were 43.0%. The table also shows that male users are more than double in comparison with female users. Another thing, which is more significant, is that the more the age limit, the more the Facebook using tendency among male and female users.

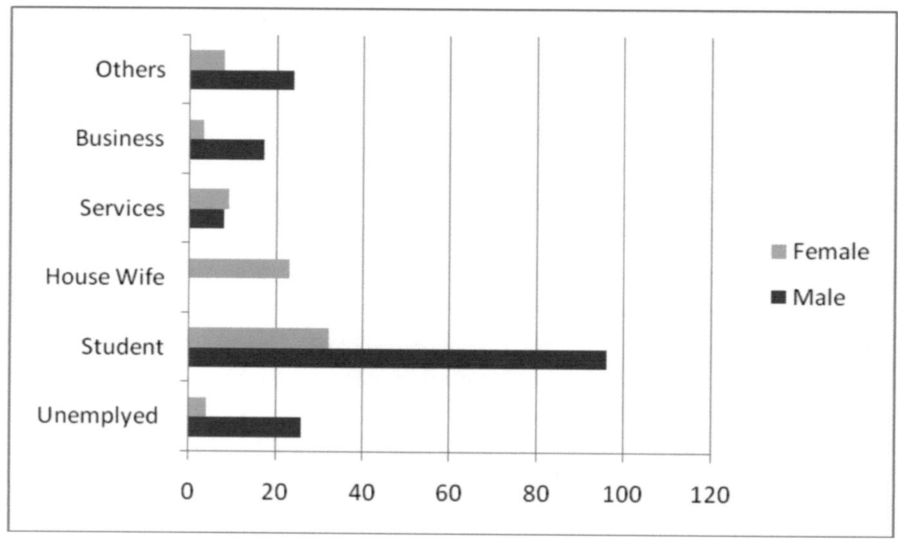

Figure 2: Professions among male and female Facebook users

The Figure 2 shows that the students are the most users among other professions. It also reveals that male users are more than female ones. Moreover, it is evident that unemployed and whose professions were not identified through the survey are using Facebook more than service holders and business persons though some male business persons use Facebook to some extent. Among the female youth, 32 were students, 23 house wife, whereas in the side of male youth, 96 were students and 26 unemployed. Among the total participants, 128 were students, 32 from different professions, 30 unemployed, 23 house wife, 20 business persons and the rest service holders.

Table 2: Participants' Identification of Climate Change Aspects

Gender		Male	Female	Total
Aspects related to Climate Change	Global Warming	43	25	25.1
	Drought	24	4	14.0
	Soil Salinity	17	8	9.9
	Cyclone	18	11	10.5
	Rise of Sea level	34	15	19.9
	Heavy Rainfall and Flood	35	16	20.5
	Total	171	79	100

The Table 2 shows that among the male youth, nos. of 43, 35, 34, 24 identified global warming, heavy rainfall and flood, rise of sea level, drought respectively, whereas, the female youth nos. of 25, 16, 15, 11 identified global warming, heavy rainfall and flood, rise of sea level, and cyclone as the aspects of climate change issues.

Table 1: Allocation and Donation for Climate Change Projects

Gender		Male 171	Female 79
Opinion about allocation and donation to Climate Change Projects	• I know about Climate Change Projects	32.2%	43.0%
	• Bangladesh Government is allocating money for Climate Change Projects	41.5%	36.7%
	• Bangladesh has already received money from donor agencies to address Climate Change	26.3%	20.3%
	• Total	100%	100%

The Table 3 portrays that 32.2% of male and 43.0% of female know about the Climate Change Projects, 41.5% of male and 36.7% of female also have the information that Bangladesh

Government allocates money for Climate Change Projects and 26.3% of male and 20.3% of female knows that Bangladesh has already received money from donor agencies to address Climate Change.

The study reveals that both male and female participants 62 and 29, respectively, said that priority should be given to Integrated Disaster Management, 50 and 21 to Capacity Building and Knowledge Management, 39 and 16 to Food Security, Health and Social Protection, and 20 and 13 to Green Energy and Low Carbon Issues to implement Climate Change Projects. The survey also reveals that 166 male youth and 77 female youth use only one Facebook accounts, whereas 5 male and 2 female youth use two Facebook accounts.

Moreover, the Table 4 clarifies that Facebook contributes to Climate Change through Likes, Comments, Sharing Photos or/and Videos in the sense that only 16 youth out of 250 either disagree or do not know the contributions of Facebook to Climate Change Projects. It is also evident that the youth who are of 19 to 29 years old are almost four-fifth of the total samples of the study.

Table 2: Facebook Contributes to Climate Change Projects

		Facebook contributes to Climate Change through Likes, Comments, Sharing Photos and/or Videos			
		Strongly Agree	Strongly Disagree	Don't Know	Total
Participants' ages	14 to 18	42	1	1	44
	19 to 23	93	1	3	97
	24 to 29	99	3	7	109
Total		234	5	11	250

The Figure 3 describes that most of the participants of the study unequivocally state that Facebook is powerhouse of knowledge, ensures civic engagement, connects youth vibrantly, a big platform to aware mass people, makes youth conscious and active, helps formulate better planning, helps set SMART goals, contributes to climate change through Likes, Comments, Sharing Photos and/or Videos. It means that Facebook has a great potential that if it is used in a designed way with a scheduled planning, it will generate great impact and influence to Climate Change Projects to ensure sustainable development.

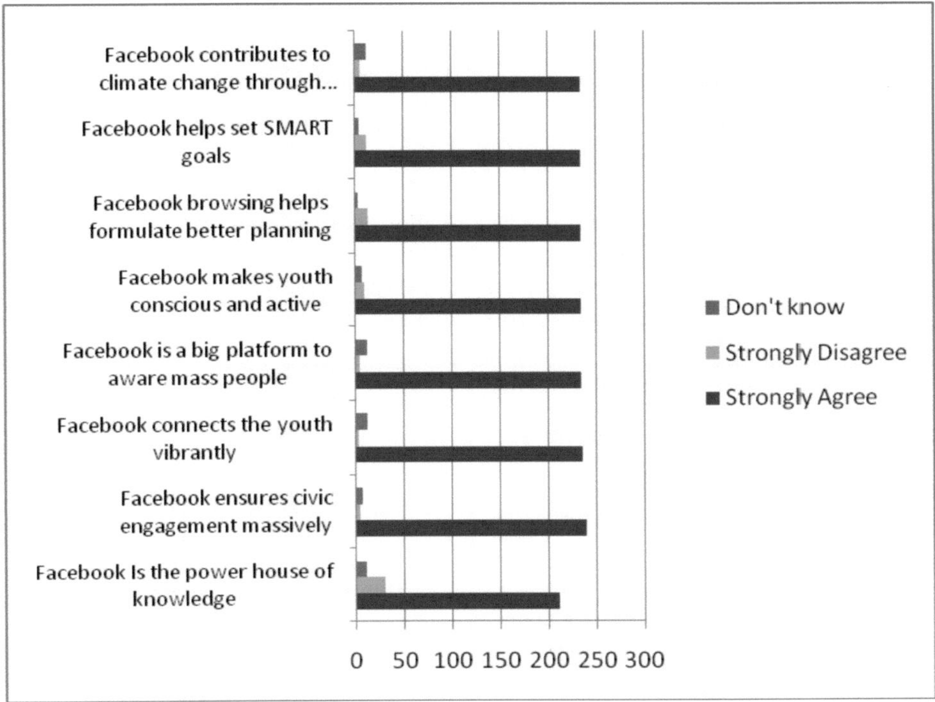

Figure 3: Facebook Impact on Youth

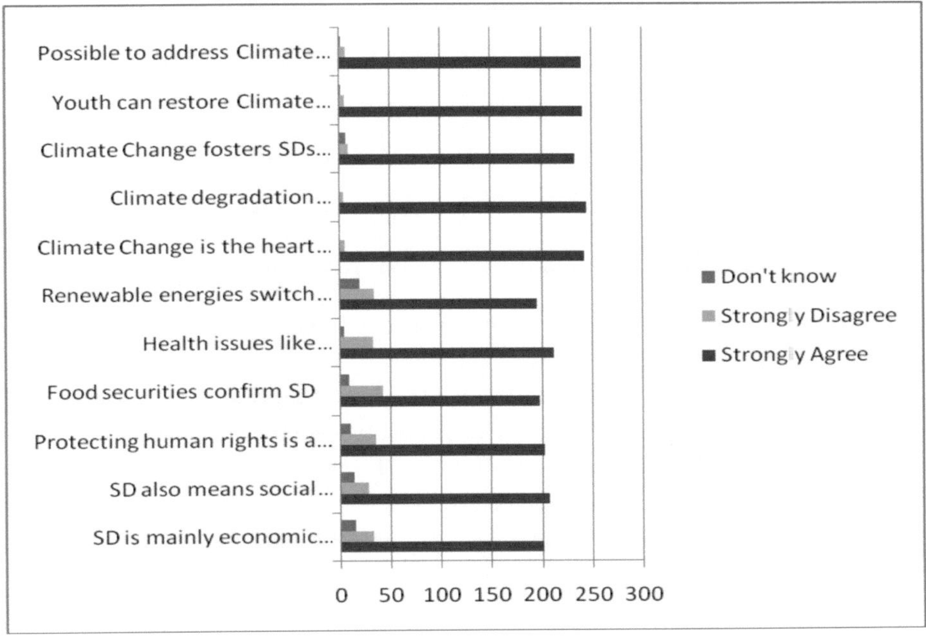

Figure 4: Facebook Impact on Youth's Opinions towards Climate Change and SD

Furthermore, the Figure 4 shows that for most of the youth Sustainable Development means economic development, social improvement, and protecting human rights. Most of them also opine that food security confirms Sustainable Development and health issues are connected with Sustainable Development. In these areas around 30 participants disagree and very few do not know the fact. But almost all of the participants gave their positive opinions on the matter that climate degradation seriously affects Sustainable Development and has some impact on Sustainable Development as well as it is possible to address the climate change projects through Facebook usage to ensure Sustainable Development.

Table 5: Correlation (it is significant at the 0.01 level -2 tailed)

A Facebook ensures civic engagement massively B Facebook makes youth conscious and active C Facebook contributes to Climate Change through Likes, Comments, Sharing Photos and/or Videos		A	B	C
A	Pearson Correlation	1	.571[**]	.717[**]
	Sig. (2-tailed)		.000	.000
	N	250	250	250
B	Pearson Correlation	.571[**]	1	.761[**]
	Sig. (2-tailed)	.000		.000
	N	250	250	250
C	Pearson Correlation	.717[**]	.761[**]	1
	Sig. (2-tailed)	.000	.000	
	N	250	250	250

The Table 5 gives a correlation matrix for three correlations: (1) Facebook ensures engagement, (2) Facebook enables consciousness and activeness, and (3) Facebook contributes to empowering. The diagonal consists of correlations of each variable with itself, always resulting in a value of 1 and the values on each side of the diagonal replicate the values on the opposite side of the diagonal. The positive correlation coefficient (.571) indicates that there is a statistically significant ($p<.001$) linear relationship between the engagement of citizen and the enabling of them by the Facebook that the more engagement is, the more enabling of people be. The same cases exist among the three correlations.

5 Conclusions

The study tried to investigate Bangladeshi youth Facebook user's engagement on Climate Change Projects through Facebook usages navigating positive role towards Sustainable Development in Bangladesh. Also, the youth participation through virtual world for fostering Sustainable Development (SD) is reflecting positive technological engagement to the society [33]. The result of the study was that Facebook usage makes the youth engaged, enabled and empowered about the Climate Change Projects in regard to Sustainable Development. As the youth learnt about Climate Change Projects, which contribute to the achieve Sustainable Development, the youth has engaged themselves with the Climate Change Projects through Facebook usage.

Moreover, this study envisaged that Facebook usage affected the youth's belief, goal setting, and planning. The youth got the platforms of interaction and participation in the context of Sustainable Development through Facebook usage. Furthermore, the study uncovered that the youth were committed, agreed and action-taking agents to Sustainable Development through Facebook usage. This study gave clear evidence that Facebook usage has played subjective role, reflective role, mutual role and societal role over the Bangladeshi youth to engage, to enable and to empower Climate Change Projects and Sustainable Development. It was found that Bangladeshi youth are the real force to adopt new communication technologies those are influential tools to address Climate Change issues for contributing the race of Sustainable Development in any situation.

6 Acknowledgements

The authors appreciate EXCEED Swindon project, German Academic Exchange Service (DAAD) and the German Federal Ministry for Economic Cooperation and Development (BMZ) for sponsorship to participate at the International Expert Workshop on *"Water on Agricultural Practices: Training the Trainers"* that took place in Rio de Janeiro, Brazil from 15 − 21 September 2019.

7 References

[1]   Zenghelis, D.: Stern Review: The economics of climate change. HM Treasury, London (2006)

[2]   de Castro, Í.B., Perina, F.C., Fillmann, G.: Organotin contamination in South American coastal areas. Environmental Monitoring and Assessment, 2012, 184(3), 1781-1799.

[3]   Smit, B., Pilifosova, O.: Adaptation to climate change in the context of sustainable development and equity. Sustainable Development, 2003, 8(9), p. 9.

[4]   Markandya, A., Halsnaes, K.: Climate change and sustainable development: prospects for developing countries. Earthscan (2002)

[5]   Bass, S., Dalal-Clayton, B.: Sustainable development strategies: a resource book. Routledge, Taylor & Francis Group, London (2012)

[6] Charles Jr, O.H., Schmidheiny, S., Watts, P.: Walking the talk: The business case for sustainable development. Routledge, Taylor & Francis Group, London (2017)

[7] White, R.: Climate change, uncertain futures and the sociology of youth. Youth Studies Australia, 2011, 30(3), p. 13.

[8] Pender, J.S.: What is climate change? and how it will effect Bangladesh. Briefing paper (final draft). Dhaka, Bangladesh: Church of Bangladesh Social Development Programme (2008)

http://www.churchofscotland.org.uk/__data/assets/pdf_file/0003/2982/climate_change_bangladesh.pdf

[9] Rasid, H., Paul, B.: Climate change in Bangladesh: Confronting impending disasters. Lexington Books., London (2013)

[10] Dankelman, I.: Gender, climate change and human security: Lessons from Bangladesh, Ghana and Senegal. Newcastle upon Tyne: Gender and Disaster Network, Radbound University, Nijmegen (2008)

[11] Nath, S.R.: Youth access to mass media in Bangladesh. Adolescents and youth in Bangladesh: Some selected issues. Research and Evaluation Division, Bangladesh Rehabilitation Assistance Committee (BRAC), Dhaka (2006)

[12] Assaad, R., Roudi-Fahimi, F.: Youth in the Middle East and North Africa: Demographic opportunity or challenge? Population Reference Bureau Washington, DC. (2007)

[13] Goldson, B.: Dictionary of youth justice. Willan, Routledge, Taylor & Francis Group, London (2013)

[14] Fien, J., Neil, C., Bentley, M.: Youth can lead the way to sustainable consumption. Journal of Education for Sustainable Development, 2008, 2(1), 51-60.

[15] Thackeray, R., Hunter, M.: Empowering youth: Use of technology in advocacy to affect social change. Journal of Computer-Mediated Communication, 2010, 15(4), 575-591.

[16] Muncie, J.: Youth and crime. SAGE Publications, London (2014)

[17] Hardi, P.: Assessing sustainable development: principles in practice. International Institute for Sustainable Development, Winnipeg, Source? (1997)

[18] Irwin, A.: Citizen science: A study of people, expertise and sustainable development. Routledge, Taylor & Francis Group, London (2002)

[19] Pawłowski, A.: How many dimensions does sustainable development have? Sustainable Development, 2008, 16(2), 81-90.

[20] Bicen, H., Cavus, N.: Social network sites usage habits of undergraduate students: Case study of Facebook. Procedia-Social and Behavioral Sciences, 2011, 28, 943-947.

[21] Bakshy, E., Messing, S., Adamic, L.A.: Exposure to ideologically diverse news and opinion on Facebook. Science, 2015, 348(6239), 1130-1132.

[22] Weedon, J., Nuland, W., Stamos, A.: Information operations and Facebook. version, 2017, 1, p. 27.

[23] Amin, M.R., Zhang, J., Yang, M.: Effects of climate change on the yield and cropping area of major food crops: A case of Bangladesh. Sustainability, 2015, 7(1), 898-915.

[24] Loo, Y.Y., Billa, L., Singh, A.: Effect of climate change on seasonal monsoon in Asia and its impact on the variability of monsoon rainfall in Southeast Asia. Geoscience Frontiers, 2015, 6(6), 817-823.

[25] Alam, M., Murray, L.A.: Facing up to climate change in South Asia. International Institute for Environment Development (IIED), London (2005)

[26] Sikder, R., Xiaoying, J.: Climate change impact and agriculture of Bangladesh. Journal of Environment and Earth Science, 2014, 4(1), 35-40.

[27] Howard,L., Bruce, L.B.: Qualitative research methods for the social sciences. Pearson, London (2001)

[28] Denzin, N.K., Lincoln, Y.S.: Strategies of qualitative inquiry. Vol. 2. SAGE Publications, London (2008)

[29] Checkoway, B.: What is youth participation? Children and youth services review, 2011, 33(2), 340-345.

[30] Buckingham, D.: Youth, identity, and digital media. MacArthur Foundation Series on Digital Media and Learning. Cambridge, MA: The MIT Press, (2007), p. 1-22.

[31] Honma, Y., Uchiyama, I.: The Relationships Between Role-Taking Ability and School Liking or School Avoidance: Rule and Moral Situations. Comprehensive Psychology, 2016, 5, p. 2165222816648079.

[32] Muuss, R.E.: Social cognition: Robert Selman's theory of role taking. Adolescence, 1982, 17(67), p. 499.

[33] Tjoa, A.M., Tjoa, S.: The role of ICT to achieve the UN Sustainable Development Goals (SDG). in: IFIP World Information Technology Forum. Springer, N.Y., (2016), p. 3-13.

IMPACTS OF WATER CRISES ON AGRICULTURE SECTOR AND GOVERNANCE CHALLENGES IN PAKISTAN

M. Mumtaz, J.A. Puppim de Oliveira

Sao Paulo School of Management, Getulio Vargas Foundation, Brazil;
Postal address: House 393, Street 59, G-10/4, Islamabad, Pakistan; mumtaz86@hotmail.com

Keywords: Adaptation, governance, Pakistan, subnational level, water scarcity

Abstract

The study develops a framework to understand the water governance challenges for agriculture sector at subnational level in Pakistan. The country is facing severe water crises and it may run dry by 2025. According to the International Monetary Fund, Pakistan is the third most water-stressed and has the world's fourth highest rate of water use. Agriculture is the major source of economy in the country. Pakistan uses 93% of its freshwater resources in agriculture. Agriculture sector contributes 24% of Gross Domestic Product, accounts for almost half of employed labor force and is the largest source of foreign exchange earnings. However, agriculture sector is under stress due to population growth, increased demands for food, ever-growing competition for water and land, climate change, and less-participatory water resources governance. It is imperative to introduce improved water management in agriculture and adaptation of agricultural systems to enhance water use performance and water productivity, particularly to face water scarcity. This study is conducted to understand the water governance for agriculture sector at subnational level by looking the case of Punjab province in Pakistan. It also identifies the key challenges in the way of effective water governance towards agriculture sector. The Punjab government has launched massive level awareness campaigns to understand the linkage between agriculture sector and water scarcity. One of the notable initiatives is the newly established national water policy of Pakistan. Some advances in irrigation system are also being incorporated. Water crises are escalating due to poor governance of water resources, lack of communication and education on water crises, and lack of investment and infrastructure development. The study proposes a framework for managing the risk of water scarcity based on preparedness rather than a crisis approach. The importance of diverse strategies to handle the forthcoming challenges associated with water resources management are emphasized, but the potential benefits depend on the appropriate multi-institutional and multi-stakeholder coordination to water resources management in a changing climate.

1 Introduction

Water is important for sustaining quality of life and ensuring sustainable development. The countries with sufficient fresh water resources have economic advantages, and it may give boost to their economic development [1]. Many developing countries lack renewable fresh water resources and facing multiple issues related to water resources [2, 3].

Pakistan is facing serious water crises. It is projected that it will become water scarce country from water stressed by 2030 [4]. According to the International Monetary Fund, Pakistan is the third most water-stressed country in the world. Pakistan has the world's fourth highest rate of water use. Due to misuse and overuse of water resources, Pakistan is facing water pollution, declination of water availability, threats to the economy, and environmental challenges.

Pakistan is an agricultural country. Pakistan's agriculture sector is a major contributor for its economy. It is indicated that agriculture sector contributes about 23.4% of the total Gross Domestic Product (GDP) in Pakistan [5]. Most of population lives in rural areas, and their livelihood is totally attached with agriculture sector [6]. It is also indicated that this sector absorbs nearly 45% labor force, and around 65% population of Pakistan is directly or indirectly attached with agriculture sector in the country. Water is very important for Pakistan due to its agrarian economy. It is reported that in Pakistan, water availability per capita has decreased from 5,260 m^3 in 1951 to 1,050 m^3 in 2008.

As water is an important driver for agricultural productivity, the sustainability of agricultural productivity is dependent on adequate and timely availability of water [3]. Water is used for enhancing agricultural productivity by irrigation practices. It is reported that Irrigation consumes about 80% of all the available water supplies in Asia [4]. Asia accounts for 79% (370 Million hectare, Mha) of all annualized irrigated areas, followed by Europe (7%) and North America (7%). About 50% of global irrigation area is found in six Asian countries: India (21.7% of the world's total irrigated area), China (19.4%), Pakistan (6.6%), Iran (2.8 %), Indonesia (1.8%) and Thailand (1.7%) [4]. In Pakistan, the Punjab province having large agricultural area is the leading province contributing for the economy of Pakistan.

Punjab is geographically located approximately at 30°000 N, 70°000 E in the semi-arid lowlands zone [7]. It is the second largest and the most populous province in Pakistan. Punjab has fertile agricultural land, which holds an extensive irrigation network and plays a leading role in the development of the economy [8]. The province accounts for 56.2% of the total cultivated area, 53% of the total agricultural GDP and 74% of the total cereal production in the country [9]. Punjab mainly contributes for agricultural sector in Pakistan. It has about 57% share of agricultural land and contributes for 53% of total Pakistan's agricultural GDP [10].

Agriculture sector is under stress due to population growth, increased demands for food, ever-growing competition for water and land, climate change, and less-participatory water resources governance. It is imperative to introduce improved water management in agriculture and adaptation of agricultural systems to enhance water use performance and water productivity, particularly to face water scarcity. Adaptation strategies in agriculture sector are an important response to climate change in Punjab province. It has been pointed out that agriculture adaptation measures can reduce losses [11]. The subnational government of Punjab is taking adaptation steps to tackle climate change. This study is conducted to understand the water governance for agriculture sector at subnational level by looking the case of Punjab province in Pakistan. The following sections discuss the role of subnational governments for effective governance, research methodology opted for this study, discussion and analysis, and the conclusion of this research.

2 Role of Subnational Governments and Adaptation Strategies

Previously, the focus remained not only on mitigation, but also top-down governance approach and the national governments were encouraged to take actions. However, presently the trend has been changed drastically especially after the Paris Agreement (PA) 2015, as the PA strongly encourages subnational/local governments and civil society organizations to act against climate change and other challenges like water scarcity. The subnational and local governments are getting many attentions in policy debate, and they are considered very instrumental to curb such challenges. Sub-national governments bring policy innovations and develop better solutions by identifying local requirements [12, 13], for instance, by local experimentations and piloting. Local governments are taking measures and innovative actions in their own as per their requirement [14, 15]. These governments are responsible for enacting laws and framing policies which are directly linked with climate change.

Subnational governments are responsible in many countries for public budget and expenditure, the effective use of these budgets can be helpful for adaptive improvements and adjustments [16]. In populated and geographical large countries, regional governments or subnational governments have reasonable administrative powers in many areas including sustainable development [17]. For example, in many countries subnational governments are responsible to implement policies like the environment, agriculture, and industry. Sub-national governments are important to raise awareness, and influence behavior and collaboration [17]. This can be achieved by launching consumer education programs, changing consumption habits, and promoting the use of green goods and services [18]. These are key aspects of adaptation strategies. These governments are effective to bring multiple stakeholders at the same point and to establish better relationships among the stakeholders at local level [19, 20].

To implement the PA and the effective use of global climate fund, it is important to understand the dynamics of subnational institutions and their collaboration with national and local level institutions [21]. These subnational level institutions are key to communicate the policy implementation challenges and opportunities between the national and local level governments. Although in the PA, the bottom-up approach and role of civil society organization are highlighted, but national governments are also an important player by a wide range of motivations such as increasing (economic) damage from climate impacts [22, 23], pressure from the public and the non-governmental organizations (NGOs) [24], learning through transnational networks [25], and (economic) competition with other countries [26]. Therefore, it is required to have a proper system of coordination for intergovernmental levels and due cooperation with international organization for effective climate adaptation governance at subnational level.

Uncertainties in climate change scenarios make it difficult to determine the precise impacts on future agricultural productivity. However, it is reported that growing food security challenges is increasing, and it will worsen the situation by 2050 if the proper actions are not taken for sustainable agriculture sector. Various studies in the literature have identified that significant losses in agriculture sector should be expected worldwide [23, 27–29]. It is widely acknowledged that policies need to provide a supportive environment that not only guides stakeholders in planning and executing adaptation interventions, but also enables farming communities to adapt to climate change [30].

Although, the local farmers communities are adapting their agriculture to the changing climate [31], but they are still vulnerable to climate change and variability [32]. Local knowledge can play a promising role to address the challenges of climate change at local scale for agriculture sector. Local knowledge is based on practice and assists farmers to make informed decisions about how to respond to environmental changes and how to improve the amount and quality of their yield [33].

Although, considerable progress has been made in developing governance system for climate change adaptation at local level but there are still implementation challenges due to lack of harmonized sectoral and local planning [34]. There are consensus that local knowledge is an important mean for effective adaptation [35, 36]. However, there is lack of evidences in literature to study local knowledge as a reflection of climate variability, its effects, and adaptations in agriculture [37]. The local knowledge of farmers has proved very useful and important in enhancing their adaptive capacity and designing climate adaptation policies [37].

It is recognized by researchers, farmers' community, and government officials around the world that adaptation to climate change is an urgent response to climate change [38]. But how this can be done most effectively? It is identified that one of the biggest challenges is the need for climate change adaptation solutions to be context specific. A one size fits all approach to policy does not work. This has led many to the conclusion that focus on local participatory approaches to adaptation planning and building adaptive capacity should be encouraged. Therefore, considering the implications of climate change to the farmers and agriculture sector in the local context is highly recognized. Agriculture has always adapted to climate with regionally specific adapted systems being observed across the world [39]. It is suggested that the adaptation policies towards agriculture sector are needed to be designed, based on how the local farmers understand climate related risks and respond to those risks [40].

In order to meet the challenges for agriculture sector effectively, the concept of the Climate Smart Agriculture (CSA) was introduced. The CSA can be defined as an approach for transforming and reorienting agricultural development under the new realities of climate change [41, 42]. Food and Agricultural Organization of the United Nations defines the CSA as *"agriculture that sustainably increases productivity, enhances resilience (adaptation), reduces/removes GHGs (mitigation) where possible, and enhances achievement of national food security and development goals"*. [41] Identified that productivity, adaptation and mitigations are critical for achieving this goal.

The adaptation in case of CSA is utilized by reducing the exposure of farmers to short-term risks and to strengthen their resilience by building their capacity to adapt and to face the long term challenges. Particular attention is given to protecting the ecosystem services, which ecosystems provide to farmers and others. These services are essential for maintaining productivity and our ability to adapt to climate changes. Here, subnational governments have an important role to establish better adaptation policies and action plans to ensure the implementation of the CSA mission.

To scale up the CSA is triggered to establish agricultural supportive policies, institutions, and financing at different level of governance [43]. The response to impacts of climate change can be observed at local level. However, these responses depend on the information and socio-economic condition of the local farmers. For example, poor farmers take measures to ensure their survival, but wealthier farmers make decisions in order to enhance their productivity and to maximize their profits [44].

3 Research Methodology

The research methodology depends on the central research question and research objectives of the study. The main objective of this study is to explore the prominent initiatives towards water governance in the province of Punjab, Pakistan. The study is conducted by employing case study as research methodology. In the first stage of this study, information from reports, policy documents, the internet, newspapers, and scientific and academic articles were collected. The field study was conducted from June to September 2018. For this study, 20 in-depth semi-structured interviews were conducted with relevant stakeholders. The stakeholders included policy experts, local government officials, think tanks working in the area, NGOs, academics, environmental protection agencies in the Punjab, civil society, and climate change activists.

4 Discussion and Analysis

The demand of water is increasing not only for domestic use due to urbanization, but also for the enhancement of agricultural productivity in order to ensure food requirements with population growth. With existing growth rate of population in Pakistan, it is projected that 250 million people will be by 2025 [45]. It is also reported that the urban population will increase to 52% by 2025, which is around 40% at the moment. Resultantly, water demand for domestic and agriculture purposes will be increased drastically. In order to fill this demand and supply gap of water resources, prominent initiates are imperative in water scare country for sustainable development. Keeping in view the existing challenges of water scarcity, government of Punjab is improving its water governance mechanisms so as the address the challenge effectively.

The provincial government is well aware of the issue. During the field study, it is noted that the different line departments related to agriculture and water are engaged. Other stakeholders are involved in establishing strategies; specifically adaptation measures for water use are communicated in various localities in the province. For instance, the small scale farmers and domestic users of water in household are educated by arranging different seminars and workshops at local scales. It was reported that water is being wasted for irrigation purposes especially by the small farmers, who are not well equipped and well aware of the use of required water for irrigation.

The government is promoting research and technological advancement so that effective strategies can be devised. For instance, some research institutions like the University of Agriculture Faisalabad and Ayub Agriculture Research Institute are continuously investigating the impacts of water scarcity on agriculture sector and set a platform for policy interventions. Likewise, local and international NGOs like the Universities for Water Network, Leadership for Environment and Development, Sustainable Development Policy Institute, and United Nations Development Programme are involved in providing policy advocacy, technical and research support to manage water crises situation in the province.

A massive level awareness campaigns have been launched throughout the province. The general public is educated about the issue of water security in Pakistan in general and in Punjab province in particular. A radio station is established, where different programs are arranged to discuss the issue of water security and possible adaptation solutions. Moreover, literature on the water security is being published in various local languages so that general public can understand the severity of the problem. One of other notable initiatives is to engage religious segment of the society. In Friday prayers sermons, the teaching of Islam about saving the water, planting trees, and taking care of environment is regularly preached by the religious leaders.

It is noted that water is being wasted due to outdated techniques of irrigation. Therefore, the government has sent many government officials to gain training from other countries in order to understand the novel and innovative techniques of irrigation so that such effective techniques can be replicated in Punjab to limit water losses.

5 Key Challenges for Water Governance in Punjab

This section is dedicated to discuss the key challenges of water governance for agriculture sector in Punjab.

Weak institutional capacity

Local institutions play an important role for effective governance at local scale. Capacity-building of local institutions is prerequisite for effective governance [46]. However, in this case of study, it is noted that local line departments are not sufficiently capable to take the challenge. This is why the water crises situation is getting wider with the passage of time. This capability building can be improved by giving them a task with proper responsibility, establishing a proper mechanism of coordination among these local intuitions and sophisticated links with national and international institutions. Although, Punjab government is trying to improve this important aspect of governance, it will take some time.

Poor irrigation infrastructure

Proper infrastructure is important for effective irrigation. It has seen in Punjab province, either infrastructure setup is missed in some localities or it is poorly managed. These infrastructure issues include leakages (reservoirs, pipes, and taps) due to damage or aging. The poorly maintained infrastructure badly affects water supply, as it is one of the major reasons for water losses during the distribution period. It is estimated that Pakistan needs to raise its storage capacity by 22 BCM (billion cubic meter) by 2025 in order to meet the projected requirements of 165 BCM. Therefore, it is of paramount importance that Pakistan gives serious attention to building new storage facilities in order to meet this challenge.

Increasing gap in supply and demand

The population of Pakistan is increasing with a rapid pace, and it is projected that it will reach to 250 million by 2025. The urbanization is one of fastest in South-Asia, and it is expected to be 52% as compared to today, which is about 40%. Due to this increase of population and urbanization, water demand for domestic, industrial, and nonagricultural will be 8% and is expected to reach 10% of the total available water resources by the year 2025 [47]. However, the water requirement for irrigation is expected to be 250 BCM in 2025 against a projected potential availability of 185 BCM from surface water resources. Even by utilizing the groundwater, the projected potential is to reach up to 190 BCM. In this scenario of "business as usual", it will pose serious implications for food shortages in the years to come and will severely hurt the national economy and livelihood of millions of people. Therefore, adaptive governance, involvement of actors, and minimization of water losses should be the top priority of subnational government.

Absence of Public Private Partnership

Public Private Partnership is instrumental for dealing the local problems. Non-state actors have a positive role to address the local issue [48]. Participatory and stakeholder involvement are critical in decision-making. In case of this study, this important concept is missing or being used not in right direction. The issue of surface water scarcity is solved by abstraction the groundwater without considering the negative implications of sucking the groundwater. It is reported that the private tube wells have increased from 10,000 in 1960 to 0.6 million in 2002 [49], and more than 1 million in 2007. The total groundwater abstraction from these tube wells is estimated at $51 \times 10^9 \, m^3$ against a recharge of $40–60 \times 10^9 \, m^3$. This overexploitation of groundwater is causing water table decline of most of the canals of Punjab province. This is another big challenge for the provincial government to monitor and to regulate such pumping and abstraction of ground water so that a balance can be maintained.

Impacts of climate change

Pakistan is ranked one of the most vulnerable countries due to climate change. The Punjab Province gets water from snowmelt during spring and summer, while in late summer monsoonal rains are the major source of water. The increase of temperature and melting of glaciers rapidly does not only cause floods, destruction of crops and infrastructure, but also decreases the water resources in long run. It is reported that climate is adverse effecting the agricultural production. It is projected by International Panel on Climate Change that South Asian countries will face with increase of monsoon by 8–24% [50, 51]. It is expected that this increase in monsoon rains will bring additional water, and cause floods and damage to the infrastructure. Pakistan needs to prepare itself for possible future climate change and its impact on Pakistan. Effective water management can be the best strategy to curb the projected climate change and its impact on agricultural economy of Pakistan.

6 Conclusions

Pakistan is ranked in the list of the countries that are most vulnerable due to climate change, and it is confronting severe water security challenges. Pakistan is facing multiple governance issues in order to put the country on the path of sustainable development. Local governments can play key role to address the wicket problems like climate change and water security. Back in 2010, on the heels of 18th constitutional amendment in Pakistan, provinces/subnational governments are empowered and decentralized system of governance was initiated. The provinces/subnational governments are responsible to deal with the issues of climate change and water security. Water scarcity in Pakistan is becoming hindrance for agricultural economy of the country. In order to address the water security and to ensure sustainable agriculture, Punjab government has taken some promising initiatives by launching massive level awareness campaigns to understand the linkage between agriculture sector and water scarcity. Adaptation strategies are promoted so that the misuse and overuse of water can be minimized especially in households' usage. One of notable initiatives is to educate local farmers to use the water for irrigation purposes. Water crises are escalating due to poor governance of water resources, lack of communication and education on water crises, and lack of investment and infrastructure development. The importance of diverse strategies to handle the forthcoming challenges associated with water resources management are emphasized, but the potential benefits depend on the appropriate multi-institutional and multi-stakeholder coordination to water resources management in a changing climate. It is recommended to enhance the coordination between user organizations, particularly between irrigation and agriculture sector. Public-private sector partnerships can also play a key role to overcome the existing challenges of water usage management. Moreover, capacity building in technical and management areas is needed for effective adaptation strategies towards water crises.

7 Acknowledgements

The author would like to thank EXCEED Swindon project and DAAD (German Academic Exchange Service) for support to participate at the Regional Workshop on "Water on Agricultural Practices: Training the Trainers" that took place in Rio de Janeiro, Brazil from 15 – 21 September, 2019.

8 References

[1]   D. Molden: Water for food water for life: A comprehensive assessment of water management in agriculture. Routledge, 2013.

[2]   M.M. Waqas, U.K. Awan, M.J.M. Cheema, I. Ahmad, M. Ahmad, S. Ali, S.H.H. Shah, A. Bakhsh, M. Iqbal: Estimation of canal water deficit using satellite remote sensing and GIS: A case study in lower chenab canal system. Journal of the Indian Society of Remote Sensing., 2019, 47(7), 1–10.

[3]   I. Ahmad, S.A. Wajid, A. Ahmad, M.J.M. Cheema, J. Judge: Optimizing irrigation and nitrogen requirements for maize through empirical modeling in semi-arid environment. Environmental Science and Pollution Research, 2019, 26(2), 1227–1237.

[4]   A. M. Bhatti, P. Suttinon, S. Nasu: Agriculture water demand management in Pakistan: a review and perspective. Society for Social Management Systems, 2009, 9(172), 1–7.

[5]   M. A. Imran, A. Ali, M. Ashfaq, S. Hassan, R. Culas, C. Ma: Impact of Climate Smart Agriculture (CSA) Practices on Cotton Production and Livelihood of Farmers in Punjab, Pakistan. Sustainability, 2018, 10(6), 1–20.

[6]   M. Hussain, M. I. Javaid, P. R. Drake: An econometric study of carbon dioxide (CO2) emissions, energy consumption, and economic growth of Pakistan. International Journal of Energy Sector Management, 2012, 6(4), 518–533.

[7]   V. Ahmed, A.A.K. Malik, S. Ramay, Z. Munawwar, A. Pervaiz: National economic and environmental development study: the case of Pakistan. Munich Personal RePEc Archive., 2011, 5–76.

[8]   M. Abid, M. Ashfaq, S. Hassan, N. Fatima: A resource use efficiency analysis of small Bt cotton farmers in Punjab, Pakistan. Pakistan Journal of Agricultural Sciences,2011, 48(1), 65–71.

[9]   PBS:Punjab Development Statistics, Punjab Bureau of Statistics Government of the Punjab Lahore. Government of Punjab, 2010.

[10]   U. Hanif, S.H. Syed, R. Ahmad, K.A. Malik, M. Nasir: Economic impact of climate change on the agricultural sector of Punjab. Pakistan Development Review, 2010, 49(4), 771–798.

[11]   S. Di Falco: Adaptation to climate change in Sub-Saharan agriculture: assessing the evidence and rethinking the drivers. European Review of Agricultural Economics, 2014, 41(3), 405–430.

[12]   J. Cauley, E.I. Im: Intervention policy analysis of skyjackings and other terrorist incidents. The American Economic Review, 1988, 78(2), 27–31.

[13]   M. Mumtaz: The National Climate Change Policy of Pakistan: An Evaluation of Its Impact on Institutional Change. Earth Systems and Environment, 2018, 2(3), 525–535.

[14]   J.A.P. de Oliveira: The implementation of climate change related policies at the subnational level: An analysis of three countries. Habitat International, 2009, 33(3), 253–259.

[15] M.M. Betsill, H. Bulkeley: Cities and the multilevel governance of global climate change. Global Governance, 2006, 12(2), 141–159.

[16] F. Barbi, L.d.C. Ferreira: Governing Climate Change Risks: Subnational Climate Policies in Brazil. Chinese Political Science Review, 2017, 2(2), 237–252.

[17] J. Setzer: Subnational and transnational climate change governance: Evidence from the state and city of São Paulo, Brazil. Fifth Urban Research Symposium, Cities and Climate Change: Responding to an Urgent Agenda, 2009, 28–30.

[18] T. Fernando: Sub-central Governments and Debt Crisis in Spain over the Period 2000–2011. Advances in Political Economy, 2013, 129-144

[19] B. Smit, O. Pilifosova: From adaptation to adaptive capacity and vulnerability reduction. Climate change, adaptive capacity and development, World Scientific. 2003, 9–28.

[20] M. Mumtaz, J.A.P. de Oliveira, S.H. Ali: Climate Change Impacts and Adaptation in Agricultural Sector: The Case of Local Responses in Punjab, Pakistan. Climate Change and Agriculture, IntechOpen, 2019, 1-15.

[21] D.W. Cash, S.C. Moser: Linking global and local scales: designing dynamic assessment and management processes. Global Environmental Change, 2000, 10(2), 109-120.

[22] E.W. Oliver, D. Thomas,T. Erla, C.K. Vincent: Understanding Contemporary Challenges to INGO Legitimacy: Integrating Top-Down and Bottom-Up. International Journal of Voluntary and Nonprofit Organizations, 2016, 27(6), 2764–2786.

[23] I. Ahmed, A. Ullah, M.H. Rehman, B. Ahmed, S.A. Wajid, A Ahmed, S. Ahmed: Climate change impacts and adaptation strategies for agronomic crops. Climate Change and Agriculture, IntechOpen, 2019, 1–15.

[24] E. Nulman: Climate change and social movements: civil society and the development of national climate change policy Basingstoke: Palgrave Macmillan, London, 2016, 1-179.

[25] A. Bauer, R. Steurer: Innovation in climate adaptation policy: Are regional partnerships catalysts or talking shops? Enviromental Politics, 2014, 23(5), 818–838.

[26] E. Massey, R. Biesbroek, D. Huitema, A. Jordan: Climate policy innovation: the adoption and diffusion of adaptation policies across Europe. Global Enviromental Change, 2014, 29, 434–443.

[27] M.H. Rahman, A. Ahmad, X. Wang, A. Wajid, W. Nasim, M. Hussain, B. Ahmad, I. Ahmed, Z. Ali, W. Ishaque, M. Awais, V. Shelia, S. Ahmad, S. Fahd, M.Alam, H. Ullah, G. Hoogenboom: Multi-model projections of future climate and climate change impacts uncertainty assessment for cotton production in Pakistan. Agricultral and Forest Meteorology, 2018, 253, 94–113.

[28] I. Ahmed, M.H. ur Rahman, S. Ahmed, J. Hussain, A. Ullah, J. Judge: Assessing the impact of climate variability on maize using simulation modeling under semi-arid environment of Punjab, Pakistan. Environmental Science and Pollution Research, 2018, 25(8) 28413–28430.

[29] A. Ullah, I. Ahmad, A. Ahmad, T. Khaliq, U. Saeed, M. H. Rahman, J. Hussain, S. Ullah, G. Hoogenboom:Assessing climate change impacts on pearl millet under arid and semi-arid environments using CSM-CERES-Millet model. Environmental Science and Pollution Research, 2019, 26(7), 6745–6757.

[30] R.J. Berman, C.H. Quinn, J. Paavola: Identifying drivers of household coping strategies to multiple climatic hazards in Western Uganda: implications for adapting to future climate change. Climate and Development, 2015, 7(1),71–84.

[31] J. Knox, T. Hess, A. Daccache, T. Wheeler: Climate change impacts on crop productivity in Africa and South Asia. Environmental Research Letters, 2012, 7(3),1-9.

[32] A.B. Heinemann, G. Hoogenboom, R.T. de Faria: Determination of spatial water requirements at county and regional levels using crop models and GIS - An example for the State of Parana, Brazil. Agricultural Water Management, 2002, 52(3), 177-196.

[33] A.J. Newsham, D.S.G. Thomas: Knowing, farming and climate change adaptation in North-Central Namibia. Global Environmental Change, 2011, 21(2), 761–770.

[34] M. Madzwamuse: Climate governance in Africa: adaptation strategy and institutions. A Synthesis Report, Heinrich Böll Stiftung, Germany, 2010.

[35] C.P. Mubaya, J. Njuki, E.P. Mutsvangwa, F.T. Mugabe, D. Nanja: Climate variability and change or multiple stressors? Farmer perceptions regarding threats to livelihoods in Zimbabwe and Zambia. Journal of Environmental Management, 2012, 102, 9–17.

[36] O. Mertz, C. Mbow, A. Reenberg, A. Diouf: Farmers' perceptions of climate change and agricultural adaptation strategies in rural Sahel. Environmental Management, 2009, 43(5), 804–816.

[37] S.A. Ogalleh, C.R. Vogl, J. Eitzinger, M. Hauser: Local perceptions and responses to climate change and variability: the case of Laikipia District, Kenya. Sustainability, 2012, 4(12), 3302–3325.

[38] B. Soubry: How Can Governments Support Adaptation to Climate Change by Small-Scale Farmers? A Case Study from the Canadian Maritime Provinces. Environmental Change Institute, 2017, 1-34.

[39] B. Ren, J. Zhang, X. Li, X. Fan, S. Dong, P. Liu, B. Zhao: Effects of waterlogging on the yield and growth of summer maize under field conditions. Canadian Journal of Plant Science, 2014, 94(1), 23–31.

[40] L. Menapace, G. Colson, R. Raffaell: Farmers' Climate Change Risk Perceptions: An Application of the Exchangeability Method. EAAE 2014 Congress on Agri-Food and Rural Innovations for Healthier Societies, 2014, 26.

[41] L. Lipper, P. Thornton, B.M. Campbell, T. Baedeker, A. Braimoh, M. Bwalya, P. Caron, A. Cattaneo, D. Garrity, K. Henry, R. Hottle, L. Jackson, A. Jarvis, W. Mann, N. McCarthy, A. Meybeck, H. Neufeldt, T. Remington, P.T. Sen, R. Sessa, R. Shula, A. Tibu, E.F. Torquebiau: Climate-smart agriculture for food security. Nature Climate Change., 2014, 4(12), 1068-1072.

[42] A. Ullah, I. Ahmad, U. Saeed, A. Ahmad, A. Mahmood, G. Hoogenbocm: Climate smart interventions of small-holder farming systems. Climate Change and Agriculture, IntechOpen., 2019, 1–16.

[43] O. Westermann, P.K. Thornton, W. Förch: Reaching more farmers: innovative approaches to scaling up climate-smart agriculture. CCAFS Working Paper no. 135. CGIAR Research Program on Climate Change, Agriculture and Food Security CCAFS, Copenhagen, Denmark., 2015, 1-110.

[44] G. Ziervogel, S. Bharwani, T.E. Downing: Adapting to climate variability: pumpkins, people and policy. Natural Resources Forum, 2006, 30(4), 294–305.

[45] M. Zakria, F. Muhammad: Forecasting the population of Pakistan using ARIMA models. Pakistan Journal of Agricultural Sciences, 2009, 46(3), 214-223.

[46] L. Ding, K.J. Wang, G.M. Jiang, D.K. Biswas, H. Xu, L.F. Li, Y.H. Li: Effects of nitrogen deficiency on photosynthetic traits of maize hybrids released in different years. Annals of Botany, 2005, 96(5), 925-930.

[47] J. Bongaarts, Z.A. Sathar, A. Mahmood : Population trends in Pakistan. Population Council Book Series, 2013, 1(1), 13-23.

[48] R.A. Beauregard: Public-private partnerships as historical chameleons: The case of the United States. Partnerships in Urban Governance, 1998, 52–70.

[49] L.F. Konikow, E. Kendy:Groundwater depletion: A global problem. Hydrogeology Journal, 2005, 13(1), 317–320.

[50] A. Amin, W. Nasim, M. Mubeen, S. Sarwa, P. Urich, A. Ahmad, A. Wajid, T. Khaliq, F. Rasul, H.M. Hammad, M.I.A. Rehmani, H. Mubarak, N. Mirza, A. Wahid, S. Ahamd, S. Fahad, A. Ullah, M.N. Khan, A. Ameen, Amanullah, B. Shahzad, S. Saud, H. Alharby, S.T. Ata-Ul-Karim, M. Adnan, F. Islam, Q.S. Ali: Regional climate assessment of precipitation and temperature in Southern Punjab (Pakistan) using SimCLIM climate model for different temporal scales. Theoretical and Applied Climatology, 2018, 131(1), 121-131.

[51] Q.Z. Chaudhry, A. Mahmood, G. Rasul, M. Afzaal: Climate change indicators of Pakistan. Pakistan Meterological Department - Technical Report No. PMD-22/2009., 2009, 1-43.

ECOSYSTEM SERVICES OF A SEASONALLY DRY TROPICAL FOREST
IN SEMI-ARID REGION OF BRAZIL

Rodolfo Souza[1], Borja Ruiz Reverter[2], José Romualdo de Sousa Lima[3], Artur Coutinho[4], Antonio Celso Dantas Antonino[1], Eduardo Soares de Souza[5]

[1]*National Observatory of Water and Carbon Dynamics in Caatinga Biome (NOWCDCB), Av. Professor Luiz Freire, 1000, Recife-PE, Brazil; rsouza@lavabit.com*

[2]*Federal University of Paraíba, Campus II, Areia-PB, Brazil;*

[3]*Federal Rural University of Pernambuco, Garanhuns-PE, Brazil;*

[4]*Federal University of Pernambuco, Caruaru-PE, Brazil;*

[5]*Federal Rural University of Pernambuco, Serra Talhada-PE, Brazil;*

Keywords: Caatinga; Carbon balance; Degradation; Ecosystem management; Grassland

Abstract

Ecosystems can provide many essential services for the society, although sometimes their valuation is ignored. The seasonally dry tropical forest in the semi-arid region of Brazil, named Caatinga, is characterized by a fine-tuned adaptation to extreme rainfall seasonality. Caatinga, like many other ecosystems, is under pressure for food production, whereas grassland is a typical replacement of native areas. The main goal of this paper is to study the carbon uptake, biomass production in Caatinga and grassland in order to quantify the level and potential revenue of environmental services in the semi-arid region of Brazil. Measurements of carbon flux, meteorological variables, and biomass in the preserved native vegetation Caatinga (CA) and in a degraded grassland (GR) in the semi-arid region of Brazil were carried out. The carbon flux was measured using the eddy covariance technique for one hydrological year. Owners chose all decisions of management areas (preserve or explore) without interference or advise. These areas have conditions that affect the price of the same product, such as carbon credit due to other not priced variables. Total rainfall during the studied period was lower than average for the region. However, carbon uptake was 2.74 t/ha in CA and 3.00 t/ha in GR. Based on current carbon credit trade value, local market and current management, CA is potentially 56% more profitable than GR. The total potential revenues in CA are 78.97 USD/ha. yr and in GR 50.74 USD/ha.yr. The lower revenue in GR is due to necessary expenses with nutrient reposition and reseeding to avoid a collapse in this area in few years. The analysis made here may be useful for policy-makers in order to preserve natural ecosystems and to reduce deforestation and land degradation.

1 Introduction

Ecosystem services are defined as the benefits that people derive from nature [1]. Ecosystems can provide many essential services for the society, e.g., carbon uptake and storage, biomass, biodiversity, water that are important for the livelihood, health, and survival, especially when the environment is preserved [2]. Even small perturbations in the ecosystem can lead to at least temporary disequilibrium and reduce quantitatively and qualitatively those services [3].

In the semi-arid region of Brazil, the native ecosystem named "Caatinga" is characterized by a fine-tuned adaptation to extreme rainfall seasonality [4]. This biome covers an area of 844,453 km^2 and has enormous endemic biodiversity, with unique characteristics that make it an exclusive Brazilian biome. One of the most striking features of the region is the severe drought periods that affect plant growth and have a substantial social and economic impact on the population. Due to the agricultural vulnerability to climate change, livestock has emerged as the main activity of the rural population, being the precursor of systematic replacement of native vegetation by grazing areas [5].

Decades of low-tech extractive agriculture, i.e., without nutrient replenishment and soil conservation management, coupled with grazing pressure and continuous extraction of forest products have resulted in intense modifications of Caatinga, including desertification and the disappearance of high tree vegetation, which is compounded by climate change. Vegetation transformations result in loss of ecosystem products and services, such as restoration of fertility, soil protection, water production, and forest products such as firewood, wood, and fodder [6].

The Caatinga contributes to regional development, as it is an energy source for industries and farming families, either in the supply of fodder to maintain livestock or offering other timber and non-timber forest products. The maintenance of seasonally dry tropical forests also suggests the regulation of processes, such as infiltration and runoff in the control of soil and subsoil water storage and erosion, respectively, in climate regulation. Also, it is support for nutrient cycling and greenhouse gas exchange as well as for their primary production.

Thus, replacing native vegetation by cultivated grass can increase the risks of degradation of the area and reduce the potential income from that area. This land-use change is even more concerning, because at least 70% of the pastures in Brazil have some degradation levels [7]. Land-use changes or any disturbances alike indeed affect the carbon sequestration by the ecosystem. Monitoring the time series of the carbon flux in terms of a net ecosystem exchange (NEE) is a key aspect for management decisions. In this work, the carbon uptake and biomass production in Caatinga and grassland as to quantify the level and potential revenue of environmental services in the semi-arid region of Brazil were studied.

2 Material and Methods

Sites description

Both experimental areas, Lagoinha (grassland) and Buenos Aires (Caatinga) farms are located at Serra Talhada city, semi-arid region of Pernambuco State, Brazil (Figure 1). According to the Köppen classification, the site is BShw. The mean annual rainfall at the experimental site is approximately 642 mm [8] with a rainy season that includes the months of December to March. The daily mean air temperature is 25.2 °C, and the potential evapotranspiration can reach 1,850 mm/yr. The soil has 17% clay, 18% silt, and 65% sand (sandy loam texture).

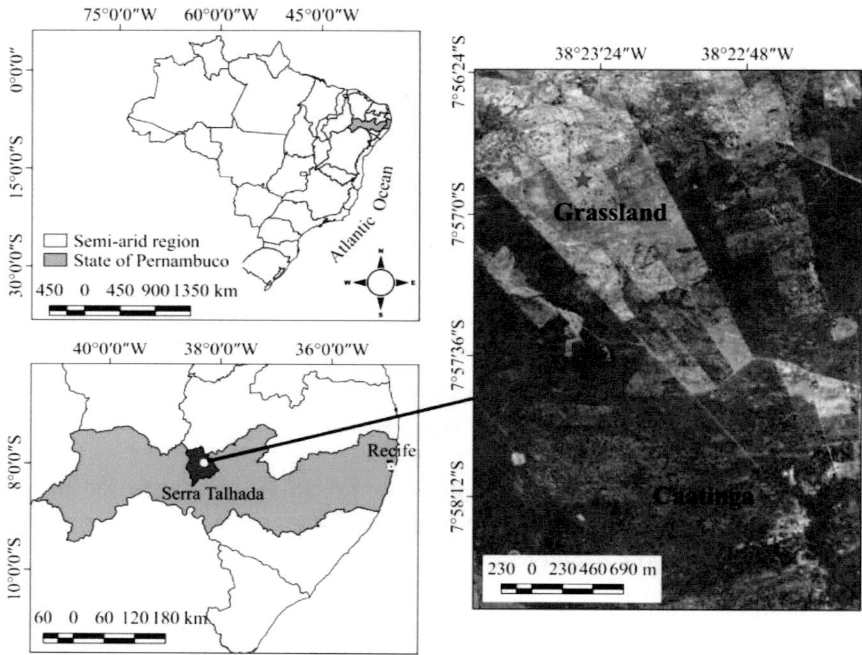

Figure 1: Map of Brazil with the delineation of the state of Pernambuco, Serra Talhada County and the Caatinga and grassland areas

The Caatinga area (7°58'5.2"S; 38°23'2.62"W) is characterized by dry, spiny and predominantly deciduous shrub/forest vegetation. This Caatinga area covers 180 ha and is composed of shrubs and tree species, most of the tree species having a height of approximately 8 m. The grassland area (7°56'50.4"S; 38°23'29"W) has 24.5 ha and was initially planted with *Urochloa mosambicensis*, which is a C4 non-native grass, but currently, most of the vegetation died due to the drought years and the management of animals and grazing.

The distance between these sites is less than 2.5 km, and in each site were installed flux towers that are part of the National Observatory of Water and Carbon Dynamics in the Caatinga Biome

(INCT-NOWCDCB). Both properties have a flat topography and were located in the city of Serra Talhada, Pernambuco State, Brazil.

The vegetation and soil features of both areas are presented in Table 1. The leaves biomass in Caatinga was measured from the litter fall of one year. For the woody biomass, the diameter of the tree at breast height is measured and the biomass estimated using allometric equations for this vegetation developed by [9].

Table 1: Vegetation and soil properties in Caatinga and grassland

	Caatinga	Grassland
Leaves biomass (t/ha)	$2.20^{\#}$	1.75
Woody biomass (t/ha)	$45.9 \pm 2.8^{\$}$	-
Soil Organic Carbon (g/kg)	18.1	8.44
Sandy (%)	71.6	66.5
Silt (%)	17.2	16.5
Clay (%)	11.1	16.7
Bulk density (g/cm^3)	1.39	1.44
Porosity (cm^3/cm^3)	0.48	0.46
Ks (mm/h)	214	36

$^{\#}$*Measured from litterfall;* $^{\$}$*Estimated from allometric equation based on [8].*

The soil organic carbon in Caatinga (CA) is about two times higher than in grassland (GR), and it is necessary to take this into account in terms of carbon quality. In respect to soil physical properties, both areas have a similar texture, but the bulk density and porosity indicate that there is some soil compaction in GR due to more intensive grazing events. The soil hydraulic conductivity in CA is about six times higher than in GR, which impacts the water infiltration and runoff directly.

The Normalized Difference Vegetation Index (NDVI) was considered as an indicator of vegetation activity with presumed links to leaf area index (LAI), cover percentage, and related effects on evapotranspiration and carbon fluxes [10, 11]. The NDVI data were obtained from moderate-resolution imaging spectroradiometer (MODIS) images with a spatial resolution of 250 m and recorded at 16-day intervals (http://www.modis.cnptia.embrapa.br).

Instrumentation and data processing

CO_2 and water vapor concentration, jointly with wind speed and sonic temperature, were measured synchronously at high frequency and, from them, the CO_2, latent, and sensible heat fluxes were calculated using the eddy covariance (EC) technique in both sites. The sensors were installed in a tower in the centers of the areas. The eddy covariance system was composed of a three-axis sonic anemometer (CSAT3, Campbell Scientific Inc., Logan, USA), and an infra-red gas analyzer (EC150, Campbell Scientific Inc., Logan, USA). Total rainfall was measured with a tipping bucket rain gauge (model TE 525 WS-L, Texas Electronics, Dallas, TX, USA). For more details about other variables and sensors, see [12]. The EC instruments were sampled at 10 Hz and low-frequency instruments for meteorological variable sampled at 30 min scale using a data logger CR1000 (Campbell Scientific Inc., Logan, USA).

The raw eddy covariance data were processed using the software AltEddy 3.90 in order to compute the 30 min fluxes and to apply WPL corrections to compensate fluctuations of air temperature and water vapor in CO_2 measurements [13]. After this step, the gap-filling and flux partitioning were performed using REddyProc [14] using the R environment [15].

Valuation of ecosystem services

To calculate the price of the ecosystem services, the products/services were considered that the owner/farmer could sell in the marketing. However, ecosystems can provide many other services such as regulation (gas, water, climate, and disturbance), soil formation, nutrient cycling, pollination, biological control, raw materials, and others [1], which could be considered to aggregate quality in traded products. In this way, for CA, the potential credit of carbon was considered, based on the net ecosystem exchange (NEE), the reserve of timber. For GR, the NEE, meat production, and the costs to replenish the soil nutrients were considered.

The price of carbon credit was considered equal to 26.00 €/t, which is equivalent to 28.82 USD [16]. The other products such as timber and meat were priced based on the local currency, Brazilian Real (BRL) and then converted to US Dollars (USD), considering that 1.0 USD is equal to 4.16 BRL, quoted at 08/28/2019 without fees and taxes that may be charged during currency conversion. The gross price for one ton of timber is about USD 17.31 and for a kilogram of lamb meat USD 3.61. Thus, the potential revenue that the farmers would receive was calculated under the two conditions: (i) keep the native vegetation (Caatinga), and (ii) manage partially degraded grassland to grow sheep.

3 Results and Discussion

The monthly variation during Oct/2014 and Sep/2015 of rain, air temperature, NDVI, and NEE is shown in Figure 2. The whole period of measurement represents a hydrological year. During this period, the total rain was 398 mm, which is about 62% of the mean rainfall (Figure 2A). This region recorded in 2012/2013 one of the most significant droughts in the last 50 years [4], with subsequent dry years as well. In the studied period, rain events were constantly recorded that contribute to a decrease in rainfall entropy [17]. However, due to a lower amount of rain, vegetation dynamics and the carbon uptake is affected.

The air temperature in CA ranged from 23.4 to 27.5 °C, while in GR it ranged from 25.4 to 29.5 °C. Monthly mean air temperature kept almost constant during the year, but in CA was about 2 °C lower than GR (Figure 2B). It was not the aim of this study to price air temperature reduction, since it is difficult to calculate the costs and especially to find people willing to pay for that [1]. However, this information is useful to assess the quality of the environment and pricing the credits of carbon.

During the dry season, the NDVI in both areas showed similar values due to reflectance of the bare soil in GR and branches in CA (Figure 2C). The increase rate of NDVI after rain events in CA was higher than in GR, and during the wet season, NDVI in CA was higher. The maximum NDVI was reached in December of 2018 because of the rains in the previous month. Then, it is possible to observe a more significant reduction of NDVI in GR than in CA. This behavior of CA is related to the ecosystem buffering capacity, which is typical for natural ecosystems such as Caatinga, specially adapted to adverse conditions like water stress and soil disturbances [18]. At the beginning of the wet season, after the first rain events, the vegetation starts to flush leaves, which can reinforce the plant to subsequent short dry spells and enhance its endurance to resist rainfall reduction [4].

The monthly balance of carbon flux varies between negative and positive values depending on water availability and rainfall distribution (Figure 2D). Negative values of NEE indicates that the ecosystem sequester carbon from the atmosphere, acting as a carbon sink. The other way around, positive values mean that carbon was released and the ecosystem acts as a source of carbon. Although the environment is semi-arid, in most of the months, NEE was negative in both areas. However, more carbon was taken up during the wet season (December to April). During June and July, NEE was positive in both CA and GR. Also, in CA positive NEE in November was recorded. The lowest NEE was -0.53 and -0.66 t/ha, respectively, for CA and GR. The highest NEE was 0.13 and 0.03 t/ha, respectively, for CA and GR.

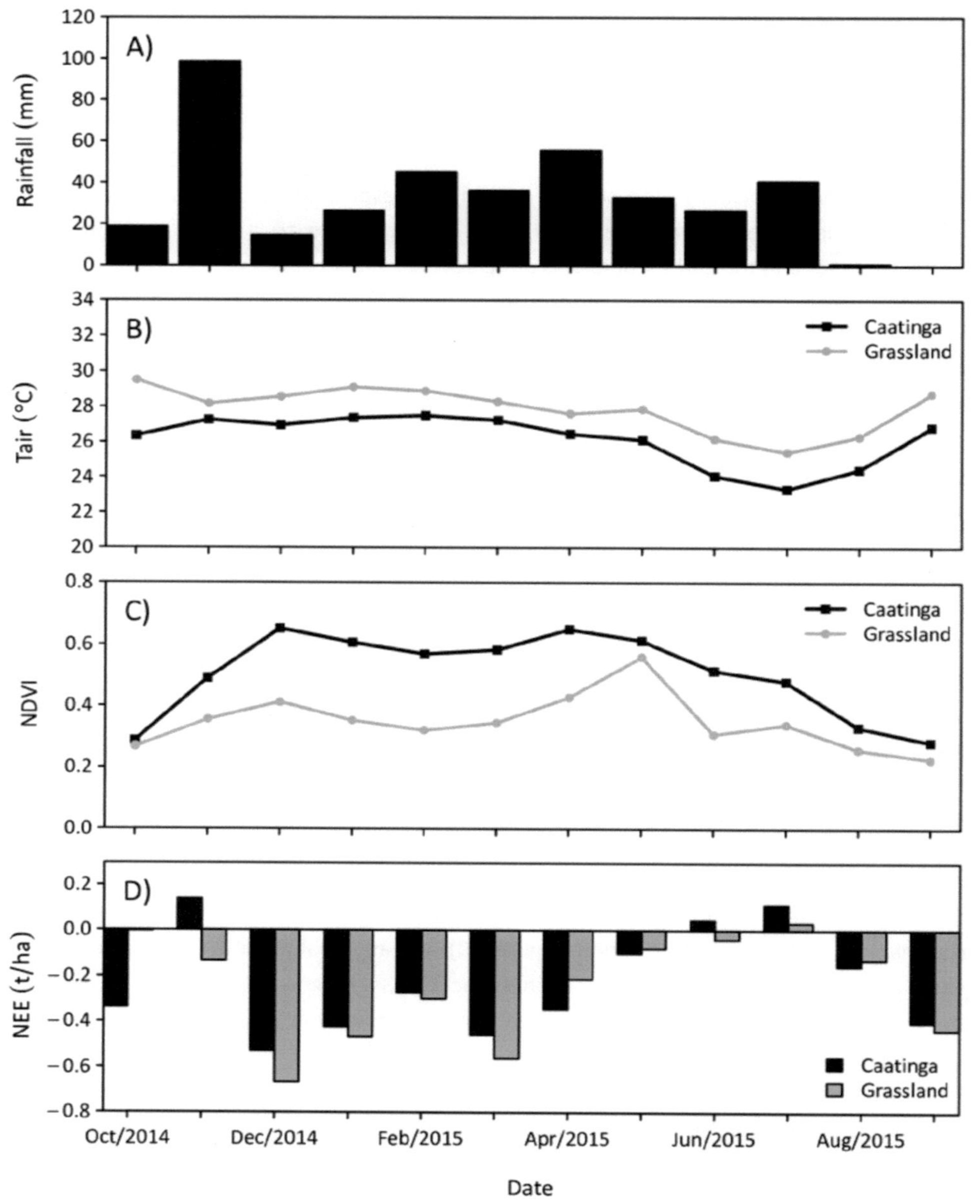

Figure 2: Monthly rainfall, mean air temperature (Tair), average NDVI and net ecosystem carbon exchange (NEE) in Caatinga and Grassland

The total carbon balance in one year was -2.74 t/ha in CA, and -3.00 t/ha in GR. Even under higher grazing pressure, water limitation and degradation conditions, grasslands tend to act as a carbon sink [19-21]. Usually, tropical grassland can uptake sometimes more carbon than other ecosystems because of the C4 photosynthesis species. The photosynthesis type of most trees and shrubs in CA is C3, but according to [22], C4 plants are more efficient in uptaking CO_2 than C3 plants, which justify the higher uptake of carbon of the GR area.

In terms of pricing, due to the current management and conditions of GR, this amount of carbon could not be traded with the same price of the uptake carbon in CA. Although the GR carbon uptake is 9.5% higher than CA, one need to take into consideration that due to different management decisions, the GR area presents higher soil degradation, nutrients, and soil organic carbon losses than the CA area. Also, the carbon stock in CA is higher than in GR, since there is an average biomass of 45.9 t/ha plus 2.20 t/ha of litter fall in CA. Besides that, there is also more carbon storage in the soil of CA (Table 1). Another aspect to consider is the difference in air temperature between areas although not being a tradable variable. For simplicity, the SOC to weight the price paid for the amount of absorbed carbon in CA and GR is considered. Thus, the potential credit of carbon in CA equal to 78.97 USD/ha.yr, and for GR equal to 40.25 USD/ha/yr are estimated.

The amount of biomass produced in GR is low because of the management (high grazing pressure) and the last droughts recorded in the region. However, this biomass could feed animals and produce roughly 30 kg/ha.yr of meat, which brings revenue regarding meat production of 108.30 USD/ha.yr. Nevertheless, in this area, about 24% of the rainfall runoffs [23] that increase soil and nutrients losses. Also, it is necessary to reseed in order to avoid higher degradation and collapse. Under these conditions, it is necessary to reseed the area and recover part of nutrients. Based on the local market, these expenses calculated can reach about 97.81 USD/ha.yr.

Table 2 summarizes all revenue and expenses for each ecosystem. The timber stock in CA was not considered in the total revenue, but it is highlighted as a potential reserve that could be managed only under the environmental license for a maximum of 80% [24]. Besides that, wood extraction releases the carbon stored in soil and vegetation, and it reduces the amount of carbon sequestered.

In the context of the farmers' own decisions and the local market and climate, if the carbon credits would be traded, keeping the natural vegetation (Caatinga) could be more profitable than grassland under current conditions. Thus, the farmer that decides to preserve could have revenue of about 56% more than the livestock farmer. There are many other ecosystem services not considered in this analysis, such as water purification, groundwater recharge, pollination, and prevention of soil erosion, which could be considered as bond services as to aggregate value in carbon uptake by Caatinga.

Table 2: Summary of potential revenues and expenses due to ecosystem services

	Caatinga	Grassland
	USD/ha.yr	
Carbon credit	78.97	40.25
Timber stock[#]	581.00	0.00
Meat	0.00	108.30
Expenses	0.00	-97.81
Total	78.97	50.74

[#]*Not considered for total revenue.*

4 Conclusions

The preserved area of Caatinga could be about 56% more profitable than grassland with the typical management of the region. Overall, a simplified study of the costs was made, however, there are other aspects regarding the biodiversity that were not considered in this study, which may potentially increase the benefits of maintaining a preserved area. These results may be useful for policy-makers in order to preserve natural ecosystems and to reduce deforestation and land degradation.

5 Acknowledgments

The authors would like to thank the EXCEED Swindon project and DAAD (German Academic Exchange Service) for support to participate at the workshop *"Water on Agricultural Practices: Training the Trainers"*, held in Rio de Janeiro, Brazil on September 15-21, 2019. They gratefully thank the Pernambuco State Research Support Foundation - FACEPE (grants: APQ-1196-5.03/15; APQ-0296-5.01/17; APQ-0498-3.07-17/NOWCDCB), the National Council for Scientific and Technological Development - CNPq (grants: 441305/2017-2; 312984/2017-0; 310537/2017-7; 307335/2017-8; 409990/2018-3; 465764/2014-2/NOWCDCB) and the Coordination for the Improvement of Higher Education Personnel - CAPES (grants: 88887.136369/2017-00/NOWCDCB; 88887.318206/2019-00/PRINT) for financial support and research fellowships. This work is part of the National Observatory of Water and Carbon Dynamics in the Caatinga Biome (NOWCDCB) supported by FACEPE articulated with CNPq and CAPES.

6 References

[1]   R. Costanza, R. D'Arge, R. de Groot, S. Farber, M. Grasso, B. Hannon, K. Limburg, S. Naeem, R.V. O'Neill, J. Paruelo, R.G. Raskin, P. Sutton, M. van den Belt: The value of the world's ecosystem services and natural capital. Nature 1997, 387 253–260.

[2]   R. de Groot, L. Brander, S. van der Ploeg, R. Costanza, F. Bernard, L. Braat, M. Christie, N. Crossman, A. Ghermandi, L. Hein, S. Hussain, P. Kumar, A. McVittie, R. Portela, L.C. Rodriguez, P. ten Brink, P. van Beukering: Global estimates of the value of ecosystems and their services in monetary units. Ecosyst. Serv. 2012, 1, 50–61.

[3]   A.M. Mapulanga, H. Naito: Effect of deforestation on access to clean drinking water. Proc. Natl. Acad. Sci. 2019, 116, 8249–8254.

[4]   R. Souza, X. Feng, A. Antonino, S. Montenegro, E. Souza, A. Porporato: Vegetation response to rainfall seasonality and interannual variability in tropical dry forests. Hydrol. Process. 2016, 30, 3583–3595.

[5]   K. Ribeiro, E.R. de Sousa-Neto, J.A. de Carvalho, J.R. de Sousa Lima, R.S.C. Menezes, P.J. Duarte-Neto, G. da Silva Guerra, J.P.H.B. Ometto: Land cover changes and greenhouse gas emissions in two different soil covers in the Brazilian Caatinga. Sci. Total Environ. 2016, 571, 1048–1057.

[6]   M. Tabarelli, I.R. Leal, F.R. Scarano, J.M.C. da Silva: Caatinga: legado, rajetória e desafios rumo à sustentabilidade, Artigos Caatinga. Cienc. Cult. 2018, 70-4, 25-39.

[7]   M.B. Dias-Filho: Diagnóstico das Pastagens no Brasil, Belém-PA, (2014). http://ainfo.cnptia.embrapa.br/digital/bitstream/item/102203/1/DOC-402.pdf.

[8]   UFCG - Departamento de Ciências Atmosférica - Universidade Federal de Campina Grande, Dados climatológicos do Estado de Pernambuco, (2015). http://www.dca.ufcg.edu.br/clima/dadospe.htm (accessed August 22, 2016).

[9]   E.V.S.B. Sampaio, G.C. Silva: Biomass equations for Brazilian semiarid caatinga plants, Acta Bot. Brasilica. 2005, 19, 935–943.

[10]   C. Aguilar, J.C. Zinnert, M.J. Polo, D.R. Young: NDVI as an indicator for changes in water availability to woody vegetation, Ecol. Indic. 2012, 23, 290–300.

[11]   B. Fu, I. Burgher: Riparian vegetation NDVI dynamics and its relationship with climate, surface water and groundwater, J. Arid Environ. 2015, 113, 59–68.

[12]   P.F. da Silva, J.R. de S. Lima, A.C.D. Antonino, R. Souza, E.S. de Souza, J.R.I. Silva, E.M. Alves: Seasonal patterns of carbon dioxide, water and energy fluxes over the Caatinga and grassland in the semi-arid region of Brazil. J. Arid Environ. 2017, 147, 71–82.

[13]   E.K. Webb, G.I. Pearman, R. Leuning: Correction of flux measurements for density effects due to heat and water vapour transfer. Q. J. R. Meteorol. Soc. 1980, 106 85–100

[14] T. Wutzler, A. Lucas-Moffat, M. Migliavacca, J. Knauer, K. Sickel, L. Šigut, O. Menzer, M. Reichstein: Basic and extensible post-processing of eddy covariance flux data with ReddyProc. Biogeosciences. 2018, 15, 5015–5030.

[15] R Core Team, R: A Language and Environment for Statistical Computing, (2019). https://www.r-project.org/.

[16] Investing.com, Carbon Emissions Futures - Dec 19 (CFI2Z9), (2019). https://br.investing.com/commodities/carbon-emissions-news.

[17] X. Feng, A. Porporato, I. Rodriguez-Iturbe: Changes in rainfall seasonality in the tropics. Nat. Clim. Chang. 2013, 3, 811–815.

[18] A.P. Kinzig, P. Ryan, M. Etienne, H. Allison, T. Elmqvist, J.P. Walker: Resilience and Regime Shifts: Assessing Cascading Effects, Ecol. Soc. 2006, 11(1), 20.

[19] C. von Randow, A.O. Manzi, B. Kruijt, P.J. de Oliveira, F.B. Zanchi, R.L. Silva, M.G. Hodnett, J.H.C. Gash, J.A. Elbers, M.J. Waterloo, F.L. Cardoso, P. Kabat: Comparative measurements and seasonal variations in energy and carbon exchange over forest and pasture in South West Amazonia. Theor. Appl. Climatol. 2004, 78(1-3), 5-26.

[20] C. Ammann, C.R. Flechard, J. Leifeld, A. Neftel, J. Fuhrer: The carbon budget of newly established temperate grassland depends on management intensity, Agric. Ecosyst. Environ. 2007, 121, 5–20.

[21] L. Gourlez de la Motte, E. Jérôme, O. Mamadou, Y. Beckers, B. Bodson, B. Heinesch, M. Aubinet: Carbon balance of an intensively grazed permanent grassland in southern Belgium. Agric. For. Meteorol. 2016, 228–229, 370–383.

[22] L. Taiz, E. Zeiger: Plant physiology, 3rd ed., England: Sinauer Associates, 2002.

[23] R.M.S. Souza, E.S. De Souza, A.C.D. Antonino, J.R.D.S. Lima: Balanço hídrico em área de pastagem no semiárido pernambucano, Rev. Bras. Eng. Agrícola e Ambient. 2015, 19, 449–455.

[24] Brasil. Ministério do Meio Ambiente, Manejo sustentável dos recursos florestais da Caatinga, Natal-RN (2008).